T0308738

Nanoelectronics

Physics, technology and applications

Online at: https://doi.org/10.1088/978-0-7503-4811-9

Nanoelectronics

Physics, technology and applications

Rutu Parekh

VLSI and Embedded Systems Group, Dhirubhai Ambani Institute of Information and Communication Technology (DA-IICT), Gandhinagar, India

Rasika Dhavse

Department of Electronics Engineering, Sardar Vallabhbhai National Institute of Technology (SVNIT), Surat, India

IOP Publishing, Bristol, UK

ISBN   978-0-7503-4811-9 (ebook)
ISBN   978-0-7503-4809-6 (print)
ISBN   978-0-7503-4812-6 (myPrint)
ISBN   978-0-7503-4810-2 (mobi)

DOI   10.1088/978-0-7503-4811-9

Version: 20231201

IOP ebooks

British Library Cataloguing-in-Publication Data: A catalogue record for this book is available from the British Library.

Published by IOP Publishing, wholly owned by The Institute of Physics, London

IOP Publishing, No.2 The Distillery, Glassfields, Avon Street, Bristol, BS2 0GR, UK

US Office: IOP Publishing, Inc., 190 North Independence Mall West, Suite 601, Philadelphia, PA 19106, USA

Dedicated to my beloved parents

Who have been my constant inspiration, throughout my life

—Rutu

To Neeraj and Aryaa—the rock-solid pillars of my life

—for their encouragement, confidence and patience

—Rasika

Contents

Preface

Philosophy and goals

Welcome to the amazing world of nanoelectronics! Nanoelectronics holds some answers as to how we can increase the capabilities of electronics devices while reducing their weight and power consumption. The International Technology Roadmap for Semiconductors (ITRS) predicts the technical capability the industry needs to continue to meet Moore's law. For several decades, planar CMOS technology has been the workhorse for integrated circuit (IC) designers. It has been scaled as per Moore's law, resulting in ever decreasing device sizes and numbers of technology nodes, the present standard being 10 nm. The ITRS predicts that the end of metal-oxide-semiconductor field-effect transistor (MOSFET) scaling will arrive in 2024 with a 7.5 nm process. Getting to 7.5 nm, however, presents some long-term difficult challenges.

The future of electronics lies with nanoelectronics. This is a completely inter-disciplinary field. Knowledge in the area of semiconductor physics, electronics, circuit design, material science, quantum physics, nanofabrication, modelling and simulation is required to understand, design and fabricate a heterogeneous IC. Nanoelectronics promises a myriad of possibilities, from high-speed computing to advanced medical treatments, and even the potential for new forms of energy production. The applications of nanoelectronics can be found in diverse fields such as quantum computing, advanced memory, wearable and flexible devices, display technology, spintronics, optoelectronics, etc. It aids in developing miniaturized portable electronics that has no counterpart in present technology. To design a circuit for such a purpose, one needs to understand the device, its physics and fabrication process, as well as its limitations.

This book provides insight into the future of electronics, emerging devices, logic and memory, sensors and system architecture, in addition to nanofabrication. In order to understand the electrical characteristics and design parameters, it is vital to know the fundamental physics behind them. That understanding is developed in the subsequent chapter contents. With this knowledge, the reader will be capable of understanding the applications and design principles of a given device or sensor. This book can be offered as a textbook for masters or undergraduate students, researchers, engineers and specialists in various fields including electron devices, solid-state physics and nanotechnology.

We would like to emphasise that, since there is no such book that bridges the gap between nanoscience, nanofabrication, nanodevices, their circuit design and appli-cations, the present book is written to address this. As far as subject depth is concerned, it cannot easily be quantified; but as experienced researchers and academicians in this area, we think we have made fair decisions and written chapters bearing in mind this aim of linking the fields of nanoelectronics at all times. It is very clear that hybrid circuit design requires knowledge in order to establish and grasp the relation between physical and electrical parameters. This book is intended in particular for engineering students, who can hopefully then begin to explore this new revolutionary technology and use it for the betterment of humanity.

Prerequisites

The prerequisites for this subject are college-level mathematics, an introduction to modern physics and knowledge of electronic circuits. We thus propose that this book is for senior-level undergraduate/graduate or beginner-level research students who wish to embark on a journey into the field of nanoelectronics. Nanoelectronics begins where conventional devices fail. As such, we will cover the limitations of the current technology and connect the same to subsequent chapters/topics. Also, it is not compulsory to teach the whole book in a single semester: the tutor can selectively decide on a topic as per their choice.

Organisation

This book is organised into nine chapters, each covering a specific area. **Chapter 1, Physical and technological limitations of nano-CMOS devices to the end of the roadmap and beyond**, focuses on the scaling trend, the "red brick wall" encountered by present nano-CMOS technology and the role of ITRS. In order to delay the absolute failure of CMOS technology, emerging devices are being developed which can be hybridised with present CMOS technology and potentially replace CMOS. A brief description of emerging and prototypical logic and memory devices is presented in brief. These devices contain the potential to revolutionise the future of electronics, and present possible advantages that have no CMOS counterpart. The chapter provides an in-depth insight into the impact of these technological options and their applications.

Chapter 2, Introduction and overview of nanoelectronics, is aimed at giving an introduction and overview of nanoelectronics and its applications. The revolution and continued evolution in this field is presented with a short history. The market potential and its requirements are further discussed, as the field is growing with ever increasing potential in the current time. This chapter also covers nanofabrication and its two approaches, top-down and a bottom-up approaches, utilising different methods. The primary goal is to establish a link between the market potential, what the technology and fabrication can offer, and the kind of applications that can be foreseen.

Chapter 3, Introduction to the quantum theory of solids, provides a brief review of a few concepts from classical Newtonian physics, and how this fails to explain phenomena at the nanoscopic scale. As is detailed within, the classical mechanical model is able to describe all observed physical phenomena; however, when it comes to the nanoscale this model breaks down, because the properties of the materials change dramatically. At this scale, the surface-to-volume ratio increases, gravitational forces become negligible and random molecular motion becomes more important. This changes the properties of the material and affects the devices that are fabricated at this scale. It is hence necessary to develop a new model that can explain such physical phenomena. This has led to a new branch of physics, i.e. quantum physics (QP) or quantum mechanics, which is able to explain the behaviour observed at nanoscale. QP enables us to explain phenomena such as the photo-electric effect, Compton effect. etc. In this chapter, we introduce quantum mechanics and how it evolved in the late 19th and early 20th centuries. Further, light's behaviour as a particle as well as a wave is described by the photoelectric effect and Young's double-slit experiment, respectively. Further, the dual nature of the particle is explained using the de Broglie principle. Schrodinger's equation is then explained, which gives the probability of finding a subatomic-scale particle at a certain position in a quantum well.

In addition to the formulation of Schrodinger's equation, its solution, the concept of a potential well, the characteristics of the wave function and its interpretation and significance, and, finally, the application of Schrodinger's equation in the structure of an atom form subsequent parts of this chapter. The equation serves as a basis with which to understand band theory. Band theory explains the difference between conductors, insulators and semiconductors. Materials are distinguished according to the differences in their band gaps. Adding impurities or varying temperature changes the band gap in semiconductors, which consequently changes the properties of the semiconductor and its electrical conduction. One more important measure to understand and exploit the properties and behaviour of materials is Fermi–Dirac statistics. It is useful to acquire the probability of electrons at a given energy level at a specific temperature. These concepts can be utilised by physicists and chip makers to make more energy-efficient devices. Hence, this chapter details the fundamentals of quantum physics and how they relate to the electrical properties of nanodevices.

An overarching goal of **chapter 4, Emerging research devices for nanocircuits**, is to survey, assess and catalogue viable emerging devices. With the evolution of the roadmap from the ITRS to the International Roadmap for Devices and Systems (IRDS), an expanded focus on systems requires that one evaluates the applicability of these novel devices to new types of systems and architectures. This goal is accomplished by (a) extending MOSFETs to the end of the roadmap, (b) employing "beyond CMOS" charge-based non-conventional field-effect transistors (FETs) and other charge-based information carrier devices, and (c) developing alternative information processing devices. Under these categories various devices, like carbon nanotubes (CNTs) and graphene-based devices, channel replacement devices, tunnel FETs, single-electron transistors, nanoelectromechanical (NEMS) systems and many more, are explained based on their physics and from a circuit application and design point of view.

With great progress being made in emerging memory technologies, the current trends and limitations are first discussed in **chapter 5, Emerging memory devices**. For the past three and a half decades of their existence, the family of semiconductor memory devices has expanded greatly and achieved higher densities, higher speeds, lower power, more functionality and lower costs. Despite its limitations, the field of conventional semiconductor memory continues to flourish, and memory device scientists are finding ways to meet these technical challenges, possibly even leading to the development of a 'universal memory' with low cost, high performance and high reliability for future electronic systems. In this chapter emerging memory devices, such as ferroelectric RAM, magnetoresistive RAM (MRAM), Mott memory, molecular memory, macromolecular memory and resistive RAM, are detailed with their physics, working and applications.

For accurate projections and evaluations of device performance in the nanoscale era, it is of utmost significance that materials, processes and devices should be modelled precisely. Indeed, modelling and simulation of nanoscale processes and devices has become an exciting topic of research today. However, when working with emerging memory or logic devices, very few commercial simulators are available. In **chapter 6, Simulation and modelling**, we give our readers a detailed account of modelling and simulation using Technology Computer-Aided Design (TCAD) tools. Multiphysics environments in which to implement MEMS/NEMS devices are outlined and methodologies demonstrating novel nanoelectronic devices with CMOS technology are also illustrated. At the end of this chapter, the reader is supplied with references to these tools for their further perusal, based on their research interests.

Synthesis at the nano level, i.e. nanofabrication, represents a gigantic leap that has brought the world of ultra-small things into reality. In fact, nano evolution is possible due to systematic progress in microfabrication, nanofabrication, thin-film synthesis, CNTs, microelectromechanical-NEMS technologies, process integration at the nanoscale and nanopackaging. All these aspects of fabrication at the nanoscale are explored in **chapter 7, Nanofabrication.**

Modern computer architecture is based on von Neumann architecture and has continuously improved, as per Moore's law. However, this scaling trend has reached its fundamental physical limit with today's architecture and materials. Much reduction in area and power demand can be achieved if a single device can be exploited as *both* a memory and logic device. **Chapter 8, Emerging architecture**, focuses on the various architectures that are being developed in the field of emerging memory devices. The architectures introduced are 3D integration of FPGAs, CMOS-based field programmable nanowire interconnects, nano-crossbar arrays, nanomagnetism-based MRAM and spin-torque transfer RAM. Furthermore, the next wave of architecture aims to mimic the human brain, which is emerging in neuromorphic and cortical architectures.

After giving a detailed account of fundamental nanoscale phenomena, devices, processes and architectures, we present the highly active and 'in demand' arena of nanosensors and transducers. This stream converges with many other streams such as chemistry, physics, biology, engineering and technology at the nanoscale and is becoming an altogether new discipline, rich with endless possibilities. Reading **chapter 9, Nanosensors and transducers**, should stimulate the readers' grey matter and help further their creativity and skills in the development of newer, smaller, lighter, cheaper, more accurate and more power efficient products.

To conclude, this book acts as an A-to-Z of topics in the field of nanoelectronics. The contents are organised such that the reader will be able to easily establish a link between device physics, systems, design, modelling, simulation and fabrication, and to gain an insight into the prospect of a particular technology. Nanoelectronics is full of exciting applications, and we are sure that it has a potential to solve the "big ticket" problems faced by mankind.

Notes to the readers

We want senior undergraduate engineering students to be equipped with knowledge of the rich and unending potential of nanoelectronics. They should know the different devices that can overcome the limitations of conventional CMOS technology, their underlying physical mechanisms, their models, fabrication techniques and the tools which can be used to model these devices. They should get an idea of the circuits and architectures that can be built in implementing these new devices. In most universities, the course curricula include all these topics, and professors and students have to refer to a number of books. Our own experience is the same. That is the very reason why we are here trying to provide a one-point solution. In this, the reader should bear in mind that this is a continuously evolving technology. Chapters can be covered and skipped as per the curriculum and need of time.

Acknowledgements

Our hearts and souls feel greatly honoured to find this opportunity to be in thankfulness before almighty for his constant innumerable blessings and also for bestowing us with enough strength and courage to complete our book.

The institute, Dhirubhai Ambani Institute of Information and Communication Technology (DA-IICT), has our appreciation for providing an atmosphere conducive to writing this book. We are indebted to students at DA-IICT, our friends and colleagues, and to one and all who directly or indirectly have lent their helping hand in this venture. This book is as much theirs as it is ours.

Authorities of Sardar Vallabhbhai National Institute of Technology, Surat (SVNIT) have been immensely supportive in this endeavour. We are grateful to them from the bottoms of our hearts.

We thank all the academicians, researchers and industry personnel for contributing to the nanoelectronics domain despite of many challenges and documenting their works meticulously. It served as great reference base.

Author biographies

Rutu Parekh

Dr Rutu Parekh received an M.Eng. in Electrical Engineering from Concordia University, Montreal, Canada, a PhD in Electrical Engineering (Nanoelectronics) from the Université de Sherbrooke, Sherbrooke, Canada, and worked as a Postdoctoral fellow at Centre of Excellence in Nanoelectronics, IIT Bombay in 2015. Her research areas are ASIC design, micro/nanoelectronics, nanofabrication, embedded systems and IOE. She has research experience at the École Polytechnique de Montréal, industrial experience with eInfochips, Ahmedabad, India and HP Karkland, Montreal, and teaching experience with Nirma University of Science and Technology, Ahmedabad. She is also associated with The Inter-University Centre for Astronomy and Astrophysics, Pune, India, as a Visiting Associate. She is currently working as an Associate Professor at DA-IICT, Gandhinagar, India. She has published a number of international journal and conference articles related to her research areas. In addition, she has been offering her services as an editorial board member for international journals and as a technical committee member in numerous international conferences. She is founder and Chair of the IEEE Nanotechnology Council Chapter, Gujarat section, India.

Rasika Dhavse

Rasika Dhavse is serving as Associate Professor in the Department of Electronics Engineering of Sardar Vallabhbhai National Institute of Technology, Surat. She pursued her doctoral degree in the field of nanocrystal-based flash memory devices. She has more than 25 years of academic experience. Her research students are working in the fields of nanoelectronics, rad-hard devices, sensors, VLSI design and technology, biomedical instrumentation, etc. In addition, she has successfully completed many government-funded research projects. She has organised numerous seminars, workshops, FDPs and STTPs on various topics related to VLSI design, semiconductor physics and technology, fabrication and characterization of MOS devices, nanoelectronics and nanotechnology, embedded systems, robotics, system automation, various tool trainings, etc. and has chaired many reputed international conferences. She has significant research papers to her credit. She is one of the founders and Vice-Chair of the IEEE Nanotechnology Council, Gujarat section, India.

Chapter 1

Physical and technological limitations of nano-CMOS devices to the end of the roadmap and beyond

Usually, discussion about nanodevices begins by outlining the limitations of metal-oxide semiconductor field-effect transistors (or MOSFETs). So, here we first introduce the two classes and working principle of MOSFETs. In the next section, we will speak about Moore's law, which states that due to technological advancements, the numbers of transistors placed on a single chip will double every 18–24 months [1]. Further, we give a qualitative account of MOSFET scaling methods and several constraints and impacts due to limitations of the technology (which are acknowledged as short-channel effects), setting back the further scaling process. We walk readers through the timeline of CMOS technology, right from this present possible failure of Moore's law due to extensive scaling issues, to finding technological solutions to elongate the timeline of CMOS scaling. We explore the underlying limitations of CMOS technology, the physical phenomena behind it, the need to overcome these limitations, and certain alternatives to work around these limitations for certain applications. We present an in-depth insight into the impact of these technological options on CMOS devices and their applications. Finally, the International Technology Roadmap for Semiconductors (ITRS) is introduced.

1.1 MOSFETs and their scaling

In the field of electronics, Julius Edgar Lilienfeld first proposed the concept of a field-effect transistor in 1925 [2]. In 1948, Walter Brattain, William Shockley and John Bardeen started a revolution by inventing the transistor: a device that can switch or amplify signals and electrical power [2]. Each and every modern electronic device uses these transistors as the primary building blocks of its circuits.

doi:10.1088/978-0-7503-4811-9ch1

A MOSFET is one type of transistor, defined as a metal-oxide semiconductor field-effect transistor. The first MOSFET was invented by Mohamed M Atalla and Dawon Kahng at Bell Labs in 1959, and first presented in 1960 [2]. A MOSFET is a voltage-controlled device with three terminals, named as the gate, drain and source. A fourth terminal substrate (or body) is primarily shorted with the source to avoid any body-bias effect on its threshold voltage. The main goal of a MOSFET is to control the flow of current in the channel region by changing the gate voltage, V_{GS}. The metal-oxide semiconductor capacitor plays a very important role in providing this functionality of MOSFET. The metallic gate and semiconductor substrate, separated by a thin oxide layer, together works as the M-O-S capacitor. The source and drain terminals are located at two opposite ends of the semiconductor surface. The gate voltage controls the conductivity between the drain and the source by enhancing a continuous path between them (i.e. a channel). The conduction current is either generated by the flow of electrons or that of holes; hence, it is a unipolar device. MOSFETs can be divided into two classes based on their types of doping elements, n-channel MOSFETs and p-channel MOSFETs.

1.1.1 n-Channel MOSFETs

Typically, n-channel MOSFETs (nMOSs) opt for a lightly doped p-type silicon substrate with two heavily doped n-type regions, i.e. a source and a drain. A thin silicon dioxide layer isolates the MOSFET channel from the gate layer. Initially, the channel between the two $n+$ regions does not exist. When the positive gate voltage, V_{GS}, is applied, 'N' channel(s) will form at the surface, which will carry the electrons from source to drain upon application of a potential difference across the drain and source. Figure 1.1(a) illustrates the structure of a nMOS [1–3].

1.1.2 p-Channel MOSFET

The structure of p-channel MOSFETs (pMOSs) is somewhat complementary to that of nMOSs. The substrate is a lightly doped n-type silicon. The source and the drain, also known as wells, are heavily doped with p-type dopant. Initially, the channel between the two p+ regions does not exist. When a negative gate voltage, V_{GS}, is applied, 'P' channel(s) will form at the surface, which will carry the holes from

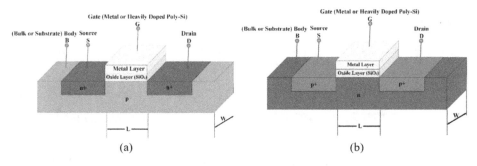

Figure 1.1. MOSFET structures: (a) n-channel, (b) p-channel.

source to drain upon application of a potential difference across the drain and source. Figure 1.1(b) illustrates the structure of a pMOS [1–3].

1.1.3 Working principle of MOSFETs

The applied voltage at the gate is either positive or negative for nMOSs and pMOSs, respectively. For an nMOS, application of positive V_{GS} repels the holes in the p-substrate downwards and generates a negative depletion region charge, which is dependent on the acceptor atoms' concentration. This also attracts and accumulates the electrons from $n+$ source and drain, creating a negatively charged channel between source and drain. At a threshold voltage V_T, the surface electron density becomes equal to the bulk hole density, and the surface is said to be completely inverted. At this point, the device is turned ON and current can flow freely in the channel when the potential difference is applied between source and drain. Similarly, in a pMOS, if a negative V_{GS} is applied to the gate, then the positively charged holes will form a channel. The transconductance and output characteristics of an nMOS are shown in figure 1.2 [1–3].

In figure 1.2(a), one can observe that $I_{DS} = 0$ at any value of V_{DS} when the gate voltage is below the threshold voltage, because the nMOS is in cut-off mode. The drain current I_{DS} will increase with increasing V_{DS} only if $V_{GS} > V_T$. In figure 1.2(b), one can observe that, if $V_{GS} > V_T$, I_{DS} will increase with increasing V_{DS}, at the beginning exhibiting linear or voltage-controlled resistor-like behaviour of the nMOS; and, after the pinch = off point, $V_{DS} = V_{GS} - V_T$, I_{DS} will remain the same for higher values of V_{DS}, causing saturation of the MOSFET. Here, the nMOS works as a constant current source. MOSFETs can be modelled by the set of equations as shown in table 1.1 [3].

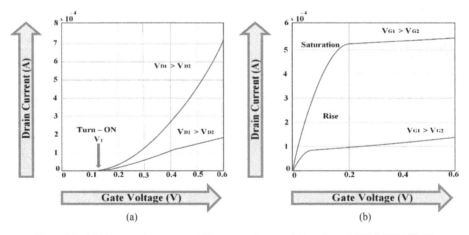

Figure 1.2. (a) Transconductance and (b) output characteristics of an nMOS MOSFET [3].

Table 1.1. Current–voltage equations of MOSFETs [3].

NMOS	Threshold voltage	$$V_{TN} = \emptyset_{GC} - 2\emptyset_F - \frac{Q_B}{C_{ox}} - \frac{Q_{ox}}{C_{ox}} - \frac{Q_I}{C_{ox}} \quad (1.1)$$ $$V_{TN0} = \emptyset_{GC} - 2\emptyset_F - \frac{Q_{B0}}{C_{ox}} - \frac{Q_{ox}}{C_{ox}} - \frac{Q_I}{C_{ox}} \quad (1.2)$$ $$V_{TN} = V_{TN0} + \gamma \left(\sqrt{|-2\emptyset_F + V_{SB}|} - \sqrt{2\emptyset_F} \right) \quad (1.3)$$ where, $\emptyset_{GC}$ is work function difference between gate and channel, $2\emptyset_F$ is surface potential (−ve), $\frac{Q_B}{C_{ox}}$ is depletion layer charge (−ve), $\frac{Q_{ox}}{C_{ox}}$ is surface charge, $\frac{Q_I}{C_{ox}}$ is implant charge, γ is the body effect coefficient (+ve), V_{SB} is source to body bias (+ve) and suffix 0 indicates respective parameter value at zero body bias.
	Cut-off region	$$V_{GS} < V_{TN}, \ I_{DS} = 0 \quad (1.4)$$
	Linear/triode region	$$V_{GS} \geq V_{TN}, V_{DS} < V_{GS} - V_{TN},$$ $$I_{DS} = \frac{\mu_n C_{ox}}{2} \frac{W}{L} (2(V_{GS} - V_{TON})V_{DS} - V_{DS}^2) \quad (1.5)$$
	Saturation region	$$V_{GS} \geq V_{TN}, \ V_{DS} > V_{GS} - V_{TN},$$ $$I_{DS} = \frac{\mu_n C_{ox}}{2} \frac{W}{L} (V_{GS} - V_{TON})^2 (1 + \gamma V_{DS}) \quad (1.6)$$ where, μ_n is surface electron mobility of NMOS.
PMOS	Threshold voltage	$$V_{TP} = \emptyset_{GC} - 2\emptyset_F - \frac{Q_B}{C_{ox}} - \frac{Q_{ox}}{C_{ox}} - \frac{Q_I}{C_{ox}} \quad (1.7)$$ $$V_{TP0} = \emptyset_{GC} - 2\emptyset_F - \frac{Q_{B0}}{C_{ox}} - \frac{Q_{ox}}{C_{ox}} - \frac{Q_I}{C_{ox}} \quad (1.8)$$ $$V_{TP} = V_{TP0} + \gamma \left(\sqrt{|-2\emptyset_F + V_{SB}|} - \sqrt{2\emptyset_F} \right) \quad (1.9)$$ where, $2\emptyset_F$ is surface potential (+ve), $\frac{Q_B}{C_{ox}}$ is depletion layer charge (+ve), γ is the body effect coefficient (−ve) and V_{SB} is source to body bias (−ve).
	Cut-off region	$$V_{GS} > V_{TP}, \ I_{DS} = 0 \quad (1.10)$$
	Linear/triode region	$$V_{GS} \leq V_{TP}, V_{DS} > V_{GS} - V_{TP},$$ $$I_{DS} = \frac{\mu_p C_{ox}}{2} \frac{W}{L} (2(V_{GS} - V_{TOP})V_{DS} - V_{DS}^2) \quad (1.11)$$ Where, μ_p is surface hole mobility of PMOS.
	Saturation region	$$V_{GS} \leq V_{TP}, \ V_{DS} < V_{GS} - V_{TP},$$ $$I_{DS} = \frac{\mu_p C_{ox}}{2} \frac{W}{L} (V_{GS} - V_{TOP})^2 (1 + \gamma V_{DS}) \quad (1.12)$$

1.1.4 Introduction and failure of Moore's law

In 1965, Gordon Moore, in his paper 'Cramming more components onto Integrated Circuits', explained that the transistor density on an integrated circuit (IC) would grow exponentially. This speculation became popularly known as 'Moore's law', and its scaling is depicted in figure 1.3.

Moore law's states that the number of transistors on ICs will double roughly every two years. To stick to this pace of technological advancement, the size of

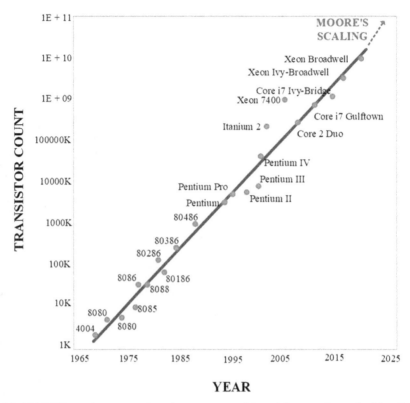

Figure 1.3. MOSFET transistor count for microprocessors. Adapted from an image by Max Roser and Hannah Ritchie/OurWorldinData (https://ourworldindata.org/uploads/2020/11/Transistor-Count-over-time.png). CC BY 4.0.

individual transistors should be reduced and their packing density should double every 18–24 months. For this, we need to miniaturise the dimensions of MOSFETs—a process known as scaling. This scaling of the MOSFET is key to continue following Moore's law. This law predicted the continuing evolution of the semiconductor industry, in which the realisation of increasingly complex devices and systems became possible. Gradual shrinking of the dimensions of the transistors below 100 nm enabled hundreds of millions of transistors to be placed on a single chip. Following Moore's law in semiconductor electronics, performance indexes such as processing speed, memory retention and efficiency were considerably enhanced [3]. Low power dissipation, smaller chip size, low cost and increased package density are additional advantages of scaling. The concepts of 'More Moore', which discusses hyper-miniaturisation, and 'More than Moore', which addresses diversification, were introduced in 2005 when the ITRS published its first white paper; these concepts are elaborated upon in figure 1.4 [4].

In order to improve the conductivity of a MOSFET by a factor (say) S, where $S > 1$, we should decrease the dimensions by S. Here, S is identified as the scaling factor. Two distinct types of scaling are (i) constant field scaling, and (ii) constant

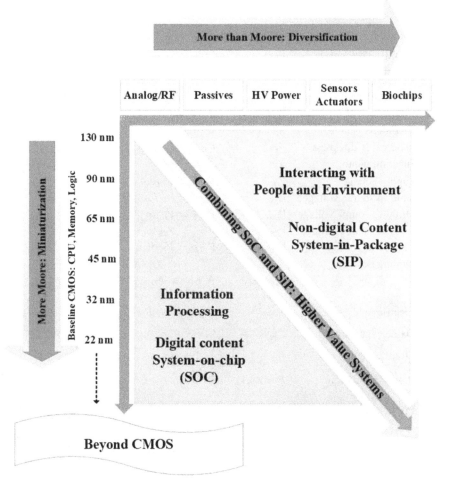

Figure 1.4. More Moore—beyond CMOS. Adapted from [4] with permission.

voltage scaling. Constant field scaling (or full scaling) retains the magnitude of the electric field inside the MOSFET, whereas the dimensions are reduced by a factor S. In this scaling type, the charge densities are gained by α because the magnitude of the electric field is kept the same. Let us say the scaling factor is $S = \alpha$. Before scaling, the power dissipated is $P = I_d \times V_{ds}$. As the power P is affected by both the drain current and source drain voltage, it will increase by α^2 [6].

In constant voltage scaling, the terminal voltage remains unchanged. Just as in full scaling, the MOSFET dimensions are reduced by S ($=\beta$). The power dissipation and power density will increase by β and β^3, respectively. The drain current is also increased by β.

The scaling-dependent changes in device dimensions and doping densities, and their effects on device parameters are listed in table 1.2 [3].

Table 1.2. Effect of scaling [3].

S. No.	MOSFET parameter	Before scaling	After scaling	
			Full scaling	Constant voltage scaling
1	Channel length	L	$L' = L/\alpha$	$L' = L/\beta$
2	Channel width	W	$W' = W/\alpha$	$W' = W/\beta$
3	Channel area	A	$A' = A/\alpha^2$	$A' = A/\beta^2$
4	Gate oxide thickness	t_{ox}	$t'_{ox} = t_{ox}/\alpha$	$t'_{ox} = t_{ox}/\beta$
5	Junction depth	X_j	$X_j' = X_j/\alpha$	$X_j' = X_j/\beta$
6	Supply voltage	V_{DD}	$V'_{DD} = V_{DD}/\alpha$	$V'_{DD} = V_{DD}$
7	Gate to source voltage	V_{GS}	$V'_{GS} = V_{GS}/\alpha$	$V'_{GS} = V_{GS}$
8	Drain to source voltage	V_{DS}	$V'_{DS} = V_{DS}/\alpha$	$V'_{DS} = V_{DS}$
9	Body to source voltage	V_{BS}	$V'_{BS} = V_{BS}/\alpha$	$V'_{BS} = V_{BS}$
10.	Threshold voltage	V_{TH}	$V'_{TH} = V_{TH}/\alpha$	$V'_{TH} = V_{TH}$
11.	Doping densities	N_A	$N'_A = \alpha . N_A$	$N_A = \beta^2 . N_A$
		N_D	$N'_D = \alpha . N_D$	$N_D = \beta^2 . N_D$
12.	Electric field	$E = \frac{V_{GS}}{t_{ox}}$	$E' = \frac{V_{GS}}{t_{ox}} = E$	$E' = \beta\frac{V_{GS}}{t_{ox}} = \beta E$
13.	Oxide capacitance	$C_{ox} = \frac{\varepsilon_{ox}}{t_{ox}}$	$C_{ox}' = \alpha\frac{\varepsilon_{ox}}{t_{ox}} = \alpha C_{ox}$	$C_{ox}' = \beta\frac{\varepsilon_{ox}}{t_{ox}} = \beta C_{ox}$
14.	Drain current	I_{DS}	$I_{DS}' = \frac{I_{DS}}{\alpha}$	$I_{DS}' = \beta I_{DS}$
15.	Power dissipation	$P = I_{DS}V_{DS}$	$P' = \frac{I_{DS}}{\alpha}\frac{V_{DS}}{\alpha} = \frac{P}{\alpha^2}$	$P' = \beta(I_{DS}V_{DS}) = \beta P$
16.	Power density	$P_d = P/A$	$P_d' = \frac{P'}{A'} = \frac{(P/\alpha^2)}{(A/\alpha^2)} = P_d$	$P_d' = \frac{P'}{A'} = \frac{(\beta P)}{\frac{A}{\beta^2}} = \beta^3 P_d$

The scaling of voltages is not preferred due to the need for complicated level-shifter arrangements. As such, constant field scaling is used only for low-power applications. Constant voltage scaling is used in high-performance applications, but an increase in power density by a factor of β^3 may cause serious reliability issues.

1.2 Limitations and showstoppers arising from CMOS scaling, and technological options for MOSFET optimisation

For a good four decades, the industry has kept pace with growth rate speculations, although the end of CMOS technology had been predicted since as early as the 1970s. The rapid growth rate seemingly approached an infinite number of transistors in an IC. In 2003, Gordon Moore revised his propositions in his paper 'No exponential is forever, but forever can be delayed!', which suggested that CMOS scaling will reach a limit but that the limit can be practically extended. Thus, it became important to seek out the material and structural design limitations that can limit CMOS scaling, hence enabling the continuation of CMOS scaling in the near term.

Scaling a transistor beyond the deep-submicron regime gives rise to many unwanted physical mechanisms, leading to the failure of their classical behaviour. Due to these mechanisms, the industry is now reaching a number of hard limits that no amount of research can overcome. These limitations are categorised into physical challenges, material challenges, thermal challenges, technological limitations and economic challenges [5–7].

(i) Physical challenges

As we go on scaling MOSFET devices, the channel length becoming ever shorter eventually leads to it being comparable to the depletion region, which contributes to short-channel effects (SCEs), including but not limited to drain-induced barrier lowering (DIBL) and sub-threshold channel currents, gate tunnel currents, gate-induced drain leakage (GIDL), junction leakage, velocity saturation, hot carrier degradation, etc. Some of these are explained below.

(a) Drain-induced barrier lowering

With increasing positive drain potential, the drain depletion region starts expanding. Normally, this would not affect the energy bands near the source region. However, in short-channel devices, it extends towards the source region, lowers the barrier between source and channel, and causes current flow between the source drain even when the transistor is in the OFF state. This is known as DIBL, and is illustrated in figure 1.5. For large drain bias voltage, the depletion region of the drain extends

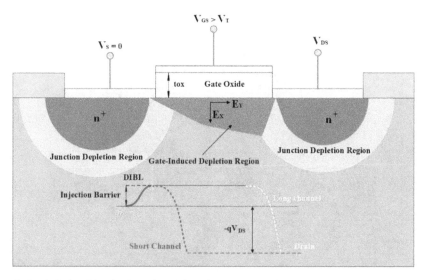

Figure 1.5. Drain-induced barrier lowering. Adapted from [3]. © The Authors. Published by McGraw-Hill Education.

towards the source and merges. This is known as 'punch through'. Punch through can be minimised by using thinner oxide, using larger substrate doping, shallower junctions and longer channels [3, 5–7].

(b) Gate tunnel currents

In MOSFET scaling, the gate oxide should be made as thin as possible so as to increase the channel conductivity and performance when the transistor is in an ON state and to reduce sub-threshold leakage when the transistor is in an OFF state. However, with very thin gate oxides the quantum-mechanical phenomenon of electron tunnelling occurs between the gate and channel, leading to increased power consumption. Insulators that have a larger dielectric constant than silicon dioxide (referred to as high-κ dielectrics), such as group IV-B metal silicates, e.g. hafnium and zirconium silicates and oxides, are being used to reduce gate leakage from the 45 nm technology node onwards [3, 5–7].

(c) Off-state leakage

The presence of a power supply when a short-channel transistor is OFF causes small drain leakage, as shown in figure 1.6. As the gate length decreases, the leakage current grows exponentially [3, 5–7]. Such off-state or sub-threshold current lead to static power dissipation, making a chip power hungry [3, 5–7].

(d) Gate-induced drain leakage

In a classical planar long-channel MOSFET, the substrate and gate are electrostatically shielded from the drain. Thus, the threshold voltage

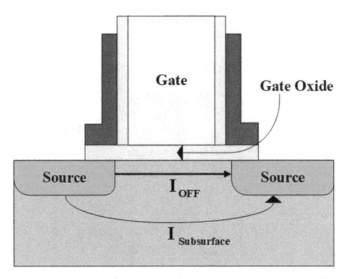

Figure 1.6. Off-state leakage.

is not dependent on drain supply. However, in the short-channel case, the drain is close enough to the gate. With a high drain supply, the $n+$ drain region under the gate becomes depleted and even inverted. This causes field crowding, resulting in avalanche multiplication and band-to-band tunnelling. Thus, minority carriers are emitted in the drain region underneath the gate and leakage current flows through the substrate [3, 5–7]. This is known as GIDL, and is illustrated in figure 1.7.

(ii) Material challenges

The failure of dielectric and wiring materials to continue providing dependable insulation and conduction, as we scale down, constitute the material challenges. The materials used in making ICs, such as silicon, silicon dioxide, aluminium, copper, etc., reach the limit of their physical capabilities like dielectric constant, carrier mobility, breakdown field strength, conductivity, etc. With the continued use of these materials, present and future scaled devices would not be able to keep up their performance [4–7].

(iii) Thermal challenges

Because the number of transistors per unit area on the chip is increasing, the total power consumption and thermal dissipation are also increasing. Because the supply voltage is not being scaled relative to the pace of the channel length, the power density is growing [4–7].

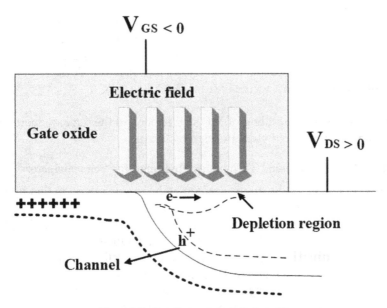

Figure 1.7. Gate-induced drain leakage [3].

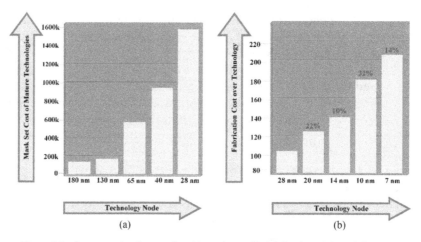

Figure 1.8. Cost vs technology nodes: (a) mask set, (b) fabrication. Adapted from [8].

(iv) Technological challenges—lithographic issues

This is considered to be the primary limitation of chip scaling. Lithographic issues pose a limit to the dimensions which can be fabricated on a chip. Research up to now has enabled lithographic technology to use the UV wavelength (193 nm), and current research is being undertaken to bring this down to the extreme-UV spectrum (13 nm). However, the dimensions of the transistor will face a hard limit: the size of the atom and molecule. The cost of fabrication has also increased and grown exponentially, which limits the profitability of increased scaling [4].

(v) Economical challenges

As aforementioned, production costs and testing costs are rising exponentially with time, as indicated in figure 1.8 and the figures given. Whilst we are advancing the technology and scaling down our semiconductor devices, we also need to simulate, test and mass produce our devices, which is what contributes the most to the cost [4–7].

Thus, various limits to the scaling of MOSFETs compel the semiconductor industry to look to novel devices and circuit design techniques.

1.2.1 ITRS: The International Technology Roadmap for Semiconductors

The ITRS was released to the public domain with the sole intention that it be used as the main common reference in the industry. It is devised to help consortia, students at institutes, and industry researchers to spur discovery in the field [4]. The ITRS consolidates a set of documents delivered by semiconductor industry professionals. These documents comprise a set of articles and reports delivered by the International Roadmap Committee (IRC), which is constituted by professional experts and representatives of semiconductor industry organisations across Europe, Taiwan, the United States, South Korea, China, and Japan between 1998 and 2015 [9]. The documents are generated with the disclaimer:

The ITRS is devised and meant for technology evaluation only and is without consideration to any commercial concern about individual commodities or devices [9].

The objectives of the ITRS include the following:

(i) To ensure cost-effective improvements in the performance of ICs, superior commodities, and applications that engage the beforementioned devices, thereby sustaining the strength and prosperity of the aforementioned organisations.

(ii) To help speed up the pace of research in the future in these domains of technology.

The documents represent the best collective opinion on the directions of research and proposed timelines extending up to about 15 years in the future for the following areas of technology:

(a) Yield enhancement: Yield is represented by the functionality and reliability of the ICs produced on wafer surfaces in the semiconductor industry. In most industries, yield has been defined as the number of products that can be sold divided by the number of products manufactured, and is an area that can potentially be made more efficient. During the manufacturing of ICs yield loss is caused by defects, faults, process variations, and design.

(b) Front-end processes: Identical ICs, known as die, are made on each wafer in a multi-step batch process. Every step adds a new layer to the wafer or modifies the existing one. These layers form the elements of the individual electronic circuits. This is known as wafer fabrication or the front-end process.

(c) Factory integration: This aspect of the ITRS focuses on integration of all the factory components needed to produce the required products efficiently in the right volumes on schedule while meeting cost targets.

(d) Test and test equipment: Semiconductor testing equipment (IC tester) and automated testing equipment (ATE) is a system for delivering electrical signals to a semiconductor device to compare output signals against expected values in its design specifications constraints.

(e) Process integration, devices and structures: This section deals with the main IC devices and structures, with overall IC process flow integration, and also with the reliability trade-offs associated with new options.

(f) Emerging research materials: Nanomaterials describe, in principle, materials of which a single unit is sized (in at least one dimension) between 1 and 1000 nm (but usually is 1–100 nm).

(g) Photolithography: Photolithography has become the most successful technique capable of producing sub-100 nm patterns.

(h) Radio frequency and analogue/mixed-signal technologies: Radio frequency (RF) and analogue/mixed-signal (AMS) technologies serve the rapidly growing advanced communications and 'More than Moore' markets. They also compose essential and critical technologies for the success of many semiconductor manufacturers, as well as the ultimate success of the future Internet of Things (IoT).

(i) IC interconnects: In ICs, interconnects are structures that connect two or more circuit elements (such as transistors) together electrically. The design and layout of interconnects on the IC is vital to its proper functioning, performance, power efficiency, reliability and fabrication yield.

(j) Assembly and packaging: The case, known as a 'package', supports the electrical contacts that connect the device to a circuit board. This process is often referred to as packaging in the IC industry; it is otherwise known as semiconductor device assembly, encapsulation or sealing.

(k) System drivers/design: The design and engineering of advanced solid-state nanoporous materials could, for example, allow for the development of novel gene-sequencing technologies that enable single-molecule detection at low cost and high speed with minimal sample preparation and instrumentation.

(l) Modelling and simulation: Semiconductor device modelling creates models for the behaviour of electrical devices based on fundamental physics, such as the doping profiles of the devices. Semiconductor process simulation is the modelling of the fabrication of semiconductor devices such as transistors.

(m) Emerging research devices: This section is motivated by the increasing difficulty of meeting all expected requirements for the rigorously scaled technologies projected for later technology nodes.

(n) Metrology: Generally, metrology denotes the methods of measuring numbers and volumes, primarily by using metrological equipment. The number of measurement points varies by semiconductor manufacturer or device.

(o) Microelectromechanical systems (MEMS): MEMS are a specialised field referring to technologies that are capable of miniaturising existing sensor, actuator or system products. Nanotechnology is a growing field that uses the unique properties of ultra-small-scale materials to an advantage.

(p) Environment, health and safety: This section addresses how nanotechnology can help in these three research areas.

Significant milestones for CMOS processor technology projected by the ITRS are shown in table 1.3.

1.2.2 Update beyond the end of the roadmap

To address the altered ecosystem of the microelectronics industry, the IRC wanted to reconstruct its constitution and goals in its annual meeting convened in Taiwan in December 2012 [9]. Accordingly, ITRS 2.0 was born in 2013–14 to fulfill the above task of the IRC. The process of reframing the ITRS was divided into 17 Technology Working Groups, which were further mapped to seven focus teams, known as IFTs [9]. These include the following:

(i) Beyond CMOS: Beyond CMOS addresses devices that are not CMOS based, but refers instead to the possible future digital electronic technologies beyond the scaling limits of CMOS. These new devices exhibit better speeds and lower densities than CMOS. Some examples of beyond CMOS devices are spin FET and spin MOSFET transistors, negative gate capacitance FETs,

Table 1.3. ITRS characteristics [9].

S. No.	Characteristics	2010	2012	2014	2016	2018	2020
1.	Metal pitch (nm)	45	32	24	18.6	15	11.9
2.	V_T (V)	0.289	0.291	0.221	0.202	0.207	0.219
		EPbulk	EPbulk	UTB FD	MG	MG	MG
3.	V_{DD} (V)	0.97	0.9	0.84	0.78	0.73	0.68
4.	Power density (W/mm^2)	0.5	0.6	0.7	0.8	0.9	1
5.	Max pin count	4900	5300	5900	6500	7200	7900
6.	Performance: on-chip (GHz)	5.88	6.82	7.91	9.18	10.65	12.36
7.	Performance: chip-to-board (Gb/s)	10	14	17	30	40	50

NEMS switches, Mott FETs, piezotronic logic transduction devices, spin-wave devices, nanomagnetic logic devices, spin torque majority logic gates, spintronics, memristors and all spin logic devices.

(ii) Outside system connectivity: This focuses on technologies based on wireless communication, including defining the type of work needed and finding the best solution. It aims at identifying technology and device requirements for enhancing intersystem communications and the corresponding research needed.

(iii) Factory integration: This focuses on new tools and processes and on producing heterogeneous integration of all these things. The specific scope of factory integration is wafer fabrication, or manufacturing in the front-end and back-end.

(iv) Heterogeneous integration: This mainly focuses on the integration of components that are manufactured separately with different technologies into a new combined component which works better than they do separately. For instance, cameras and microphones are two components that are manufactured differently.

(v) More Moore: The More Moore focus team provides electrical, physical and reliability requirements for memory and logic technologies to assist More Moore scaling for cloud (IoT and server), mobility and big data applications. Shrinking of CMOS is the main objective.

(vi) Heterogeneous component: This chiefly focuses on the various components that create heterogeneous systems such as in power generation, sensing devices and MEMS. These components cannot certainly scale down as per Moore's law.

(vii) System integration: System integration is a topic that concentrates on the design of and knowledge pertaining to the integration of heterogeneous blocks.

The above IFTs include elements from the ITRS along with many **new** elements, like novel charge-based and non-charge-based devices, wireless connectivity, heterogeneous integration, etc [9]. It is needless to say that the purpose of ITRS 2.0 was not limited to CMOS devices and their technologies but served rather to address a newer approach to system integration, as well as the topics of traversing the means of communications from conductors to wireless and then to optical fiber, in addition to continuing to search for non-electron-based technologies [9]. The last version of ITRS was published in 2013, which as mentioned aimed to provide a roadmap for the next 15 years, that is, up to 2028. But, as we now know, Moore's law was by then already reaching its last days, so researchers decided to generate a new roadmap, named the International Roadmap for Devices and Systems (IRDS). The IRDS, which was introduced in 2016, is thus the effective successor of the ITRS. Its goals include the following [11]:

(a) To find new technology for devices and systems and provide a roadmap for the next 15 years.
(b) To determine needs and challenges, and to find new opportunities for change.
(c) To encourage people in these aims around the world by conducting seminars, workshops or through IEEE conferences.

1.2.3 The show must go on!

According to many predictions, if the trend continues as per IRDS guidelines, CMOS scaling will no longer be effective. Practical speculations restrict the technology to a node size of 14 nm; and, even before this restriction is reached, there are extreme barriers to overcome. So, what technological options might be used to optimise MOSFETS and elongate the CMOS scaling period? The next section shall provide some solutions to successfully (however theoretically) overcome these constraints and optimise MOSFET structures.

(i) Improvements and technological aspects of MOSFET optimisation
   The optimisation of MOSFET structures to extend the boundaries of CMOS scaling takes a two-faceted approach: (a) use of new materials in existing structures, and (b) use of new structures.
   (a) Use of new materials
      The problems faced by the chips can be handled by adjusting the relevant critical properties. These sensitive properties are low-resistivity conductors, low-k dielectrics and strained silicon. It has been a common practice to use tungsten and copper to minimise device and interconnect resistance. Additionally, metal gate electrodes offer many advantages over doped-polysilicon gates. The polysilicon depletion effect (PDE) of doped-polysilicon gates effectively increases gate oxide thickness at inversion by a couple of nanometres, resulting in degradation of the gate capacitance. For sub-50 nm CMOS nodes, the gate oxide thickness is typically less than 1.5 nm. The PDE is quite dominant in this regime.

Additionally, boron penetration in the thin oxide underneath the doped poly-gate degrades its quality; and further, there is a practical limit on oxide scaling due to plausible gate leakage. A metal gate offers no depletion, very low gate resistance, no boron penetration and compatibility with high-κ. Strained silicon crystal has been the star of technology scaling for high-performance requirements. In this, the silicon lattice is subjected to physical strain (mostly by adding a layer of another lattice having a larger lattice size on top of the silicon) to improve carrier mobility and cut down the device resistance as well as several other critical properties. The disadvantage of this technique is in the fabrication aspect, as it does not fit with many technologies that are new in the market. High-κ dielectrics address the issue of off-state power consumption and more particularly gate leakage due to tunnelling currents. As dielectric thickness ultimately restricts the gate length, the dielectric thickness will be the first to reach atomic dimensions. Practically, the length of the gate has to be 40 times in comparison to the dielectric thickness so that it can properly control the SCEs. The proposed solution is to find and use a material with higher dielectric constant than SiO_2 that will allow us to reduce the effective dielectric thickness without affecting tunnelling [9].

(b) Use of new structures

Proposals for changing the structures of the transistor have also been made. The two main proposed structural modifications are silicon-on-insulator (SOI) and double-gate (subsequently multi-gate and gate-all-around) complementary metal-oxide semiconductors (DGCMOSs). Normally, a transistor is fabricated by connecting its body to a substrate, but in SOI we first bury an oxide on top of the substrate and the transistor is fabricated on its pinnacle. This isolates the frame electrically from its environment and a positive body bias exists (for nMOS, of course), reducing the device threshold voltage and increasing performance. An insulating layer is used to achieve a highly resistant element with zero junction area capacitance, and no reverse body effect by stacked circuits. Schematics of a bulk MOSFET and a SOI MOSFET are shown in figure 1.9.

DGCMOSs utilise the fundamental concept of adding an extra gate to boost coupling among the gate and the channel. This scheme has been hailed as the 'perfect structure for scalability'. A cross section of a DG-SOI MOSFET is shown in figure 1.10 [9].

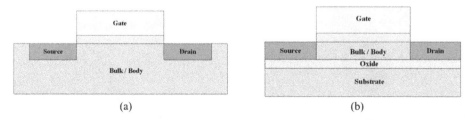

(a) (b)

Figure 1.9. FET schematics: (a) bulk, (b) SOI.

The principal objectives of DG-SOI MOSFETs are to reduce SCEs while significantly maintaining good electrical characteristics. Less gate leakage and sub-threshold current and high ON current are the benefits conferred by DG-MOSFETs; however, realising their structure poses some difficulties. When using traditional fabrication techniques, adding a second gate below the device's body results in troublesome alignment issues. This demands a complex process, higher gate capacitance and higher source to drain series resistance. The abovementioned problems are addressed by using fin-shaped FETs (or FinFETs). Working beyond the 45 nm node, FinFETs have done wonders. These are categorically different from planar MOSFETs: they offer low leakage with high driving capability, supply voltage scaling and increased intrinsic gain. A three-dimensional profile of a FinFET is illustrated in figure 1.11.

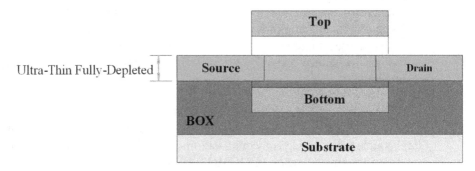

Figure 1.10. Double-gate SOI MOSFET.

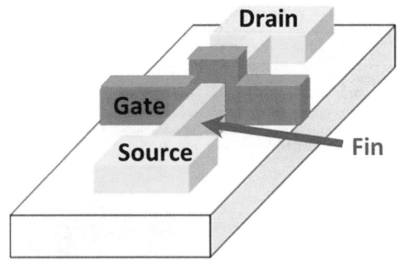

Figure 1.11. FinFET structure. (This Doublegate FinFET.PNG image has been obtained by the author(s) from the Wikimedia website where it was made available under a CC BY-SA 3.0 licence. It is included within this book on that basis. It is attributed to Irene Ringworm.)

FinFETs demonstrate high parasitic resistance and quantised device width. An important point to note is that SCEs can be controlled only if the width of the body is taken to be a quarter of the gate length. As the gate length is the smallest dimension fabricated, it is a difficult challenge to overcome. Although FinFETs have proved their worth, research in this area is still in progress. Better gate control can be obtained in further variants of FinFETs such as multi-gate or gate-all-around MOSFETs.

(ii) Novel devices

Novel and emerging logic devices that extend MOSFETs to the end of the roadmap, in addition to unconventional FETs and non-FETs are illustrated in figure 1.12 [11].

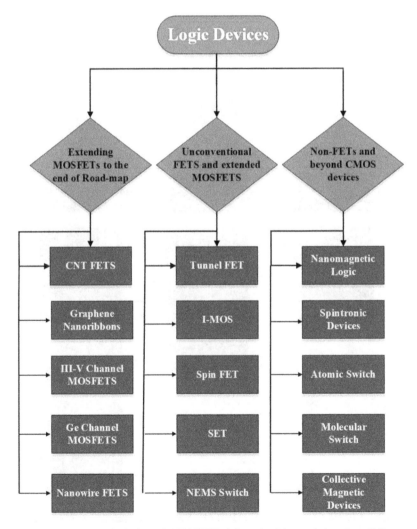

Figure 1.12. Logic devices. © 2021 IEEE. Adapted, with permission, from [11].

On similar grounds, memory devices are categorised into three classes: baseline, prototypical and emerging. These are illustrated in figure 1.13 [11].

A comparison of the major emerging devices based on their traits is briefly outlined in table 1.4.

In summary, the emergence of novel nano-regime devices and advancements in fabrication technology down to such a tiny scale are saviours for the semiconductor

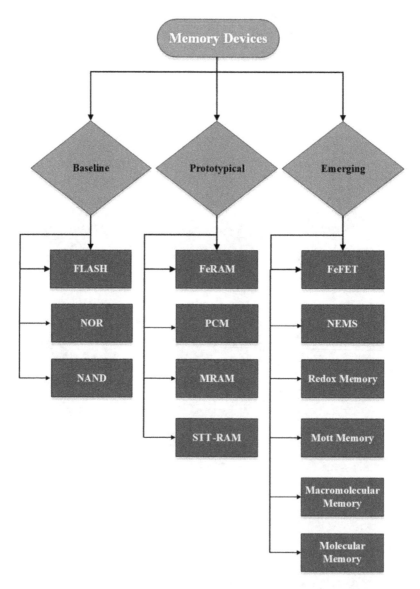

Figure 1.13. Memory devices. © 2021 IEEE. Adapted, with permission, from [11].

Table 1.4. Comparison of different device technologies [11].

Sr No.	Technology	Pluses	Minuses
1.	CNTs/graphene/ nanotubes	High conductivity	Fabrication issues
2.	Ge channel devices	High mobility, high performance	Carrier scattering
3.	Nanowire devices	High aspect ratio leading to better channel control, remarkable optical properties	Poor current drive
4.	RTD/tunnel FET	Apt for RF applications	Device matching across the wafer
5.	Spintronics	Scalability, low power, high performance	Perfection in spin control mechanism
6.	Single-electron transistors	Ultra-low power	Room-temperature operation, lower drive current
7.	Fe-RAM	Does not need charge pump, guaranteed data reliability	Limited performance
8.	PCM	Fast switching, scalability	Thermal sensitivity, fabrication
9.	MRAM	Refresh not required	Less packing density
10.	Fe FET	Low voltage operation	Material choice, leakage, charge injection
11.	NEMS	Efficient, small and cheaper systems	Reliability and long-term stability
12.	Mott memory	Fast switching, ultra-low power operation	Instability
13.	Molecular electronics	Can overcome long interconnect issues	Stability problem

industry. Though complete replacement of CMOS technology is a far-fetched proposal, newer technologies can coexist with the existing technology and the lifetime of CMOS can be extended. With newer materials, devices, instruments and processes, it is possible to commence on the development of exclusive, smart, portable, power/energy efficient, high-performance and cheaper products.

Questions

1. State Moore's law. Does it hold in today's age?
2. What is the ITRS? State its significance.
3. State the different types of scaling. Explain the physics behind them along with their applications.
4. What are the limitations of CMOS scaling? Explain briefly.
5. What is the future of microelectronics/very-large-scale integration (VLSI) technology?

References

[1] Sze S M 2002 *Semiconductor Devices, Physics and Technology* 2nd edn (New York: Wiley)

[2] Streetman B 2015 *Solid State Electronic Devices* (London: Pearson) 7th edn

[3] Kang and Leblebici 2003 *CMOS Digital Integrated Circuits: Analysis and Design* 3rd edn (New Delhi: Tata McGraw-Hill)

[4] *2005 International Technology Roadmap for Semiconductors (ITRS) by Semiconductor Industry Association* 2005 (https://semiconductors.org/resources/2005-international-technology-roadmap-for-semiconductors-itrs)

[5] Haron N Z and Hamdioui S 2008 Why is CMOS scaling coming to an END? *3rd Int. Design and Test Workshop, Monastir* pp 98–103

[6] Iwai H 2004 CMOS Scaling for sub-90 nm to sub-l0 nm *Proc. 17th Int. Conf. on VLSI Design (VLSID04)* pp 30–5

[7] Rairigh D 2005 *Limits of CMOS Technology Scaling and Technologies Beyond-CMOS* (Piscataway, NJ: IEEE)

[8] Ferguson J 2017 Assessing the true cost of node transitions (https://techdesignforums.com/practice/technique/assessing-the-true-cost-of-node-transitions)

[9] *Introduction to International Technology Roadmap for Semiconductor 2.0* 2015 (http://itrs2.net)

[10] Hiraki K 2012 *Speculative Aspects of High-Speed Processor Design* (Tokyo: The University of Tokyo)

[11] IEEE 2021 *International Roadmap for Devices and Systems (IRDS)* (Piscataway, NJ: IEEE) (irds.ieee.org)

IOP Publishing

Nanoelectronics
Physics, technology and applications
Rutu Parekh and Rasika Dhavse

Chapter 2

Introduction and overview of nanoelectronics

Nanoelectronics is a king-sized topic about miniaturised things. It deals with the precise manipulation of ultra-small electronic components/systems to provide compact, power-efficient, fast, accurate and inexpensive technological solutions to various complex functionalities. Nanoelectronics is exciting, inspiring and challenging. In this chapter, we present a bird's-eye view of nanoelectronics. A short history, its revolution and continued evolution, and some market requirements are summarised. Nanoscale fabrication and its two approaches, top-down and bottom-up, are also introduced.

2.1 Introduction

The concept of nanotechnology was first coined by Richard P Feynman at the American Physical Society in 1959. From his famous presentation 'There's Plenty of Room at the Bottom', he suggested that by manipulating individual atoms we can make materials of exciting and desired properties [1]. This laid the foundations for the emergence of nanotechnology and nanomaterials. Professor Norio Taniguchi, of Tokyo University of Science, first developed a scanning tunnelling microscope (STM) that could examine atoms at the nanoscale (ranging from 1 to 100 nm) and opened up the gates to the nano world [2]. Gradually, increasing demand for ever more miniaturised materials in technological fields led to the evolution of nanotechnology.

2.1.1 Nanotechnology

A nanometre is one billionth of a metre. To give some idea of its scale, the thickness of paper is about a hundred thousand nanometres; the diameter of a single gold atom is about a third of a nanometre. Objects whose dimensions are between 1 and 100 nm are known as nanoscale objects.

Nanotechnology and nanoscience involve the use of these incredibly tiny objects across various science fields, e.g. physics, chemistry, biology, materials science and

doi:10.1088/978-0-7503-4811-9ch2

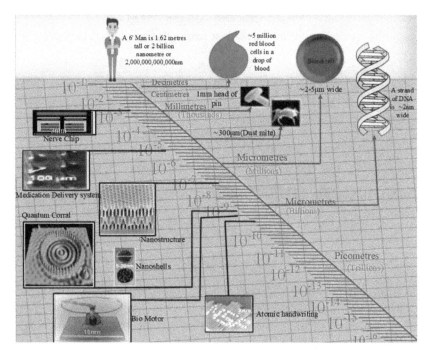

Figure 2.1. Nanoscale and nanotechnology size comparison. Copyright © 2011, James F. Leary, All rights reserved.

engineering. Various nanoscale biological, physical and chemical systems when observed exhibit totally different properties than those of their bulk counterparts. Figure 2.1 represents a size comparison of nanoscopic materials with bulk counterparts.

The following provides some reasons why nanotechnology has changed the whole dynamics of materials science:

(i) Nanoscale technology offers novel (as well as some unexpected) properties.
(ii) At the nanoscale we can design a material to have specific properties as per our requirements, rather than its normal behaviour at the microscale or macroscale.
(iii) Small materials are constantly moving so that they can arrange themselves to form interesting and useful objects, like computer chips, or solutions, such as in curing cancer.

Nanotechnology is constantly evolving, expanding and giving birth to various new technological applications. Its essence lies in the development of creative fabrication processes for novel devices, and the formulation of newer materials and chemicals. Current-generation equipment can be replaced with better resolution, more power efficient and increasingly compact ones using nanotechnology. We can also derive benefits in terms of the lower consumption of materials and energy. Figure 2.2 illustrates the application fields of nanotechnology.

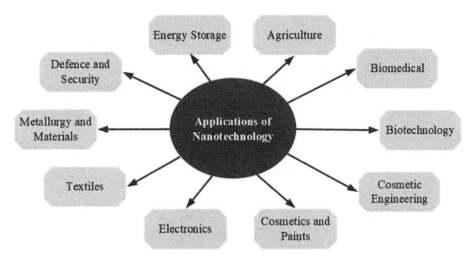

Figure 2.2. Nanotechnology application and reference to nanoelectronics.

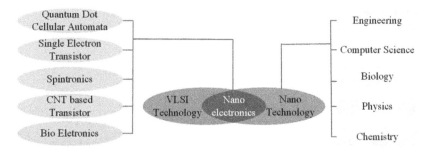

Figure 2.3. Overlap in application of nanotechnology and nanoelectronics with conventional very-large-scale integration (VLSI) technology.

2.1.2 Nanoelectronics

The use of electronic devices has become continuously prevalent over the past few decades. The need for miniaturising these devices has led to the inclusion of nanotechnology in the electronics field, as well. The resulting field developed from the combination of nanotechnology and electronics is known as 'nanoelectronics'.

Figure 2.3 shows the overlap of nanotechnology and conventional very-large-scale integration (or VLSI) technology, indicating the emergence of nanoelectronics. It deals with the integration of purely electronic devices and circuits at miniaturised scale. As previously defined, these nanodevices are often only a few nanometres in size (1–100 nm). Nanoelectronic components have different properties than non-nanoelectronic components due to device-level quantum-mechanical effects and interatomic interactions. At the nanoscale, manipulation of 'other-than-charge' electron properties can be carried out to obtain a desired functionality. The use of nanodevices in memory chips (data storage), logic chips, sensors, transducers, displays, molecular electronics, wearable electronics and many other electronic systems can lead to a brighter future because of some of the extraordinary properties of these nanodevices.

In 1947, the first transistors were built; their size was around 1 cm. Now, the smallest prototyped transistor is 7 nm long—1.4 million times smaller than the first transistor. Because of 7 nm transistor manufacturing techniques, today we can make processors in which there are billions of transistors, with these processors further being integrated on a small single footprint.

A brief account of selected developments that eventually led to nanoelectronic devices over the last few decades is given in table 2.1.

Table 2.1. Revolutionary developments and continued evolution in nanoelectronics.

S. No.	Year	Milestone	Remarks
1	1925	First use of the term 'nanometre' in the study of gold particles with sizes down to 10 nm (nanocrystals)	By Richard Adolf Zsigmondy, a Hungarian chemist known for his works in colloids
2	1959	'There's Plenty of Room at the Bottom'	Famous talk given by Richard Feynman
3	1959	First silicon pressure sensor demonstrated	By Kulite
4	1960	Nanolayer-based metal-semiconductor transistors first demonstrated	Initially proposed and demonstrated by A Rose, and later in 1962 by L Geppert, M Atalla and D Kahng
5	1967	Anisotropic deep silicon etching	By H A Waggener *et al*
6	1967	The concept of a double-gate thin-film transistor (TFT) proposed	By H R Farrah (Bendix Corporation) and R F Steinberg
7	1968	Resonant gate transistor patented	In 1964, H Nathanson from Westinghouse produced the first batch-fabricated microelectromechanical (MEMS) device by using a surface micromachining process
8	1970s	Bulk etched silicon wafers used as pressure sensors	By using a bulk micromachining process
9	1974	Atomic layer deposition	Developed and patented by T Suntola and co-workers in Finland
10	1979	Micromachined ink-jet nozzle	By HP, using a bulk micromachining process.

11	1981	Invention of scanning tunnelling microscope	By G Binnig and H Rohrer at IBM Zurich Research Laboratory
12	1982	Silicon used as a structural MEMS material	By K Petersen
13	1982	LIGA process invented	By KFK, Germany
14	1982	Disposable blood pressure transducer	By Honeywell
15	1983	Integrated pressure sensor	By Honeywell
16	1983	Infinitesimal machinery	Another famous talk by Richard Feynman
17	1985	Sensonor crash sensor introduced	Popularised in airbags nowadays
18	1985	Discovery of the 'Buckyball' fullerene	By H Kroto, R Smalley and R Curl
19	1986	Invention of the atomic force microscope	Application of nanoelectromechanical (NEMS) devices demonstrated by IBM
20	1987	First 10 nm MOSFET using tungsten-gate technology	Demonstrated by an Iranian engineer, Bijan Davari, and his team in IBM-led research
21	1988	Batch-fabricated pressure sensors via silicon wafer bonding	By NovaSensor
22	1989	First FinFET transistor, also known as a 'DELTA' transistor (depleted lean-channel transistor), demonstrated	Developed by the Hitachi Central Research Laboratory, which later became a basis for FinFETs
23	1991	Discovery of carbon nanotubes (CNTs)	By S Iijima of NEC using an arc-discharge technique
24	1996	Mass production of CNTs of uniform diameter	R Smalley developed technique based on laser ablation
25	1998	Demonstration of CNTFET	By Dekker, Tans, Verschueren and Alwyn
26	1998	17 nm n-channel FinFET devices fabricated successfully	By Hisamoto of Berkley along with TSMC's C Hu
27	1999	P-channel FinFET (sub-50 nm) and FinFET-based CMOS technology developed	By Hisamoto of Berkley along with TSMC's C Hu
28	2001	15 nm FinFET process development	By Hisamoto of Berkley along with TSMC's C Hu
29	2002	Development of 10 nm FinFET process	By Yu and Chang of Berkley along with TSMC's C Hu

(*Continued*)

Table 2.1. (*Continued*)

S. No.	Year	Milestone	Remarks
30	2004	High-κ/metal-gate FinFET invented	By A Agarwal and M Ameen
31	2004	Investigation of graphene	By A Geim and K Novoselov at the University of Manchester
32	2006	3 nm gate-all-around (GAA) FinFET technology developed	Thanks to Korean researchers from the Korea Advanced Institute of Science and Technology (KAIST) and the National Nano Fab Center
33	2010	3D ICs	Used widely in NAND flash memory and mobile phones
34	2013	Commening of commercial production of 10 nm FinFET-based semiconductor chips	By Samsung
35	2019	Extreme-UV lithography with a resolution of 5 nm	By TSMC
36	2010–	2020	Proliferation of BioMEMS, NEMS and optical MEMS
37	2021	Commercialisation of 3 nm GAAFET process.	By Samsung

These inventions expanded Moore's law, as illustrated in figure 2.4. In this figure, 'More Moore' refers to the expansion of the CMOS platform by conventional dimensional and functional scaling. This CMOS platform can be further extended by a 'More than Moore' approach, where diverse functionalities like sensors, interfacing circuits, power-management modules, etc., can be combined with 'More Moore' components. 'Beyond CMOS' technologies refer to new information processing devices and architectures. The heterogeneous integration of Beyond CMOS (new technologies) and More than Moore (existing technologies) into conventional More Moore scaling will extend the CMOS platform's functionality to form ultimately an 'Extended CMOS'.

2.2 Market requirements for nanoelectronics

For a very long time, electronics has been used to dealing with electrons. Nanoelectronics is playing a role whereby electronics will now deal with the single electron. Smaller dimensions and new physical mechanisms are giving rise to an unparalleled market for nanoelectronic devices and gadgets. Right from materials to devices, devices to circuits, circuits to standard cells, standard cells to modules, modules to systems: nanoelectronics can do wonders. As well as conventional electron charge, many properties such as the magnetic properties of electrons, spin

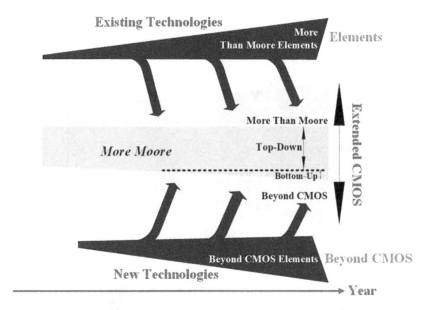

Figure 2.4. Evolution of so-called Extended CMOS. Adapted from [3] with permission.

and orbit properties, molecular orientations, optical properties, etc., can be exploited at the nanoscale. It is no surprise that market interest has risen for novel nanomaterial-based applications in sensors, low-power and low-energy systems, portable consumer electronics, remote operating gadgets, high-performance computing, Internet-of-Things components, biomedical systems, in addition to expanding transmission speed between integrated circuits (ICs), reducing power utilisation, and so on. Furthermore, nanoelectronics innovation is important in fabricating new gadgets, organising engineering and structuring new assembly procedures with minimal effort and time.

Nanoelectronics has a huge market scope for novel transistors, ICs, memory devices, mobile wireless devices, displays (e.g. LEDs), wearable health monitoring, waterproof electronic coatings, spintronics, optoelectronics, molecular electronics, as well as applications in energy, etc [4]. Figure 2.5 indicates typical Beyond CMOS nanoelectronic devices that can change the course of progress in the nanoelectronic industry.

Many of these devices are still in their infancy, and it will take a couple of decades for the technology to mature. Nonetheless, they have already demonstrated their worth, and the semiconductor industry is eagerly anticipating them as the drivers of the future.

2.3 Nanofabrication

Nanofabrication can be seen as a technology for making systems, components and/or devices in the nanometre range. In this, there is no difference in the definition to distinguish it from previous microfabrication technology. Yet, a typical Pentium chip occupies around 1 square inch of area and comprises approximately 50 million

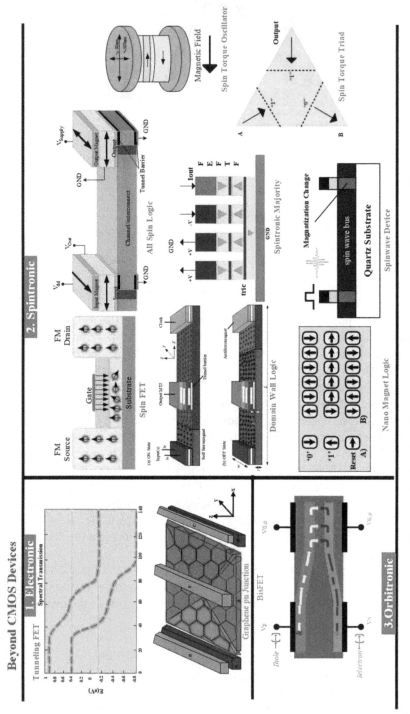

Figure 2.5. Devices determining the market in nanoelectronics. Adapted from [5].

transistors. In comparison, an Intel i7 processor has 1.4 billion transistors in around the same area. Now, such i7 processors utilise a 22 nm transistor. Future nano-electronics requires something much smaller than this. It is thus obvious that new techniques, new tools and newer manufacturing technologies are needed to fabricate objects at the nanoscale. Accordingly, top-down and bottom-up forms, as shown in figure 2.6, compose the two chief nanofabrication techniques [6]. In the top-down approach of nanofabrication, as the name suggests, we build nanomaterials/nano-devices from the big bulk state of the materials by cutting, etching, milling, chipping, etc. This is achieved by scaling down big raw substrates to desired nanomaterials, thus the naming 'top-down'. Similarly, in the bottom-up approach, just as the name suggests, we build the nanomaterial atom by atom, gradually constructing from the bottom-up the desired nanomaterial.

(i) Top-down approach

Mostly, the top-down approach depends on lithography. In this most common patterning practice the bulk material is taken and the important needed part of it is protected by a mask while its exposed part is removed by etching. The many available lithography techniques variously involve photo-lithography, immersion lithography, lithography involving e-beams and X-rays, as well as other techniques that include nano-imprints, scanning probes, etc.

(ii) Bottom-up approach

Bottom-up manufacturing uses physical and chemical procedures at a nanoscale level to arrange basic units into bigger structures. The bottom-up approach imitates natural biological or chemical processes, wherein

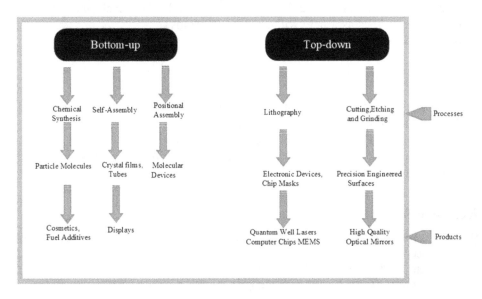

Figure 2.6. The two main approaches to nanofabrication.

individual atoms heap up in a steady progression on a substrate to frame atoms. The most typical methods are self-assembly methods like chemical vapour deposition, atomic layer deposition, vapour phase reactions, physical deposition, molecular organic deposition, probe microscopy, etc.

However, these techniques are resolution sensitive and do not guarantee consistency in production. Exhaustive technology computer-aided design (or TCAD) and Monte Carlo simulations are required to finalise the process steps. Fabrication plants (colloquially known in the industry as 'fabs') are a bit reluctant to adopt a technology without having substantial and promising proofs regarding the functionality and reliability of nanoelectronic devices and processes. A noteworthy point is that many industries need to ensure adequate cash flow for funding of research and development in this field. However, governments should also understand that, currently, the semiconductor industry is going through a giant transition phase. And this phase requires huge amounts of capital. Billions of dollars' worth of funds are required in order to ethically explore the effects of nanoelectronic materials and devices on the environment, human beings, the economy and wider society.

In summary, nanoelectronics has tremendous influence on our health, wealth and growth. A computer processor built using nanoelectronics will be more powerful than today's processors that are built using semiconductor fabrication techniques. Nanoelectronics constitute a paradigm-shifting phenomenon. It rests on a rock-solid foundation of theoretical and practical scientific breakthroughs, and promises countless new possibilities for realising processes of small structures. Scientists and researchers will always look for novel emerging ideas; and fabs will always evaluate them from a mass-production point of view. Thus, there will always be a conflict between idealistic researchers and market-orthodox fabs. However, one thing can be sure: nanoelectronics as a field will emerge victorious.

It has to, doesn't it?

Questions

1. What is nanotechnology? State its implications for electronics.
2. Explain (a) nanoelectronics, (b) nanofabrication.
3. List some examples of nanodevices and their significance.
4. State some of the applications of nanoelectronics that have no counterpart in present technology.

References

[1] Feynman R P 1992 There's plenty of room at the bottom [data storage] *J. Microelectromech. Syst.* **1** 60–6
[2] Taniguchi N 1974 On the basic concept of 'Nano-Technology' *Proc. Int. Conf. Proc. Eng. Tokyo, Part II* (Japan Society of Precision Engineering)

[3] Semiconductor Industry Association 2011 2011 International Technology Roadmap for Semiconductors (ITRS) (https://www.semiconductors.org/resources/2011-international-technology-roadmap-for-semiconductors-itrs/)

[4] Zhang G Q and van Roosmalen A J 2009 The changing landscape of micro/nanoelectronics ed G Zhang and A Roosmalen *More than Moore* (Boston, MA: Springer)

[5] Belthangady C 2014 *Quantum-assisted magnetometry with NV centers in diamond* https://nanohub.org/resources/20560

[6] Hanson G W 2009 *Fundamentals of Nanoelectronics* (New York: Pearson Education) 1st edn

IOP Publishing

Nanoelectronics
Physics, technology and applications
Rutu Parekh and Rasika Dhavse

Chapter 3

Introduction to the quantum theory of solids

Present-day classical mechanics is the outcome of numerous developments from experiment and reasoning over centuries. From a pin to a satellite, every object can be clearly described with the help of classical mechanics, which was formalised and presented by Isaac Newton in 1686. Newton described what he saw. Hence, in classical mechanics, waves and particles are understood as two completely distinct entities. A wave is continuous and extends in space, whereas a particle is discrete and has very little or no extent in space. One can know the present state and predict the future of an object with utmost precision, making classical mechanics a deterministic theory. One can easily 'observe' the laws of classical mechanics and describe them mathematically. In this way, everyday practical experiences match with the theoretical concepts.

Howsoever established it may be, classical mechanics ceases to apply when one discusses atomic and subatomic particles. Newtonian laws fail to explain quantum phenomena such as electron behaviour, atomic structure, atomic spectra, thermal radiation, the photoelectric effect, Young's double-slit experiment, etc. A broader and more universal concept of quantum mechanics was thus devised in the early 20th century, when scientists focused their efforts on reworking physics for ultra-small entities. One such phenomenon at the ultra-small scale, wave–particle duality, is an immensely significant notion in quantum mechanics. In terms of scope, quantum mechanics encompasses classical mechanics. Classical mechanics is applicable to bigger and top-level entities, whereas quantum mechanics is built from the bottom up. In a phrase, you would use Newtonian physics to describe a basketball and quantum physics to describe a Buckyball!

This chapter is hence dedicated to the quantum nature of matter.

doi:10.1088/978-0-7503-4811-9ch3

3.1 Classical particles, classical waves and quantum particles

3.1.1 Classical particles

A classical particle [1] can be defined as a point like object such as a ball or a bullet, of mass m and position $r(t)$ in a three-dimensional system, at some given time t. Here, the classical particle obeys classical Newtonian mechanics, which are based on the Newton's laws of motion, dependent on the forces acting on the particle and the motion of the particle. The classical physics for a classical particle can be identified if at a certain time t we can precisely express the velocity of particle and position of the particle:

$$r(t) = a_x x(t) + a_y y(t) + a_z z(t). \tag{3.1}$$

Equation (3.1) defines the position of the particle in three-dimensional space, and its first and second derivatives yield its velocity, $\vec{v}$, and the acceleration, $\vec{a}$, at a certain time, t. According to Newton's second law, the force acting on the particle is decided by its momentum, $\vec{P}$, as per equation (3.2):

$$\vec{F} = \frac{d\vec{P}}{dt}. \tag{3.2}$$

Also, we know that the force is

$$\vec{F} = m\vec{a}. \tag{3.3}$$

Comparing equations (3.2) and (3.3), we get

$$\frac{d\vec{P}}{dt} = m\vec{a}. \tag{3.4}$$

Now, acceleration is the first derivative of velocity. So,

$$\frac{d\vec{v}}{dt} = \vec{a}. \tag{3.5}$$

Combining equations (3.4) and (3.5), we get

$$\frac{d\vec{P}}{dt} = m\frac{d\vec{v}}{dt}. \tag{3.6}$$

Since the mass m is constant, we can write

$$\frac{d\vec{P}}{dt} = \frac{d(m\vec{v})}{dt}. \tag{3.7}$$

So, it is apparent that the momentum of a classical particle is

$$\vec{P} = m\vec{v}. \tag{3.8}$$

3.1.2 Energy of classical free particles

In physics, a classical free particle is a particle that is present in a region where the potential energy does not vary or is not bound by any external force. Without the loss of generality, we can arbitrarily set the potential energy for such region to zero [1]. According to the law of conservation of energy, the total energy of such a particle is the summation of its potential energy and kinetic energy [2]. Yet as mentioned, here the potential energy is arbitrarily set to zero, so we have

$$\text{Total energy} = \text{kinetic energy} = \frac{1}{2}mv^2. \tag{3.9}$$

Using equations (3.8) and (3.9):

$$\text{Total energy of a free classical particle} = \frac{p^2}{2m}. \tag{3.10}$$

3.1.3 Classical waves

Classical waves are the manifestation of the propagation of radiation. Waves have a fundamental importance in almost every branch of physics. For example, light and sound energy propagates in the form of waves. Waves carry energy from one location to another. A simple example of a wave is when we drop a stone in a still pond of water: the consequent disturbance in the water spreading in all directions travels in the form of waves [1]. Thus, we can define a wave as a disturbance in some physical system that is periodic in nature (i.e. repeats itself within some fixed interval) in both space and time. According to classical wave theory:

(a) The wave intensity is equal to the energy incident per unit area per unit time.
(b) The energy conveyed in a wave is proportional to the square of its amplitude.

3.1.4 Mathematical interpretation of a wave

In classical physics, the most elementary one-dimensional wave is a harmonic function, as shown in figure 3.1, with amplitude A [1].

The position of any point in the wave at a distance x at some given time t is given by $y(x, t)$, as

$$y(x, t) = A \cos(\omega t - kx), \tag{3.11}$$

where

- A stands for amplitude, where the amplitude is defined as the maximum distance between the highest point of disturbance and the equilibrium point of the wave;
- k stands for wave number;
- X represents the direction in which the wave is travelling;
- t is the coordinate for time;

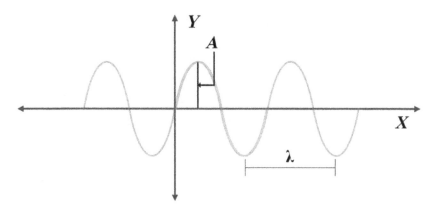

Figure 3.1. A simple harmonic wave.

- ω stands for angular frequency; and
- λ stands for wavelength.

A classical wave obeys the classical Newtonian laws of motion. We can transform the wave equation (3.11) to a more general equation by adding an imaginary term as follows:

$$y(x,\ t) = A \cos (\omega t - kx) - iA \sin (\omega t - kx).$$
$$\text{Then, using } e^{-i\theta} = \cos (\theta) - i \sin(\theta), \qquad (3.12)$$
$$\text{we get } y(x,\ t) = Ae^{-i(\omega t - kx)}.$$

Thus, equation (3.12) represents a one-dimensional wave equation for a wave travelling along the x direction with respect to time t.

3.2 Quantum particles and principles of quantum mechanics

The scientific understanding regarding particles that prevailed in the 18th and early 19th centuries underwent a serious reassessment when various scientific experiments began to show that light can manifest both a particle-like nature (known as photons) as well as a wave-like nature. In the early 19th century, Thomas Young, who had studied diffraction and the interference of light, suggested that light shows a wave-like nature, which supported the wave theory propounded by Christian Huygens and opposed the corpuscular theory of Isaac Newton. Following this, several influential experiments indicated the dual nature exhibited by light, such as the photoelectric effect, blackbody radiation, the dual-slit experiment, Compton scattering of X-rays, Heisenberg's uncertainty principle, etc. Thus arose a new idea, that of wave–particle duality, reflecting that light can act both as a particle and a wave at the quantum scale [2]. Let us look at some of these experimental effects in brief:

(i) Photoelectric effect

The photoelectric effect shows the particle nature of light. When a light is incident on a metal surface, the energy contained within the light gets

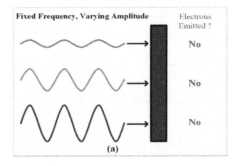 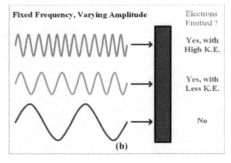

Figure 3.2. Photoelectric effect, the number of electrons emitted when (a) increasing the amplitude, (b) changing the frequency.

transferred to the electrons that are present in the metal, resulting in their emission. However, this transfer of energy happens in discrete packets, which are also known as quanta of energy. Classical physics held that the outermost electrons were weakly bound with the attractive electric force towards the nucleus of an atom. Hence, these electrons could be freed and emitted out when there was an increase in the light intensity.

As shown in figure 3.2, it was observed that no electrons were emitted when the amplitude (intensity) of the incident light was increased; rather, it was dependent on the frequency of light, which had to exceed a threshold frequency. Moreover, the number of electrons increased with the intensity, but the energy was still the same. Thus, classical physics failed to explain this effect.

The solution:

Quantum theory provides an explanation for this problem in the fact that light behaves like a particle. These light particles should have enough energy in order to free the electron and emit it. Further, when we increase the intensity of the incident light, the number of light particles that are emitted increases but there is no increase in the energy of the particles, which gives an explanation of the phenomenon observed earlier. Quantum theory refers to these light particles as **photons,** whose energy is dependent on their frequency, given by [3]

$$E = h\upsilon, \qquad (3.13)$$

where h stands for Planck's constant, and υ stands for the frequency of the incident light. This was contradictory to the assumption of light acting as a wave, an enigma that was solved by Einstein in 1905 by describing light as being composed of said photons rather than a continuous wave. The kinetic energy of the emitted electron is given by

$$KE = E_{\text{photon}} - E_{\text{binding}} = h(\upsilon - \upsilon_0). \qquad (3.14)$$

(ii) Blackbody radiation

Scientists refer to a blackbody as a material that emits light on heating, absorbs all the light incident on it and radiates nothing. An approximation of a blackbody can be given as a pure absorber of electromagnetic radiation that does not depend on the frequency or the angle of the incident wave, such as in the case of a hole that when light is shone into it, the light undergoes continuous internal reflection till it is absorbed or a small amount escapes. On heating, the hole would glow because of the acceleration and deceleration of electrons on the surface of the material, which are perturbed thermally [4]. This body has homogeneous and isotropic radiation; an illustration is shown in figure 3.3.

When a metal object is heated it emits thermal radiation due to the high temperature. The yellow glow visible to the human eye is because the frequency of radiation emitted falls in the visible light spectrum. Extending this principle, it is then true to say that everything in the Universe emits thermal radiation, however in a different intensity range, of a larger wavelength that cannot be detected by the human eye. An infrared camera can be used to detect light of a higher wavelength, falling in the infrared range, and so can be used to detect such thermal radiation.

Objects emit electromagnetic radiation when they are heated, and this radiation depends on the temperature of the object. As shown in figure 3.4, if the temperature of the object is comparatively low, like in the example of a horseshoe forged by a blacksmith, it appears red in colour, while in an object with a much higher temperature, like the Sun, it appears yellow or white in colour.

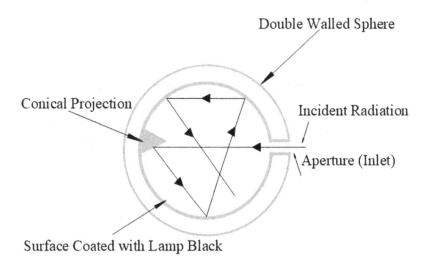

Figure 3.3. A blackbody consisting of a double-wall cavity that is blackened inside.

Figure 3.4. The first image shows a horseshoe at low temperature, while the second image shows the surface of the Sun at higher temperature. (Left: This Hot Horseshoe (stevefe).jpg image has been obtained by the author(s) from the Wikimedia website where it was made available under a CC BY 2.0 licence. It is included within this book on that basis. It is attributed to Steve Ford Elliott. Right: © Getty Images/iStockphoto/ Thinkstock Images.)

There are two laws relating to blackbody radiation. The **Stefan–Boltzmann law** states that the intensity of the radiation emitted by a body is independent of the nature of the body and depends only on the temperature (and varies as T^4). **Wien's displacement law** states that the temperature of a body depends on the frequency of the emitted radiation. The relation of the temperature and frequency is given as $v_{max} \propto T$, where the proportionality constant is 5.789×10^{10} Hz K^{-1}. Lord Rayleigh and J H Jeans introduced an equation for the distribution of radiation in a classical blackbody. The relation of the frequency with the temperature of the body was given as [5]

$$d\rho(v, T) = \rho_v(T)dv = \frac{8\pi K_B T}{c^3}v^2 dv. \qquad (3.15)$$

Experiments were performed to verify this Rayleigh–Jeans law, and it was found that the experimental data deviated from the expected result. Figure 3.5 shows spectra for the light emitted by a blackbody at 3000 K, 4000 K and 5000 K. We can see that, during relatively low temperatures, the radiation that is released is mostly of a wavelength that is longer than 700 nm, which lies in the infrared spectrum. As the temperature of the body increases, the peak frequency changes to a shorter wavelength. According to the given law, the frequency should start to diverge at higher temperatures; but experiment resulted in the convergence of the frequency with increment of the temperature. This behaviour is known as the ultraviolet catastrophe.

The solution:

The main reason for this failure of the Rayleigh–Jeans law is that the energy is taken to be a continuous variable, and that it can take any

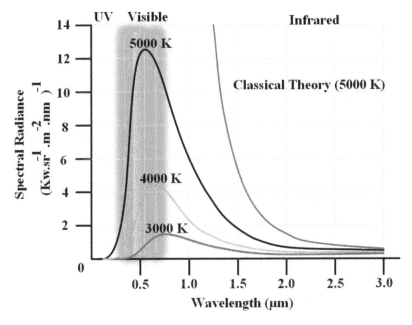

Figure 3.5. Relation between spectral radiation and corresponding emitted wavelength. (This Black body.svg has been obtained by the author(s) from the Wikimedia website, where it is stated to have been released into the public domain. It is included within this book on that basis.)

continuous value. Max Planck explained this apparent error in the experimental behaviour of energy by describing energy rather as a form of discrete packets, whereas Rayleigh and Jean had described it as continuous. Planck, then, quantised the energy in the form $E = nhv$, where n is an integer and v is the frequency. Planck explained the emission of the electron in which energy is described as a packet of hv. Planck's distribution is given as follows [5]:

$$d\rho(v,\ T) = \rho_\lambda(T)d\lambda = \frac{2hc^2}{\lambda^5}\frac{1}{e^{\frac{hc}{\lambda K_{BT}}} - 1}d\lambda. \tag{3.16}$$

(iii) Dual-slit experiment

Light interference when it passes through a screen with several slits can only be explained by the wave nature of the light. This can be satisfyingly adduced by the famous dual-slit experiment. In a dual-slit experiment [1, 2], a monochromatic light is allowed to be incident on a detector screen through two slits. When the detector output is observed over time, the resulting intensity pattern shows the interference effect, as depicted in figure 3.6(b). It was concluded that the observed pattern corresponds to the pattern expected based on wave theory. This observation showed that photons are not classical particles, as thought earlier and depicted in figure 3.6(a). This is a significant substantiation of the incompleteness of classical physics.

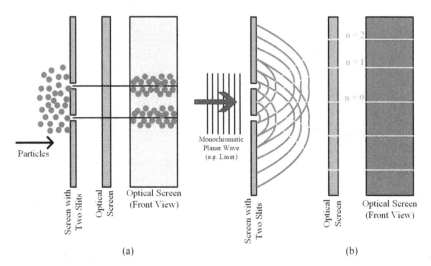

Figure 3.6. Dual-slit experiment: (a) light in the form of a particle, (b) light in the form of a wave. (These Two-Slit Experiment Light.svg and Two-Slit Experiment Particles.svg images have been obtained by the author(s) from the Wikimedia website, where it is stated to have been released into the public domain. It is included within this book on that basis.)

(iv) Compton scattering of X-rays

Compton scattering is a phenomenon wherein electrons collide inelastically with photons. The difference of the wavelength of the reflected photon (λ') and the incident photon (λ) is given by a formula, which is known as a Compton shift [1, 2]:

$$\lambda' - \lambda = \frac{h}{mc}(1 - \cos\theta). \tag{3.17}$$

The fundamental difference between Compton scattering and the photo-electric effect is that there is no scattered reflected photon in the latter, as shown in figure 3.7. In the photoelectric effect, the energy of the incident photon is absorbed completely, while in the case of Compton scattering, the energy of the incoming photon is absorbed only partially, with the remaining energy transformed to a scattered photon.

(v) Heisenberg's uncertainty principle

Another experiment, conducted by Werner Heisenberg, affirmed the inference of light having wave-like properties. This principle of the experiment relates to the measurement of the position and momentum of a particle with certainty. In this, the light scattered after illuminating a particle is captured by a microscope lens, as shown in figure 3.8 [1, 2].

Due to the wave-like properties of light, the position of the particle can only be estimated with an uncertainty of [1, 2, 7]

$$\Delta x \approx \frac{\lambda}{\sin\alpha}, \tag{3.18}$$

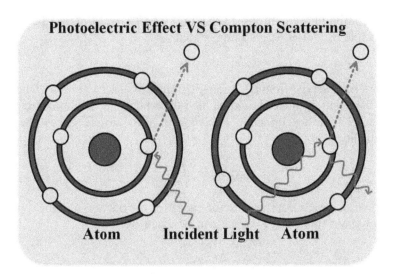

Figure 3.7. Photon emission in the photoelectric effect vs the Compton scattering effect. [6] John Wiley & Sons. Copyright © 2004 WILEY-VCH Verlag GmbH & Co. KGaA.

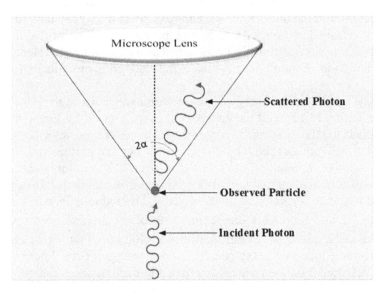

Figure 3.8. Heisenberg's experiment. [7] John Wiley & Sons. © 2013.

where the angle subtended by the lens at the particle is denoted as 2α. The sideways momentum of the scattered photon is given by

$$\Delta p \approx \frac{h}{\lambda}\sin \alpha. \qquad (3.19)$$

On combining equations (3.18) and (3.19), we obtain the uncertainty principle of Heisenberg's fame, stated as

$$\Delta x \, \Delta p \approx h. \tag{3.20}$$

It can be observed from equation (3.20) that in order to attain higher accuracy for the position, there should be high uncertainty in the momentum, and vice versa. The exact principle applying to the particle obeys the following inequality:

$$\Delta x \, \Delta p \geqslant \frac{h}{4\pi}. \tag{3.21}$$

Similarly, we cannot measure the energy of the particle and the time at which it possesses this energy, which can be given by

$$\Delta E \, \Delta t \geqslant \frac{h}{4\pi}. \tag{3.22}$$

From equations (3.21) and (3.22), one can see that, for an atom, the orbit in which the electron resides cannot be determined with absolute precision. However, one can determine the probability that an electron is at a given point around the nucleus. This means that we can only say that an electron is at a certain location with a certain probability. This probability density function of an electron can be modelled as a wave function. Hence, a region with high probability density means the electron is more likely to be there.

De Broglie combined the energy of an electromagnetic wave given by Planck, as shown in equation (3.13), and the energy of a particle given by Einstein ($E = mc^2$, where m stands for the mass of the object, and c stands for the velocity of light). He assumed that these entities show both a wave and particle nature, and from this he derived a wave-like representation of particles, given by $\lambda_e = h/mv$ (where λ_e stands for de Broglie's wavelength, h is Planck's constant, m stands for the mass of the object, and v stands for velocity of the object). The reduced Planck's constant is given by $\hbar = \frac{h}{2\pi}$, and represents the quantisation of the angular momentum of the electron. In 1926, physicist Erwin Schrodinger inserted this formulation by de Broglie into the equation for the conservation of energy (Total Energy = Kinetic Energy + Potential Energy), which describes the behaviour of electrons. We can predict the behaviour of electrons and other quantum particles in various situations using Schrodinger's equation. One can obtain the electronic energy band structure band gap/energy levels and the behaviour of the dopants by solving the Schrodinger equation in momentum space (essentially, Schrodinger's equation in terms of momentum). Further, details of various impurity densities and other crystalline defects can be well described by the Schrodinger equation.

Subatomic particles or quantum particles are classed as particles that are much smaller than the atom. Particles at this scale are further characterised into

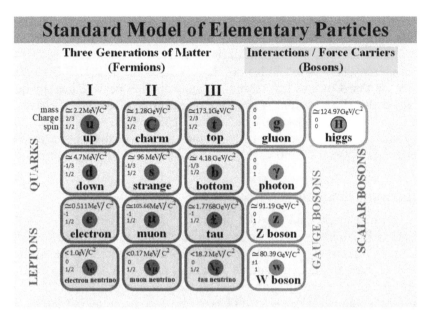

Figure 3.9. Different types of quantum particles. (This Standard Model of Elementary Particles.svg image has been obtained by the author(s) from the Wikimedia website where it was made available under a CC BY 3.0 licence. It is included within this book on that basis. It is attributed to MissMJ.)

elementary particles and composite particles. Figure 3.9 shows the standard model for quantum particles.

Electrons are one of the fundamental, indivisible building blocks of the Universe, and they are the first standard model elementary particle ever to be discovered. Electrons are bound to the atom's nucleus by electromagnetism. The nucleus is made up of protons and neutrons. These protons and neutrons are further made up of small, indivisible particles known as quarks. A proton contains two up quarks and one down quark, while a neutron contains two down quarks and one up quark. A very powerful force, carried by particles known as gluons, is responsible for holding the nucleus together. High-energy experiments have revealed that there are actually six quarks, down and up, strange and charm, and bottom and top. In addition, heavier siblings of the electron, namely muon and tau particles, have been discovered to exist. These heavy particles are only produced for very brief moments in high-energy collisions and are not seen in everyday life because they decay very rapidly into lighter particles. Such decay involves the exchange of weak-force-carrying particles, known as Z and W bosons, which, unlike photons, have mass. There are additional particles in the standard model known as neutrinos, which interact with other particles through weak force only. Matter and antimatter particles are produced in pairs, and they annihilate each other when they meet. Finally, the most talked about particle of the standard model is the so-called Higgs boson: a quantum ripple in the background energy field of the Universe. Interacting with this field is how all the fundamental matter particles acquire mass.

As established, photons and electrons both behave as particles and waves. This cannot be understood with classical mechanics. For this to be comprehended, we have to utilise concepts from quantum mechanics. The quantum world is probabilistic, while the classical world is not (i.e. it is deterministic) [1, 2]. An effect that cannot be explained by classical mechanics and can only be explained by quantum mechanics is called a quantum effect. Due to quantum effects, the behaviour of a particle changes at the nanoscale, which in turn changes materials' optical, thermal and electrical properties. For example, platinum, an inert element at the macroscale, becomes a catalyst at the nanoscale.

3.3 Quantum tunnelling

Let us take an example, and say that, in the classical world, we have a ball at the bottom of a hill. Now, if we want to move this ball over to the other side of the hill, we need to provide enough energy to the ball so that it can climb up the hill and cross its peak. However, in the quantum world, things change drastically.

Let us take another example, and say that there is an electron and there is a thin potential barrier, and that this electron has to cross the potential barrier. Even if the electron does not have enough energy to cross the potential barrier, it can sometimes cross the barrier. This is called quantum tunnelling, and is depicted in figure 3.10. Here, instead of treating the electron as a particle, we treat it as a wave. Thus, when the electron wave collides with the potential barrier, because the wave does not have enough kinetic energy, a certain fraction of the wave is reflected back. If the barrier is thin enough, then the electron can cross over the barrier. Then, there is a very small but non-zero probability that the electron will be present at the other side of the barrier [1, 2]. When we

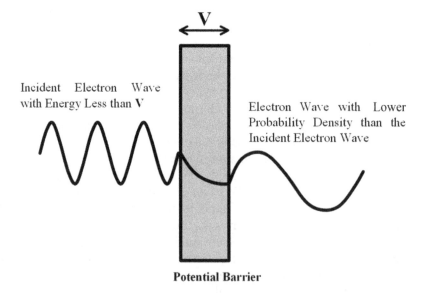

Potential Barrier

Figure 3.10. Electron wave passing through a potential barrier.

solve Schrodinger's equation for the abovementioned scenario, we get an exponentially decaying probability density function on the other side of the barrier.

3.4 Quantum confinement

When materials are miniaturised to a very small scale, such as that comparable to the de Broglie wavelength of an electron, the electron becomes confined to a particular energy level. This phenomenon is called quantum confinement. At the macroscale, the properties of a given material do not change with its size. However, at the nanoscale, due to quantum effects, every variation in size matters. This phenomenon is particularly important for semiconductors. One can tune a given material's band gap in a quantum structure by altering its size. A typical example would be of optical absorption/emission from a semiconductor. Differently sized nanocrystals/quantum dots lead to different energy band gaps even for the same material, as depicted in figure 3.11. This enables absorption/emission of different wavelengths of a solar spectrum, leading to improvement in efficiency.

3.5 Schrodinger's wave equation—meaning, boundary conditions and applications

Let us summarise a couple of concepts in a qualitative manner before visiting Schrodinger's wave equation. Through the prior work of Planck and Einstein, it was proven that energy can be quantised and that light exhibits a wave–particle duality. Following this, de Broglie extended this duality to include matter, as well, meaning that all matter has a wavelength, from a tiny electron to our whole human body, to a massive star. However, an object's wavelength is inversely proportional to its mass,

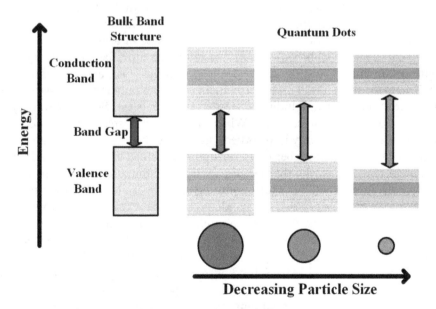

Figure 3.11. Effect of particle size on band gap energy.

so objects larger than the molecular scale have a wavelength that is so tiny as to be completely negligible.

Yet an electron is incredibly small, so small that its wavelength is indeed relevant, being around the size of an atom, so accordingly we will treat electrons as both particles and waves from now on [1, 2]. Therefore, we must next discuss the wave nature of the electron. So, what kind of wave might this be? We can regard an electron in an atom as a standing wave, just like the kind we learned about in classical physics, except that rather than being something like a plucked guitar string, an electron is a circular standing wave surrounding the nucleus. If we understand this, it becomes immediately apparent why the quantisation of energy applies to the electron, because any circular standing wave must have an integer number of wavelengths in order to exist; further, it is given that an increasing number of wavelengths means more energy carried by the wave.

Due to these wave characteristics of an electron, we will deal with uncertainty about its position. If we see it as a wave, then the electron could be anywhere within that wave. This kind of behaviour we will define in terms of probability and wave function. The wave function for any electron or subatomic particle is not as important as the square magnitude of the wave function because it depicts the probability of having an electron or subatomic particle at that place. In quantum physics, a vector in a cube system (also recognised as an endless prospective well or an endless rectangular well) defines a free electron to travel in a tiny room encircled by impenetrable obstacles. For instance, in conventional schemes a particle caught inside a big container can migrate within the cabinet at any velocity and is no more probable to be discovered in one place than another. Similarly, it can never have zero energy, so the particle can never be at a 'standstill'. Moreover, based on its energy level, it is more probable to be discovered at certain locations than at others. As such, at certain locations the electrons can never be identified. There is hence a probability distribution of the electron being located at various locations within the box. We can here see Bohr's model for the hydrogen atom begin to emerge, as we imagine a standing wave with one wavelength, and then two, and then three and so forth. This is the reason why an electron in an atom can only inhabit a discrete set of energy levels: the circular standing wave that represents the electron can only have an integer number of wavelengths. When an electron is struck by a photon of a particular energy, this energy is absorbed, promoting the electron to a higher energy state and increasing the number of wavelengths contained within the standing wave (see figure 3.12). This is why the electron shifts to inhabit a higher energy level, and this is what is fundamentally occurring during electron excitation. Furthermore, it is the constructive interference of these standing waves that explains how orbital overlap results in covalent bonding; thus, we can attain a new clarity in our understanding of chemistry because of modern physics. Once it was realised that electrons exhibit wave behaviour, the physics community set out to find a mathematical model that could describe this behaviour. Erwin Schrodinger achieved this goal in 1925, when he developed his Schrodinger equation, which incorporated the de Broglie relation [1]. This provided a version of quantum mechanics based on waves. Schrodinger's equation is also useful in understanding the structure of the atom. For instance, we

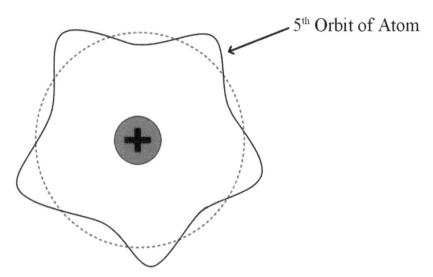

5th Orbit of Atom

Figure 3.12. Orbit of an atom.

can understand the energy quantisation and possible energy levels for the hydrogen atom. A hydrogen atom has only a single electron moving around a single proton. For multi-electron atoms, this equation requires certain assumptions and estimations, which increase its difficulty; but for the hydrogen atom, Schrodinger's equation can accurately define its characteristics. Accordingly, in this section we will examine Schrodinger's equation and its significance, the meanings of the equation and their implications, its solution and what it represents, and so on.

3.5.1 Schrodinger's equation

(i) Time-dependent Schrodinger's equation

From the theory of waves, we can say that any wave that propagates in space with time can be modelled by equation (3.12). Here, we consider the wave nature of particles so that we can model the wave function as per equation (3.23) [1, 2]:

$$\psi(x, t) = A e^{i(kx - \omega t)}. \tag{3.23}$$

Here, we are adopting the above wave function as a suitable wave function for a free particle of momentum

$$p = \hbar k \text{ and energy, } E = \hbar\omega. \tag{3.24}$$

Using $E = \frac{p^2}{2m} = \frac{\hbar^2 k^2}{2m}$, the second-order differential equation to describe the kinetic energy of the particle is

$$\frac{\partial^2 \psi}{\partial x^2} = -k^2 \psi, \tag{3.25}$$

which can be written as

$$-\frac{\hbar^2}{2m}\frac{\partial^2\psi}{\partial x^2} = \frac{p^2}{2m}\psi. \tag{3.26}$$

Similarly, by partially differentiating equation (3.23), we can write

$$\frac{\partial\psi}{\partial t} = -i\omega\psi. \tag{3.27}$$

A small manipulation in the energy denoted by equation (3.24) using equation (3.27) gives the potential energy as

$$E\psi = \hbar\omega\psi = i\hbar\frac{\partial\psi}{\partial t}. \tag{3.28}$$

The total energy of the particle can be calculated by summing equations (3.26) and (3.28) as

$$E\psi = \frac{p^2}{2m}\psi + V(x)\psi, \tag{3.29}$$

$$-\frac{\hbar^2}{2m}\frac{\partial^2\psi}{\partial x^2} + V(x)\psi = i\hbar\frac{\partial\psi}{\partial t}. \tag{3.30}$$

The above equation (3.30) is what is known as the time-dependent Schrodinger's equation. By analysing and solving this equation, we can understand the basics of quantum mechanics. In general, the solution of the time-dependent equation will describe the dynamic behaviour of a particle, which is similar to Newton's classical physics. As will be recalled, through Newton's formula we can evaluate the location of the object as a function of time, but Schrodinger's equation solution shows us how the probability of a particle being in a particular space differs as a function of time.

(ii) Time-independent Schrodinger's equation

The time-dependent Schrodinger's equation depends on space as well as time; but, in most cases, we are interested in the probability distribution of having an electron within an atom to understand the structure of the atom. Hence, we need only the space parameter. The time-dependent Schrodinger's equation is also difficult to solve for quantum systems. Thus, another version of Schrodinger's wave equation exists that depends solely on space. This is known as the time-independent Schrodinger's equation, and is discussed as follows.

The time dependency enters into the wave function by a complex exponential factor. This means that, to remove this time dependence, we have to take a wave function to the Schrodinger wave equation in exponential form, as given below:

$$\psi(x, t) = \psi(x)\ e^{-\frac{iEt}{\hbar}}, \tag{3.31}$$

where the time and space dependences of the whole wave function are retained as separate factors. Next, we have to verify that this method leads us to derive an equation for $\psi(x)$, which is the space part of this wave function. If we replace the ψ of equation (3.30) with equation (3.31), and use partial derivatives, we get

$$-\frac{\hbar^2}{2m}\frac{\partial^2\psi(x)}{\partial x^2}\ e^{-\frac{iEt}{\hbar}} + V(x)\psi(x)\ e^{-\frac{iEt}{\hbar}} = i\hbar.\ -\frac{iE}{\hbar\ e^{-\frac{iEt}{\hbar}}}\psi(x) = E\psi(x)\ e^{-\frac{iEt}{\hbar}}. \quad (3.32)$$

Therefore,

$$-\frac{\hbar^2}{2m}\frac{\partial^2\psi(x)}{\partial x^2} + V(x)\psi(x) = E\psi(x). \quad (3.33)$$

Equation (3.33) is a time-independent form of Schrodinger's equation. Here, E is a depiction of the total energy of a particle. If we want to calculate the wave function of an electron at any particular value of E in a particular existing potential well of $V(x)$, then we have to substitute the values of E and $V(x)$ and calculate the wave function from equation (3.33). We will have different wave functions for different values of E. For a particular value of E, this wave function can be depicted as $\psi_E(x)$.

However, these wave functions might not be valid. A wave function has to be finite and single valued as per the the definition of a function. For a wave function to be valid, it has to follow two conditions: (i) the wave function has to be normalised (the normalisation condition), and (ii) the wave function and its derivative must be continuous (the continuity condition). Generally, these are referred to as boundary conditions [8] and are described as follows.

(iii) The normalisation condition

The wave function is a complex quantity. If we multiply it with its complex conjugate then we will obtain a square of its magnitude. This gives us the probability distribution of a particle over space at that particular time instant. As such, this sum of all probabilities over the space has to be 1 [8], and so the integration of the probability density function of the particle over the space has to be 1. This constitutes our normalisation condition for the particle, which can be mathematically stated as

$$\int_{-\infty}^{+\infty} |\psi(x,\ t)|^2\ dx = 1. \quad (3.34)$$

The above equation depicts that as x tends to infinity, $\psi_E(x)$ has to tend to zero. Only then does the normalisation condition hold true. This also implies the convergence of the wave function and the probabilistic interpretation of the wave function.

(iv) The continuous condition

If the potential is discontinuous in some way, then Schrodinger's equation can find that they are discontinuous. However, in the real world, discontinuity has no probability (as this would suggest an infinite force acting on the particle). Therefore, we always find continuous wave functions for continuous potentials. We thus consider an additional condition that the wave function itself and its spatial derivative must be continuous [1, 2]. Mathematically, for two regions, 1 and 2, this can be stated as

$$\psi_1(p) = \psi^2(p) \ \text{ and } \ \frac{\partial \Psi_1(p)}{\partial x} = \frac{\partial \Psi_2(p)}{\partial x}. \tag{3.35}$$

Hereafter, we will concentrate on the time-independent Schrodinger's equation.

3.5.2 The solution of Schrodinger's equation and potential well

In quantum mechanics, one of the first-used models of a quantum system is the so-called particle in a box model (also known as the infinite potential well model). This defines a particle that is free to move in a small space surrounded by impermeable barriers, meaning that the potential energy of the particle is zero. This is a simple model that does not require many approximations and can be solved analytically. It is generally used to explain the difference between classical and quantum systems, energy quantisation, etc. In addition, the model can be further extended or modified to study complex quantum systems like atoms and molecules.

Let us say we have one such particle with mass m within a region $0 < x < L$, as illustrated in figure 3.13. Infinite potential walls bound this region. An infinite potential well can be described by equation (3.36) [1, 2, 7]:

$$V(x) = \begin{cases} 0 & 0 < x < L \\ \infty & x \leqslant 0, \ x \geqslant L \end{cases}. \tag{3.36}$$

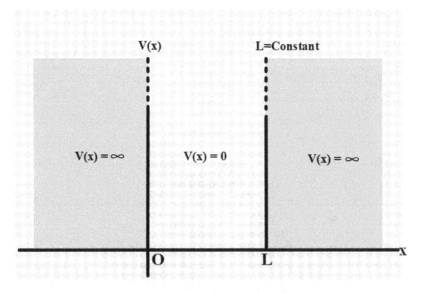

Figure 3.13. Infinite potential well.

Electrons in a metal block or gas in a large container describe such systems precisely. In these examples, the potential on the electron as it goes to the edge of the metal block is infinite [1, 2, 8]. This does not allow the electron to leave the surface of the block. The same can be considered for a gas molecule at the bottom of a container. This shows that if a particle has a lower energy than the height of a potential well, it cannot escape that potential well.

To analyse the particle in a well problem, we divide the system into three regions, I, II and III, as shown in figure 3.13. In regions II and III, the potential energy is infinity. The particle is confined inside of the well, and there is no probability of the particle being present at these places. As such, the wave function in regions II and III will be zero. Thus, we must apply this restriction, i.e. the boundary condition [1, 2, 8]:

$$\psi(0) = \psi(L) = 0. \tag{3.37}$$

In region I, the potential energy is null. So, equation (3.36) becomes

$$-\frac{\hbar^2}{2m}\frac{d^2\psi(x)}{dx^2} = E\psi(x). \tag{3.38}$$

Substituting $k = \sqrt{\frac{2mE}{\hbar^2}}$, we obtain $\frac{d^2\psi(x)}{dx^2} + k^2\psi(x) = 0$.

This is a second-order differential equation with a general solution as

$$\psi(x) = A \sin\ (kx) + B \cos\ (kx). \tag{3.39}$$

Now, we will impose the boundary condition; noting that ψ is continuous, we obtain the following:

At $x = 0$, $\psi(0) = 0$, leading to $B = 0$.

At $x = L$, $\psi(L) = A \sin(kL) = 0$. Hence, $\sin(kL) = 0$, where $kL = n\pi$, $n = 0, \pm1, \pm2, ...$ Thus, now we have

$$k_n = \frac{n\pi}{L}, \text{ where } n = 1, 2, ... \tag{3.40}$$

This leads to

$$E_n = \frac{\hbar^2 k_n^2}{2m} = \frac{\hbar^2 n^2 \pi^2}{2mL^2}, \text{ where quantization level } n = 1, 2, ... \tag{3.41}$$

After considering this boundary condition and the normalisation condition, the final solution is given by [1, 2, 8]

$$\psi_n(x) = \begin{cases} \sqrt{\frac{2}{L}} \sin(n\pi x/L) & 0 < x < L \\ 0 & x < 0, x < L \end{cases}. \tag{3.42}$$

The sinusoidal wave function depicted by equation (3.42) is the solution of Schrodinger's equation of an infinite potential well model. It indicates a sinusoidal wave behaviour between two limiting points [1, 2]. If we were to join these two end points together, we would obtain an atomic structure, as illustrated in figure 3.12. For now, we can visualise a particle that shows various wave functions corresponding to various energy levels; that is, we can identify valid wave functions that represent allowed and quantised (for discrete values of n) energy levels [1, 2]. From the quantised energy equation, we can also find the wavelength of that particle for that particular energy level. The integer multiple of this wavelength has to be fit within the perimeter of the orbit corresponding to that particular energy level [1, 2]. This also gives insights to the nature of the particle–wave duality: it is how Schrödinger's equation explains various quantum phenomena under the umbrella of quantum physics.

3.5.3 Interpretation of wave function in terms of probability density function

The square of the magnitude of the wave function gives a probability density function of a particle over a space at that point in time. Integration of the probability density function (example shown in figure 3.14) from point A to point B in space at a particular instant of time yields the probability of finding that particle between A and B as

$$P(A \leqslant x \leqslant B) = \int_{A}^{B} |\psi(x)|^2 \, dx. \qquad (3.43)$$

Figure 3.15 shows wave functions and corresponding probability density functions for different values of n. We can observe that electrons can be found at some places with very high probability, while at other places their probability is zero. Further, there are multiple locations where the probability of finding electron is the same. This is a phenomenon that cannot be explained using classical physics—hence the need for quantum physics to understand such behaviour. Infinite potential wells are hypothetical and of little interest as they imply no escape of electrons, meaning

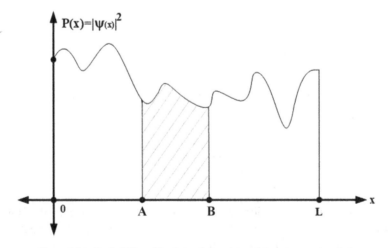

Figure 3.14. Probability of having subatomic particle between A and B.

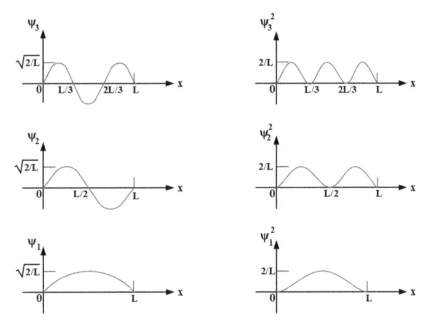

Figure 3.15. Wave functions and probability density functions for various values of n.

no current, and thus no potential for development in real applications. In the real world, we usually deal with finite potential wells. In a finite potential well, as the name would suggest, the height of the wall is finite, as shown in figure 3.16. Any particle with an energy greater than this wall height can leave the system. As such, a finite potential well can be mathematically modelled as equation (3.44):

$$V(x) = \begin{cases} 0 & 0 < x < L \\ V & x \leqslant 0, x \geqslant L \end{cases}. \tag{3.44}$$

These potential wells are of utmost interest to device engineers and material scientists, because in semiconductors it is possible to manufacture such wells and to verify our calculations and accordingly tune band gaps and energies. They provide true and more accurate models of atoms. Using a potential well, we can cause an electron to escape its atom by supplying it with a certain amount of energy.

3.5.4 Application of Schrodinger's equation in atomic structure

In the previous section, we have seen the solution and interpretation of the wave function. Solution of Schrodinger's equation gives us a set of wave functions and energy levels for those wave functions. This leads us to study the structure of an atom. The possible values of quantised energy are known as 'orbits' of an atom. An electron in any atom can only stay in specific orbits. These orbits also contain 'orbitals'. An orbital is a description of electron spatial density within an orbit. The

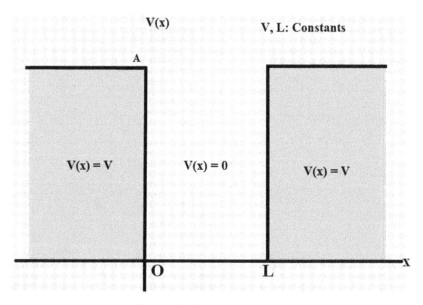

Figure 3.16. Finite potential well.

probability of electrons being found within orbitals is very high as compared to forbidden spaces within the atomic structure. These orbitals are described by a set of quantum numbers, which show the position and state of the electrons within an atom.

Orbitals in an atom are represented by numerals and letters. The main quantum number, which gives the possible quantised energy levels, is called the 'principal quantum number' and is denoted by the symbol n. As aforementioned, this serves as a depiction of the energy levels at which the orbitals can reside. The second quantum number is called the 'azimuth quantum number' and is symbolised by l. This is a description of the angular momentum of an electron. For a particular value of n, the possible values of l are 1 to $n-1$. These various values of l give orbitals with various shapes, which represent the probability density function of an electron in that atom. Values of l are further denoted by orbital symbols s, p, d, f, etc., for $l = 0$, 1, 2, 3, ..., respectively. The third quantum number is known as the 'magnetic quantum number'. This gives the orientation of the orbitals with respect to the magnetic field. This orientation describes a three-dimensional orientation of orbitals. The magnetic quantum number is given with reference to a specific angular quantum number as m_l, where m_l takes values from $-l$ to $+l$. The fourth quantum number is known as the 'spin quantum number'. Two electrons residing in the same orbit do not possess the same energy because they do not have the same spin. This quantum number describes the magnetic field of an electron, which affects its energy. The spin quantum number can have only two possible values, ½ or $-$½. These quantum numbers are briefly summarised in table 3.1.

Table 3.1. Quantum numbers [2].

S. No.	Name	Symbol	Values	Property
1	Principal	n	Positive integers (1, 2, 3,..., n)	Energy levels
2	Angular momentum	l	Positive integers (1, 2,..., $n-1$)	Orbital shape (probability distribution)
3	Magnetic	m_l	Integers from $-l$ to $+l$	Orbital orientation
4	Spin	m_s	$+\frac{1}{2}$ or $-\frac{1}{2}$	Direction of electron spin

3.6 Significance of the band theory of solids

At a temperature of 0 K, solids are in their lowest-energy state. This is known as the ground state of a material. The valence band is defined in this state. It can be fully or partially occupied. The empty band just above this is known as the conduction band. Electrons present in the conduction band account for the conductivity of the solid. The band gap separates the valence and conduction bands. Now let us compare the band gap in semiconductors, insulators and conductors (figure 3.17).

The electrons of an atom always start by occupying the lowermost available energy state. The lowermost energy state represents the innermost orbital, and electrons residing in that state are closer and hence highly confined to the nucleus. The outermost electrons are the valence electrons, which usually participate in physical processes. In semiconductors, a partially filled valence band and small band gap leads to the liberation of valence electrons from the valence to conduction bands upon the application of a thermal/optical/electrical (or any other form) of external stimulus. This will further cause a current flow. A silicon atom has 14 electrons with an electronic configuration of $1s^2 2s^2 2p^6 3s^2 3p^2$. All the energy orbitals (or sublevels) through 3s are full and the 3p orbital is partially filled as there are only four electrons in it. During lattice formation, electrons from the 3s and 3p orbitals hybridise to form a sp orbital with an occupied lower sp hybrid orbital and an empty higher sp hybrid orbital. They are separated by a small energy band gap.

In conductors (such as metals), the valence band and conduction band usually happen to form in the same band. There are no forbidden energy levels in between. Very small energy (thermal or otherwise) is required to convert valence electrons into conduction electrons, and the conduction electrons belong to the whole lattice, aiding high conductivity.

Materials having electrons exactly equal to the number of energy states in a valence band remain stable. They exhibit a huge energy band gap and require an extremely large external stimulus to liberate an electron from the valence band to the conduction band. Swapping of electrons may happen in the lattice due to some energy, but this does not contribute to any current as there is no net electron motion.

To summarise:

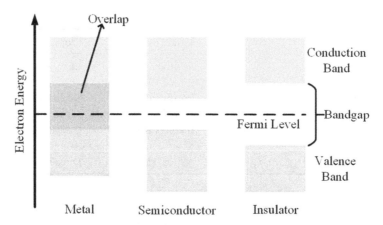

Figure 3.17. Band gap in conductors, semiconductors and insulators.

(i) In semiconductors, the extent of the forbidden energy gap is small. As such, a small supply of energy excites the electron and lifts it up from the valence band to the conduction band, leaving a hole behind; as in, e.g., germanium, silicon, gallium arsenide, etc.
(ii) In conductors, the valence band and conduction band overlap; as in, e.g., gold, aluminium, copper, etc.
(iii) In insulators, there is a large gap between the two bands; as in, e.g., rubber, wood, carbon, etc.

Based on the energy–momentum profiles of free electrons in semiconductor crystals, they are categorised into direct band gap and indirect band gap semiconductors. Refer to the *E–k* diagrams shown in figure 3.18. If in an *E–k* diagram, the minima of the conduction band and the maxima of the valence band exist at same values of k, then it is known as a direct semiconductor; otherwise, it is known as an indirect semiconductor. An electron would always prefer to be at the minima and the hole to be at the maxima. Ideally, an electron would be promoted at same k value. However, phonons (a type of quasi-particle) due to lattice vibrations impart a change in the momentum and, hence, the k value is different for the minima of the conduction band and the maxima of the valence band. This process is thermal in nature. In a direct semiconductor, when the electron drops from the conduction band to the valence band, there is a change in the energy but no change in its momentum. This is possible only when photons are emitted out; as they have no mass, their momentum is zero. A defect-driven recombination is also possible in a direct semiconductor. Most optics-related devices use direct semiconductors. In indirect semiconductors, when the electron drops from the conduction band to the valence band a defect will capture it midway, and a phonon is emitted and thermal energy released in the form of lattice vibration. There is no photon emission. Examples of direct band gaps include GaAs, ZnO, etc., and examples of indirect gaps include Si and Ge.

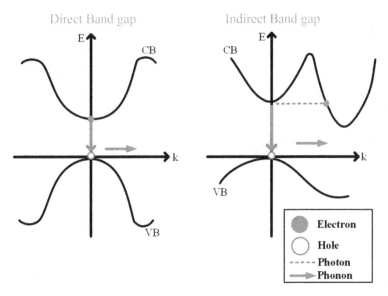

Figure 3.18. Direct and indirect band gap semiconductors.

3.7 Factors affecting the energy band gap

The energy band gap is a material constant. However, it can be manipulated by controlling particle size, operating temperature and the composition of certain semiconductor alloys. This process is popularly known as band gap engineering.

(i) Dependence of energy band gap on particle size

Quantum mechanics explains energy quantisation and orbital manifestation at the atomic level. When a huge number of atoms come close together, their orbitals couple with those of adjacent atoms and lead to the formation of energy bands in lattice structures. However, when a small number of atoms or molecules combine to form nano-sized particles, the number of coupling orbitals decreases. This leads to an increase in the band gap (due to the small dimensions of the particle) and splitting of energy states (instead of band formation). The greater the extent of the forbidden gap, the higher the electron confinement. This is illustrated in figure 3.19 (same as figure 3.11).

(ii) Dependence of energy band gap on operating temperature

The band gap of a material can also be affected by virtue of thermal energy. This effect of the material is quantified by a linear expansion coefficient. An increment in thermal energy leads to higher atomic vibrations. These higher-amplitude vibrations cause an increment in the interatomic distance, which further increases the average kinetic energy of an electron. According to the law of conservation of energy, an increase in kinetic energy leads to a decrease in potential energy. Hence, the band gap of

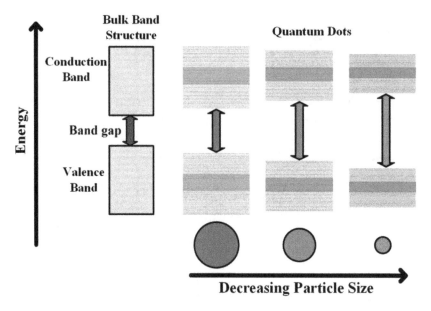

Figure 3.19. Band gap diagram illustrating variation of semiconducting materials as a function of particle size. Colour variation with decreasing particle size signifies blue shift.

a semiconductor can be reduced. The temperature dependence of the energy band gap has been experimentally determined yielding the following expression for E_g as a function of temperature T:

$$E_g(T) = E_g(0) - \frac{\alpha\,T^2}{T + \beta}, \tag{3.45}$$

where $E_g(0)$ is the energy gap at 0 K and α, β are fitting parameters. These fitting parameters are listed for germanium, silicon and gallium arsenide in table 3.2.

(iii) Dependence of energy band gap on doping

By controlling the composition of alloys or constructing layered materials with alternating compositions, the band gap of a material can be altered. Variation in the material concentration affects its bonding nature significantly. This modifies the lattice constant, carrier mobility, and so on, thereby modifying the band gap of the resulting metal alloy.

Let us take an example where zinc oxide (ZnO) is doped with copper (Cu) and manganese (Mn) separately. It is observed that the band gaps of $Zn_xCu_{1-x}O$ and $Zn_xMn_{1-x}O$ are smaller than that of ZnO.

As we can see from table 3.3, $Zn_{0.99}Cu_{0.01}O$ has the smallest band gap because of the higher content of copper in it (the band gap of CuO is between 1.2 and 1.3 eV). $Zn_{1-x}Cu_xO$ and $Zn_{1-x}Mn_xO$ are doped compounds so they have smaller band gaps in comparison to a pure ZnO

Table 3.2. Fitting parameters for thermal dependency of energy band gap of common semiconductors [2].

S. No.	Fitting parameter	Germanium	Silicon	Gallium arsenide
1	$E_g(0)$ eV	0.7437	1.166	1.519
2	α eV/K	4.77×10^{-4}	4.73×10^{-4}	5.41×10^{-4}
3	β K	235	636	204

Table 3.3. Band gap energies of ZnO compounds.

S. No.	Compounds	Band gap energies (eV)
1	ZnO	3.33
2	Zn0.99Cu0.01O	3.288
3	Zn0.99Mn0.01O	3.32

compound. Due to this effect, the valance band maxima of the doped ZnO compounds shift downwards compared to the pure ZnO example. It is possible to synthesise materials with a desired band gap using advanced processes like molecular beam epitaxy.

In summary, band gaps can be engineered. At the nanoscale, with an increase in the band gap, there is a decrease in conductivity. In semiconductors, this is of utmost significance because it opens door to endless possibilities for new materials and novel devices leading to better, smarter, more accurate, faster and more power-efficient electronic/optical applications.

3.8 Fermi statistics and electrical conduction in solids

Before the invention of Fermi–Dirac statistics, some of the properties of electrons and their behaviour under certain conditions were very difficult to understand. For example, at room temperature, the number of electrons that contribute to the heat capacity of a metal is much smaller than the number of electrons that take part in electronic current. Another poorly understood phenomenon was the generation of current in metals when applying high electric field being not at all dependent on temperature. One of the prior models which encountered such difficulties was the Drude model (1900) [1, 2]. As per classical statistical theory, all electrons were understood to be equivalent in all terms, meaning it was held that when a metal is heated, the total heat is due to the sum of contributions from all electrons independently and is a multiple of the Boltzmann constant. Fermi–Dirac statistics was first introduced in 1926 [1, 2], independently by Paul Dirac and Enrico Fermi. Before understanding Fermi–Dirac statistics, we must understand some of the fundamental concepts with which they built their new statistics.

The most significant term in this regard is a fermion. If any particle takes the following set of properties, then it is referred to as a Fermi particle, or fermion, and accordingly follows Fermi–Dirac statistics, e.g. electrons, neutrinos, quarks and protons. In the following, we will first look into the meaning of each of these properties. We will then explore Fermi statistics, followed by an examination of its implications and uses in designing electronic systems.

- They have a spin of a half integral of h (Planck's constant). This is why they are called half-spin particles.
- They obey Pauli's exclusion principle.
- All particles are assumed to be identical and indistinguishable.
- There has to be no interactions between two particles.

3.8.1 Spin of a particle

The spin of a particle has different interpretations depending on whether we are considering classical mechanics or quantum mechanics. Spin is one type of angular momentum (rotation) of a particle. Basically, there are two types of rotation (angular momentum):

(i) Orbital angular momentum, which is referred to as the revolutionary motion of a mass around the centre of another mass, for example the revolution of the Earth around the Sun (classical mechanics) and the revolution of an electron around a nucleus (quantum mechanics).
(ii) Spin angular moment, which is defined as the rotation of a centre of mass on its own axis, for example Earth's rotation around its own axis (classical mechanics) and an electron's rotation around itself (quantum mechanics).

When we say that 'a particle has a spin', we mean the quantum mechanical wave function that describes the particle has some particular direction in which it oscillates. This phenomenon is shown in figure 3.20.

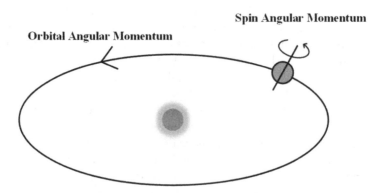

Figure 3.20. Example of two kinds of angular momentum.

3.8.2 Pauli's exclusion rule

In 1925, Austrian-Swiss physicist Wolfgang Pauli explained that no two electrons with the same spin value can exist in the same energy level and have identical quantum numbers. The spin of same-energy-level electrons must be opposite. This is known as Pauli's exclusion rule.

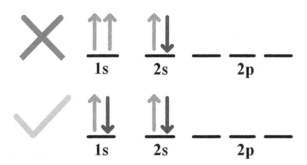

Figure 3.21. Wrong and correct configuration of electrons.

For example, let us consider an atom with four electrons. As shown in figure 3.21, the upper configuration shows an energy level consisting of electrons with the same spin, which is not possible. The lower configuration is the correct configuration, in which every energy level contains electrons with opposite spin [1, 2].

3.8.3 Fermi level

The Fermi level describes the highest layer of electron energy levels at absolute zero temperature (0 K). Since electrons are fermions and they obey Pauli's principle, that no two electrons can stay in the same energy state, at 0 K temperature they sit in the minimum energy state in the valence band. The conduction band remains empty. Thus, we get a threshold energy level above which we cannot find any electron. This is known as the Fermi level. The Fermi level can be calculated using the total electrons in a particular energy band and the electrons' probability distribution function.

Since the Fermi level is an energy level in the band diagram, sometimes it is also referred to as 'Fermi energy'. Table 3.4 shows some metal elements and their respective Fermi energies (in eV). We can see that some metals have low Fermi energy, and hence good conductivity, and vice versa. In other words, the probability of finding an electron above the Fermi level is zero, that below the Fermi level is 1, and that at the Fermi level is 0.5. Thus, in other words, the 'Fermi level is the energy level which has 50% probability of being occupied by an electron. It is the maximum energy level that an electron can have at 0 K.' The Fermi level as it applies in metals, semiconductors and insulators is given in figure 3.22 [1, 2, 7, 8].

In metals, there is no gap between the valence and conduction bands; they overlap. Consequently, here the Fermi energy gives us information about the speed of the electrons. Electrons nearer to the Fermi level can be excited and can be made

Table 3.4. Fermi level of some example metals [8].

Element	Cu	Ag	Au	Be	Hg	Al	Ga
Fermi energy (eV)	7.00	5.49	5.53	14.3	7.13	11.7	10.4

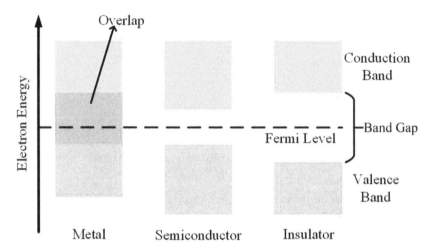

Figure 3.22. Fermi level in metals, semiconductors and insulators.

to participate in conduction. At the Fermi level, the electron energy E_F will be converted to kinetic energy when it starts to move [1, 2, 7, 8]. Thus, we can write

$$\frac{1}{2}mv^2 = E_F, \tag{3.46}$$

$$v = \sqrt{\frac{2E_F}{m}}. \tag{3.47}$$

This is called Fermi velocity. One of the useful characteristics of the Fermi level in a metal is that the density of free electrons can be implied from the Fermi energy.

In semiconductors, there is a band gap between the valence and conduction bands; there is no overlap. Thus, the Fermi level makes more sense when applied to semiconductors. This is shown in figure 3.23. For undoped semiconductors, the Fermi level lies at the exact middle point of the valence band and conduction band because the number of holes and electrons are the same. However, for doped semiconductors we have two different cases. N-type semiconductors are normally doped with a material that has a high number of free electrons. Since the number of electrons is greater, when the semiconductor stabilises at a temperature of 0 K, there is a very high number of electrons compared to holes. Thus, they fill some of the positions in the conduction band. Because of this, the Fermi level shifts towards the

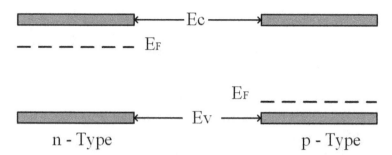

Figure 3.23. Different Fermi levels in n-type and p-type semiconductors.

conduction band. P-type semiconductors are doped with a material that has a low number of free electrons. Since the number of electrons is smaller, they cannot even fill the valence band fully. Thus, some levels in the valence band remain empty, and eventually the Fermi level comes closer to the valence band. This is again demonstrated in figure 3.23.

3.8.4 Fermi–Dirac distribution

Using a Fermi–Dirac distribution, we can only obtain a probability of the existence of particles at a given energy level. We cannot gather any information about the total number of states available at a particular energy level. This distribution can be described by a mathematical expression that denotes the probability density function for some particle, such as an electron, to obtain a fixed energy state. In electronics, the conductivity of materials is one of the biggest factors considered when designing nanodevices. As per the band theory of solids, the conductivity is dependent on the number of free electrons in the material. As mentioned, free electrons lie in the conduction band; thus we are interested in knowing how many electrons there are in this conduction band. This can be ascertained by using a Fermi–Dirac distribution. A Fermi–Dirac distribution function basically yields the density of electrons in particular energy states.

The normalisation factor for the Fermi–Dirac probability density function is temperature dependent and is given by $e^{-\frac{E_F}{kT}}$. At 0 K temperature, the probability of finding the electrons is 1 for energies less than E_F and zero for energies greater than E_F. This means that all the levels above the Fermi energy are not filled, but all the levels below the Fermi energy level are filled. All the particles below the Fermi energy level follow the Pauli exclusion principle. The probability of finding an electron at energy level E eV is thus given by [1, 2, 7, 8]

$$f(E) = \frac{1}{e^{(E-E_f)/kT} + 1}.\tag{3.48}$$

Thus, in a semiconductor, for an energy level lying 0.01 eV below the Fermi level, the probability of this level being occupied by an electron is $\frac{1}{e^{-0.01eV/0.026eV} + 1}$, i.e. 0.595, or 59.5%. For some energy level E much higher than the Fermi level, the exponential term becomes zero and hence the chance of obtaining the occupied energy state above the Fermi level is zero. For some energy level E less than the Fermi level, the function value signifies that every energy level up to the Fermi level E_F will be occupied. Figure 3.24 shows the characteristics of Fermi energy at different temperature values of $T = 0$ K, $T = 300$ K and $T = 2500$ K. At $T = 0$ K, the graph has step-like behaviour. For $T \neq 0$ K, when $E = E_F$, then the value of the Fermi–Dirac function will be independent of temperature. When $E < E_F$, then the exponent will be negative and the value of function will increase with increasing energy. The value of the function will start from 0.5 and increase and approach 1 as the energy decreases. When $E > E_F$, then the exponent will be positive and the value of function will decrease with increasing energy. The value of the function will start from 0.5 and decrease and approach zero as the energy increases.

An electron has to be close to the Fermi level by an amount kT for its possible boosting from the valence band to the conduction band. From table 3.3, the Fermi energy of gold is 5.33 eV. So, for a gold crystal operating at 300 K, any electron which is located at $5.33 - 0.026 = 5.504$ eV can become a conduction-band electron. Fermi–Dirac statistics helps to predict the behaviour of free electrons inside materials. At small scale, a small difference in conductivity can affect applications and their operating temperature range significantly. Hence, finding proper materials is very important. Fermi–Dirac statistics helps to select the most appropriate material for fabrication of nano-level devices. Another good example of such an application is in sensors like charged coupled devices, as are used in cameras. From Fermi–Dirac statistics, we can get information about how much light is required to produce a proper charge in the circuit and attain the desired results. Using

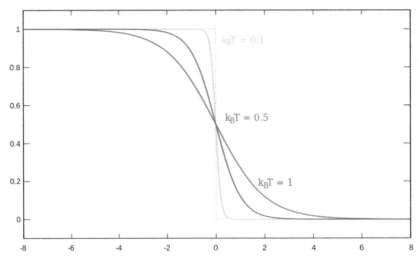

Figure 3.24. Probability distribution function vs energy difference $(E - E_F)$. (This Fermi dirac distr.svg image has been obtained by the author(s) from the Wikimedia website where it was made available under a CC BY-SA 3.0 licence. It is included within this book on that basis. It is attributed to Vulpecula.)

Fermi–Dirac statistics, we can also find magnetic susceptibility, thermal conductivity, superconductivity, etc.

3.8.5 The density of state for solids

In a small-scale material, electron energies are quantised and distributed across many levels and sublevels. In bulk lattices, there are so many sublevels that it is difficult to differentiate between successive ones. This leads to the manifestation of energy bands. For this same reason that we cannot distinguish between them, we discuss them in the form of the density of states per unit volume per unit energy increment instead of an individual level or sublevel.

Based on quantum mechanics, for a solid cube with characteristic dimension D, the total number of electron states in an energy increment E is given by [1, 2]

$$N_s = \left(\frac{8\pi}{3}\right)(2mE)^{\frac{3}{2}}\frac{D^3}{h^3},\tag{3.49}$$

where m is the mass of the electron (9.31×10^{-31} kg) and h is Planck's constant (6.626×10^{-34} m^2 kg s^{-1}). Then, the number of energy states per unit volume is given by

$$n_s = \left(\frac{8\pi}{3}\right)\frac{(2mE)^{\frac{3}{2}}}{h^3}.\tag{3.50}$$

Derivation of this equation with respect to energy increment dE leads to the density of states function, described per unit energy per unit volume as

$$D_s(E) = \frac{8\sqrt{2}\,\pi m^{\frac{3}{2}}}{h^3}\sqrt{E}.\tag{3.51}$$

Thus, it can be seen that the density of states is proportional to the square root of the energy. For a cubic micrometre of gold at 3 eV, the electron energy states per unit energy can be calculated from equation (3.51) as 7.36×10^{46} J^{-1} m^{-3}. Multiplying this by the volume of the gold cube, we can obtain 11.8×10^9 per eV number of energy states. At 2 eV, the same can be calculated as 6.009×10^{46} J^{-1} m^{-3} and 9.61×10^9 energy states per eV, respectively. Many of the electrical, magnetic and conducting properties of materials depend upon the density of states in a material.

3.8.6 Electron density in a conductor

Multiplication of the Fermi–Dirac function and density of states function gives us the number of free electrons per unit volume per unit energy increment [1, 2, 7, 8]:

$$f(E)D_s(E) = \frac{1}{e^{(E-E_f)/kT} + 1}\frac{8\sqrt{2}\,\pi m^{\frac{3}{2}}}{h^3}\sqrt{E}.\tag{3.52}$$

The total number of free electrons per unit volume can be calculated by integrating equation (3.52) with respect to energy from $-\infty$ to ∞. However, the limits of integration can be modified to 0 to E_F for a conductor operating at 0 K. Thus, the volume density of free electrons in a conductor can be given by

$$n_e = \int_0^{E_F} \frac{1}{e^{(E-E_f)/kT} + 1} \frac{8\sqrt{2}\,\pi m^{\frac{3}{2}}}{h^3} \sqrt{E}\, dE = \frac{8\sqrt{2}\,\pi m^{\frac{3}{2}}}{h^3} \int_0^{E_F} \frac{1}{e^{(E-E_f)/kT} + 1} \sqrt{E}\, dE$$

$$= \frac{8\sqrt{2}\,\pi m^{\frac{3}{2}}}{h^3} \left(\frac{2}{3} E_F^{\frac{3}{2}} \right). \tag{3.53}$$

At the nanoscale, our task is making solid materials exhibit more atom-like behaviour. Equation (3.53) can be rewritten to yield the Fermi energy of a conductor as

$$E_F = \frac{h^2}{2m} \left(\frac{3n_e}{8\pi} \right)^{\frac{2}{3}}. \tag{3.54}$$

The rest of the terms being constants, we see that the Fermi energy of the conductor depends on its electron density. So, by definition of electron density and equation (3.54), we deduce that

$$E_F \propto \frac{N_E}{D^3}. \tag{3.55}$$

Equation (3.55) is simple but misleading if taken by its straightforward meaning. Therefore, a deeper understanding of this concept is essential. One must understand that the free electron density as well as Fermi energy of a conductor are material constants: they will not vary with the size of the material. The discrete energy states in a nanomaterial will spread over the same energy increment as the extent of the energy band in a bigger bulk solid. For small-scale materials, however, the energy states will be widely spaced. This gap between successive energy states is a function of the characteristic dimension D. Also, it is a nonlinear function of energy value, meaning the energy gaps will not be equal for all energy levels. Thus, we can only estimate the average value of spacing between energy states as

$$\Delta E \approx \frac{E_F}{N_E}. \tag{3.56}$$

Thus, we can conclude that the spacing between energy states is inversely proportional to the volume of the bulk solid:

$$\Delta E \propto \frac{1}{D^3}. \tag{3.57}$$

As per quantum mechanics, one isolated atom has one energy state in its sublevel. Two identical atoms brought together to form one molecule will have two energy states, a particle of three atoms will have three splits in the energy level, and likewise. Thus, by manipulating the number of atoms—and hence the characteristic dimension—of a particle, it is possible to tune to various band gaps. This concept is very much exploited in quantum-confinement-based nanostructures and their applications.

In summary, with the help of quantum mechanics, we can engineer 'artificial atoms' with tunable band gaps at the nanoscale. Quantum mechanics has gained vital significance in the domain of nanotechnology as it is the only way of understanding and modelling small systems. By estimating the additional amount of energy required to liberate electrons near the Fermi level, we can ensure development of useful novel devices.

Questions

1. Why is it necessary to study quantum theory/mechanics to understand the field of nanoelectronics?
2. What is a quantum particle? Explain the principles/effects that explain its behaviour.
3. What is effect of quantum confinement? How does quantum confinement work in nanomaterials?
4. State the significance of Schrodinger's wave equation. Relate how it helps in studying the physical and electrical properties of a nanodevice.
5. What is meant by a Fermi–Dirac distribution? What is the significance of the Fermi–Dirac distribution function?
6. What is the density of states? Can you relate it to Fermi energy? How does it affect the electrical property of a nanodevice?

References

[1] Rogers P and Adams 2008 *Nanotechnology: Understanding Small Systems* (Boca Raton, FL: CRC Press, Tayler and Francis Group)
[2] Hanson G W 2009 *Fundamentals of Nanoelectronics* (New York: Pearson Education) 1st edn
[3] Ferry D 2001 *Quantum Mechanics: An Introduction for Device Physicists and Electrical Engineers* (Boca Raton, FL: CRC Press)
[4] Griffiths D 1985 *Introduction to Quantum Mechanics* 3rd edn (Upper Saddle River, NJ: Prentice-Hall) p 07458
[5] Bergman T L, Lavine A S and Incropera F P 2011 *Fundamentals of Heat and Mass Transfer* 7th edn (New York: Wiley)
[6] Griffiths D J 1987 *Introduction to Elementary Particles* (New York: Wiley) pp 11–51
[7] Phillips A C 2013 *Introduction to Quantum Mechanics* (New York: Wiley) pp 11–17
[8] Streetman B 2015 *Solid State Electronic Devices* (New York: Pearson) 7th edn

IOP Publishing

Nanoelectronics
Physics, technology and applications
Rutu Parekh and Rasika Dhavse

Chapter 4

Emerging research devices for nanocircuits

Over the past six decades, metal-oxide-semiconductor field-effect transistors (MOSFETs) have been the workhorse of the semiconductor electronic industry. However, after continued shrinking the technology has matured and reached saturation, which has led to a myriad of physical and economic issues, as we saw in chapters 1 and 2. It is time to look at the matter from a novel nanoelectronics perspective. Novel nanoelectronics-based devices' circuits and systems are not only evaluated based on their power-performance-area (commonly referred to as PPA) aspects but their implementations have to additionally clear the demands of mass production, uniformity, consistency, reliability, durability and scalability too. Based on this purpose, nanoelectronics devices are broadly categorised into logic devices and memory devices. In this chapter, we illustrate some of the most promising logic devices—whose future is bright, beyond doubt.

Extending MOSFETs to the end of the semiconductor roadmap has been a gradual but no less astounding journey. All efforts have been targeted at improving the scalability and gate control of MOSFET-like device architecture. Devices like SiGe transistors, silicon on insulator technology, Fin-FETs, high-κ metal gates, double-gate/multi-gate FETs, gate-all-around (GAA) transistors, etc., along with circuit/system/process-level strategies have ensured appropriate technology advancement. A great deal of literature has been published on these topics. In 1959, American physicist Richard P Feynman gave a lecture at Caltech, named 'There's Plenty of Room at the Bottom: An Invitation to Enter a New Field of Physics'. In this talk, he considered the manipulation of matter at an atomic scale and raised several extreme properties that had not been noted theretofore. His talk led to many great inventions in the field of nanotechnology. In this chapter, we focus on a number of emerging nanoscale devices based on radically different conduction mechanisms/materials and carrier properties. The International Roadmap for Devices and Systems (IRDS) organises emerging nanodevices into three categories, as listed in table 4.1 and as briefly described in chapter 1.

Table 4.1. Emerging nanodevices.

MOSFETs (extending the channel of MOSFETs to the end of the roadmap)	Unconventional FETS (charge-based extended CMOS devices)	Non-FETs (non-chargebased 'Beyond CMOS' devices)
Unconventional geometry devices	SpinFETs	Molecular switches
III–V channel MOSFETs	SETs	Atomic switches
Ge channel MOSFETs	Tunnel FETs	Collective magnetic devices
Nanowire FETs	I-MOS	Spin transfer torque logic
CNTFETs	NEMS switches	Moving domain wall devices
Graphene nanoribbons FET	Negative Cg MOSFETs	Pseudo-spintronic devices
	Excitonic FETs	Nanomagnetic (M:QCA) devices
	Mott FETs	

Extending the channel of MOSFETs to the end of the roadmap

The nanodevices in this first category are MOSFET structures that use field-effect functionality to control the flow of current. They may utilise alternate channel material. This in turn offers high-mobility, modified geometry for better performance and channel control in addition to high charge carrier velocity. The rich physics offered by materials like strained Ge, SiGe, III–V compound semiconductors, carbon nanotubes (CNTs) and graphene are introduced. Such extensions to the technology allow the development of new information-processing devices. The technology is, however, confronted by several very difficult materials, process and device challenges.

Extended CMOS (charge-based devices)

The set of devices under this category offers co-integration with the CMOS platform. In order to extend and enhance the functionality of ultimately scaled CMOS, hybrid integrated systems with nanoelectronic devices are devised. These devices offer lower power dissipation as they provide a steeper subthreshold turn-on current–voltage characteristic ($S < 60$ mV/dec). The tunnel FET (TFET) is one such device, where the channel current is modulated via a source-channel tunnelling process. However, there are challenges with TFETs, such as their low ON current, development of device modelling and a hypersensitivity to structural and material variations.

Another device in this category is the impact ionisation MOSFET (or IMOS). In this device the channel current is modulated by generation of channel charge via impact ionisation. There is an increase in the device drivability and scalability because of its higher impact ionisation generation rates and reduced breakdown voltages. However, the practicality of high-density integration is a challenge for these devices. The next device is the negative gate capacitance (Cg) MOSFET. It is obtained by replacing the insulator of a MOSFET gate stack with a ferroelectric insulator. This acts like a step-up voltage transformer that will amplify the gate voltage. With this, it enables low-voltage and low-power operation because of the subthreshold slope, lower than the classical limit of 60 mV/dec. The attendant challenges are that the negative capacitance may not be enough to realise a high-performance switch and the response time of the polarisation in the ferroelectric gate dielectric.

Next in this category are spin transistors, divided into two types, spin-FETs and spin-MOSFETs, each having a different operating principle. The source and drain of both types of devices are composed of a ferromagnetic material. The switching operation in spin-FET is achieved by spin precession or dephasing of spin-polarised charge carriers injected into the channel. In SpinMOSFET, relative magnetisation configurations of the source and drain are used to modify the output. Despite this, SpinMOSFET transistors have not been accomplished; the key challenge is injecting a high percentage of spin-polarised electrons from a half-metal source into the channel, and in how these devices interconnect. In addition, nanoelectromechanical (NEMS) switches face a major challenge in volt level operation at GHz switching speeds. Finally, the last device mentioned here, the single-electron transistor (SET), can potentially deliver high device density and power efficiency at good speeds. Furthermore, the logic driving capability of SETs with CMOS-comparable output voltage and GHz frequency at room temperature has been investigated by simulation and analytical approach.

Beyond CMOS (non-charge-based devices)

Logic devices under this category exploit novel mechanisms to achieve reduced switching energy. They explore non-charge-based state variables like individual spins, collective spin polarisation or collective spin oscillation. These new devices are suitable for more specialised tasks such as associative processing, communication and multi-valued logic, and utilise ferromagnetic elements for their nonvolatility, radiation hardness and error tolerance. They are different from Si CMOS technology in device structure, materials, logic representation and architecture; and they demand architectures other than classical von Neumann. Some examples of this category of device are collective spin devices, moving domain wall devices, atomic switches, molecular devices, pseudo-spintronic devices, bilayer pseudo-spin field-effect transistors (or BiSFETs) and nanomagnetic logic devices. Each of these devices offers a common advantage of reduced power dissipation, but each poses several challenges.

Many of these potential devices are detailed below with their underlying physics and applications.

4.1 Channel-replacement devices

In the high-performance and energy-efficient domain of integrated circuits (ICs), we adopt two methods [1]: (i) replacing the bulk of the material with a material which has a higher mobility than silicon, like indium antimonide (InSb), gallium antimonide (GaSb), n-type-germanium or lead telluride (PbTe); (ii) using strained silicon channels or nanowire FETs, where the nanowire is made up of zinc sulphide (ZnS), cadmium sulphide (CdS), zinc selenide (ZnSe), cadmium selenide (CdSe), etc.

Employing strained silicon channels is becoming difficult as we approach atomic dimensions. The major challenges faced here are high leakage current and heat dissipation. An obvious remedy is to use silicon technology with improved channel mobility. The electron and hole mobility for the potential channel-replacement devices are shown in figure 4.1. Some of the proposed channel-replacement materials and devices are graphene, CNTFETs, TFETs, NWFETs, p-type III–V channel-replacement devices and n-type Ge channel-replacement devices.

4.2 Graphene

Believe it or not, graphene is a magical material, with the tremendous potential to make everything work in the world of solid-state devices. It is the single most popular nanoscale invention in the recent perspective. Its parent element, carbon, is

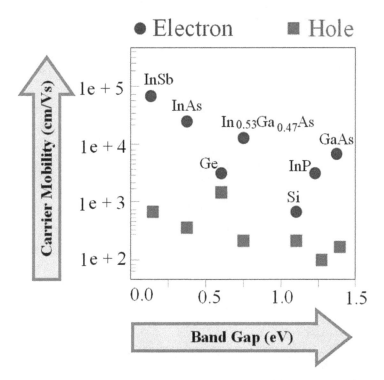

Figure 4.1. Trade-off between lattice constant and mobility. Adapted from [1]. Copyright (2018), with permission from Elsevier.

abundant in nature, in all living things and many nonliving things. Carbon is usually known for being found in two forms; that is, diamond, one of the hardest known materials, and graphite, one of the softest known materials at present. A third variant of carbon, graphene, is found as a single-layer graphite structure—with some miraculous properties [2].

Canadian physicist Prof. P R Wallace theoretically postulated the single-layer structure of graphite in 1947. The experimental discovery was made by Hanns-Peter Boehm and his co-workers in 1962. The term 'graphene' was introduced in 1986, a combination of the word graphite and the suffix '-ene', denoting its similarity to the class of polycyclic aromatic hydrocarbons. Graphene is a sheet of carbon that exhibits sp^2 hybridisation. The carbon atoms are densely packed in a honeycomb-like structure, as illustrated in figure 4.2. This feature of graphene makes its unique at this scale and makes it useful in a number of applications.

Graphene is strong at the level of its 2D structure. However, it has weak bonding between separated layers. Because of this characteristic, graphene is stronger than diamond. It is also a most stretchable and pliable material, showing up to 20% elasticity. It is also stiffer than diamond, which means it has more tensile strength;

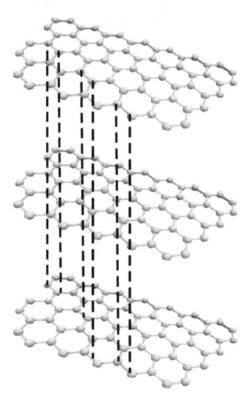

Figure 4.2. Monolayers of graphene: the intralayer covalent bonds are very strong and the interlayer van der Waals forces are very weak. (This Graphen.jpg image has been obtained by the author(s) from the Wikimedia website where it was made available under a CC BY-SA 3.0 licence. It is included within this book on that basis. It is attributed to AlexanderAlUS.)

because of this, in the future it may even be able to replace steel. Graphene is one of the strongest materials discovered. As compared to steel, it is 200 times stronger. It is further stronger than diamond. It has Young's modulus of nearly 1.3 TPa. Furthermore, graphene has greater conductivity than copper; it has a resistivity on the order of 10^{-6} Ω cm, and a high intrinsic electron mobility of 200 000 cm^2 V^{-1} s^{-1}. As a result, it can replace copper in future applications. It has the largest mean free path at room temperature so it can be used as an interconnect. It also has the largest current density at room temperature. Graphene shows extreme thermal conductivity, with a thermal conductivity higher than steel, of ~2000–4000 W m^{-1} K^{-1}. Because of its small dimensions, it also has the largest surface area, of around 3000 m^2 g^{-1}. This means graphene can be used as a future material to incorporate a greater number of components onto it. Monolayer graphene absorbs $\pi\alpha = 2.3\%$ of white light, where α is the fine-structure constant. This means graphene transmits 97.7% of white light, and so it is also a good optical transmitter. We can even say that graphene is a transparent material. Moreover, it is impermeable to helium.

There are many different methods for graphene synthesis [1, 2]. In general, techniques for graphene synthesis are divided into two types: top-down techniques and bottom-up techniques. Top-down methods involve mechanical exfoliation, chemical exfoliation and chemical synthesis. Following its invention, researchers tried to make graphene using exfoliation techniques. Further mechanical exfoliation methods involve the use of adhesive tape and the tip of an atomic force microscope (AFM). In the micromechanical exfoliation process, pencil is rubbed onto a surface and then, by using sticky tape, graphene sheets can be extracted. One of the biggest advantages of this method is that it is cheap and does not require any special equipment. However, use of this peeling technique is not suitable for large-scale production. The chemical synthesis techniques include bath sonication and reduced graphene oxide technique. In contrast, the bottom-up methods comprise three main approaches for graphene synthesis, namely, pyrolysis, epitaxial growth and chemical vapour deposition (CVD). By epitaxial growth on electrically insulating surfaces such as silicon carbide, we can generate graphene on a large scale. This process requires high temperatures, however, so the cost of manufacture is higher than other methods. CVD involves thermal- and plasma-based processes [1].

Graphene today is used in single-molecule gas detection, graphene-based transistors, ICs, transparent conducting electrodes, ultra-capacitors, graphene biodevices and display screens.

4.3 Fullerenes and carbon nanotubes

Fullerenes and CNTs exist in interstellar dust and geological formations on Earth. These are hollow structures and made entirely of carbon. When carbon (1s2 2s2 2p2) undergoes sp^2 hybridisation at ambient temperature, anisotropic formation of a planar (2D) graphite structure takes place. Diamond is formed under high temperatures and pressures as sp^3 hybridisation leads to cubic (3D) structure formation. A combination of sp^2 and sp^3 hybridisation under specific conditions leads to cylindrical fullerene formation. These fullerenes can be spherical, elliptical or tubular, wherein every atom is a surface atom. Carbon fullerenes like Buckminsterfullerenes (also known as Buckyballs) with C60 composition (containing 60 carbon atoms) were first discovered in 1985 in Texas by Croto, Curl and Smalley; an example is shown in figure 4.3. The cage-like structure of the fullerenes has the potential to act as a molecular storage enclosure for other substances, such as drugs or other elements [2]. If potassium vapours are diffused into such a structure, the potassium atoms occupy the empty spaces between the carbon atoms and form the compound K3C60. Whereas C60 is an insulator, K3C60 is conductive in nature and, below 18K, acts as a superconductor. It is possible to manufacture fullerenes of various compositions. Larger fullerenes up to C540 have been synthesised.

CNTs were discovered by Japanese physicist Sumio Iijima. He presented a paper in 1991, which led to unmatched considerations at the atomic scale in carbon matter. This also boosted the pace of research in nanotechnology. CNTs can be considered as cylindrical fullerenes [2]. They are usually hollow, around 1–1.5 nm in diameter and a few hundred nm long. There are two types of CNTs: single-walled nanotubes (SWNTs) and multi-walled nanotubes (MWNTs), as illustrated in figure 4.4. Both differ according to their geometrical properties. Two distinct zones with various physical and chemical characteristics make up a SWNT. The tube's walls and end caps are its first and second components, respectively. The end cap structure is related to or generated from a smaller fullerene, such as C60. They are chemically stable and inert. They have good thermal conductivity. CNTs can act as both thermal conductors and thermal insulators. Based on their orientations, CNTs can be metallic or semiconducting in nature. The electrical conductivity of metallic CNTs can be up to 6 times higher magnitude than that of copper. Because of their structure, CNTs exhibit ballistic conduction. CNTs have high tensile strength and Young's modulus due to their sp^2 molecular structure. CNTs have very high

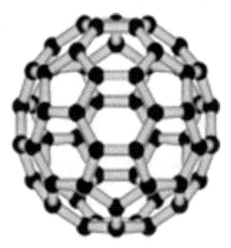

Figure 4.3. An example Buckminsterfullerene (C60): 60 carbon atoms, 700 pm diameter. (This C60-rods.png image has been obtained by the author(s) from the Wikimedia website where it was made available under a CC BY-SA 3.0 licence. It is included within this book on that basis. It is attributed to Rob Hooft.)

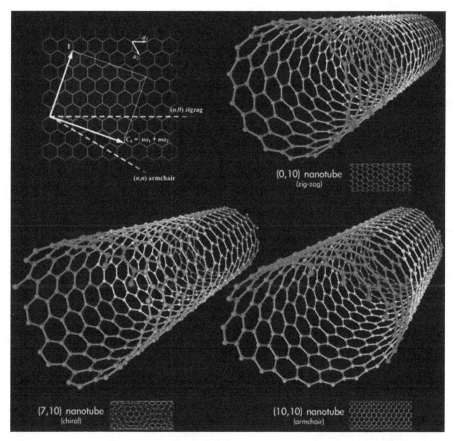

Figure 4.4. Types of CNTs and formation of various CNT structures. (This Types of Carbon Nanotubes.png image has been obtained by the author(s) from the Wikimedia website where it was made available under a CC BY-SA 3.0 licence. It is included within this book on that basis. It is attributed to Mstroeck.)

stiffness, which means they can resist a high amount of deformation in response to applied force [3]. MWNTs too possess very high thermal conductivity, and their Young's modulus ranges between 1000 and 3000 GPa—which is the highest ever measured. This is at least 10 times higher than steel, whilst weighing 6 times less. They are extremely flexible and resilient, chemically inert and yet can accommodate hydrogen up to about 100 times their own volume. In addition, tubes can be grown coaxially; these are referred to as MWNTs. They have a diameter of around 10 nm. The formation of a CNT can be understood by visualising a rolled-up sheet of graphite. Any finite-sized graphene layer will have dangling bonds. These dangling bonds correspond to high-energy states. During nanotube formation, the dangling bonds are eliminated and the bonds' strain energy is increased, leading to energy reduction of the overall system and structure stabilisation.

The structure of a SWNT can be represented based on the 'honeycomb lattice' structure of 2D graphene. The lattice growth of graphene is a bit unconventional in that the unit vectors a_1 and a_2 are not perpendicular. When a graphene sheet of a particular size and direction is wrapped, the SWNT structure is produced. We can only roll in a specific set of directions to create a closed cylinder because the outcome will be a symmetric cylinder. One of the two atoms in the graphene sheet is selected to act as the origin. The sheet is rolled up until the two atoms meet. The chiral vector is a vector that points from the first atom to the second, and its length is equal to the circumference of the nanotube [2–4]. The chiral vector and the nanotube axis are at right angles to each other. The optical activity, mechanical strength and electrical conductivity of SWNTs with various chiral vectors varies. A pair of indices (n, m) along the unit vectors a1 and a2 represent how the graphene sheet is wrapped. Nanotubes are known as zigzag nanotubes if the graphene sheet is wrapped in such a way that for all end atoms $m = 0$, armchair nanotubes if $n = m$ or chiral nanotubes in all other cases. A formula can be used to determine a nanotube's diameter as [4]

$$d = \frac{\sqrt{3} \times lcc \times \sqrt{n^2 + nm + m^2}}{\pi} = \frac{B}{\pi}, \tag{4.1}$$

where lcc is the carbon–carbon bond length = 1.46 Å and B is the length of the chiral vector.

If the detailed geometry of the lattice, as shown in figure 4.5, is analysed, it is easy to see that

$$\sin \theta = \frac{h}{na_1}, \tag{4.2}$$

and

$$\sin [180° - (120° - \theta)] = \frac{h}{ma_2}. \tag{4.3}$$

This can be rewritten using the trigonometric identity $\sin(\alpha - \beta) = \sin \alpha \cos\beta - \cos\alpha \sin \beta$, and we can use the fact that $a_1 = a_2 = \sqrt{3} \; lcc$ to get

$$\tan \theta = m\frac{\sqrt{3}}{(2n + m)}. \tag{4.4}$$

If $(2n + m)$ is a multiple of 3, then the CNT acts as a metal; otherwise, it acts as a semiconductor. The total number of hexagons composed by the CNT is calculated as $2(m^2 + mn + n^2)/(2n + m)$.

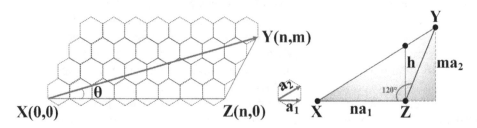

Figure 4.5. Rolling-up of a CNT.

Armchair-type CNTs behave like a metal with a conductivity approximately 50 times that of copper. As such, they are popularly used as nanoscale wires. In this, it is possible to achieve a current density of 10^7 A cm^{-2}. Moreover, in theory, if SWNT ropes are used in place of a single tube, then we will be able to achieve much higher stable current densities—as high as 10^{13} A cm^{-2}. However, there may be some defects in SWNTs. If such CNTs then combine together, they may form transistor-like devices. The same does not hold for MWNTs as the conductivity is quite complex in such cases. In addition, interwall reactions within MWNTs were found to redistribute the current nonuniformly across individual tubes. The resistivity and conductivity of SWNT ropes can be measured by placing electrodes at various parts of the CNTs. Subsequently, this means that the most conductive carbon fibers known are SWNT ropes. Alternatively, zigzag and chiral nanotubes can behave like semiconductors. They are being researched for use as transistors in the ever-shrinking circuitry of computers.

4.3.1 Synthesis and fabrication of carbon nanotubes

It is not exactly known whether the growth mechanism of a CNT is a tip-growth or a root-growth mechanism. The growth technique consists of three steps, as shown in figure 4.6 [2]. First, to form nanotubes and fullerenes, a metal precursor is placed on a substrate. Then, C_2 is formed on the surface of the metal catalyst particle; a rod-like carbon forms rapidly from this metastable carbide particle. Observation based on transmission electron microscope reveals a slow graphitisation of its wall. The

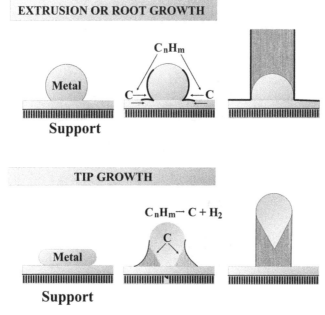

Figure 4.6. Growth of CNTs. Adapted from [75]. CC BY 4.0.

exact atmospheric conditions depend on the technique used. For all techniques used, the growth mechanism of the nanotubes seems to be the same. However, in order to employ them for technological tasks, it is exceedingly difficult to manage their size, direction and structure.

CNTs were synthesised for the first time in 1991, at NEC laboratories in Tsukuba, Japan, by Iijima. He was researching the material that had been deposited on the cathodes during the synthesis of fullerenes using arc evaporation. It was found that a variety of closed graphitic structures, including nanoparticles and nanotubes, were present in the cathodic deposit's central core. Currently, the three major methods for economical fabrication of CNTs are CVD, arc discharge and laser ablation, which will detail below [2–4].

(I) **Chemical vapour deposition**

CVD synthesis is realised by putting a carbon source into the gas phase. A plasma or a resistively heated coil transfers energy to the gaseous carbon molecule. Metal catalyst particles like iron, nickel and cobalt or some combination thereof are used as a substrate. The substrate is heated to approximately 700 °C. Two types of gases are bled into the chemical reactor, one a carbon-containing gas and the other a process gas. Ammonia, nitrogen and hydrogen are some commonly used process gases and acetylene, ethylene, methane or ethanol are carbon-containing gases. The energy source is used to 'crack' the molecule into reactive atomic carbon. CNTs will grow around the metal catalyst, as shown in figure 4.7.

On the nanometre scale, CVD makes exceptional alignment and positioning control possible. By using this method, the diameter and rate of

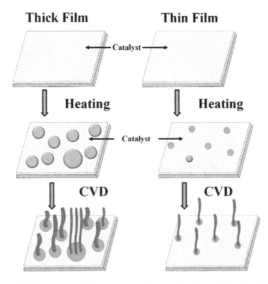

Figure 4.7. Fabrication of CNTs using CVD. Adapted from [5]. Copyright (c) 2007 The Japan Society of Applied Physics. All rights reserved.

growth of the nanotubes can be controlled. It is essentially a two-step procedure that begins with the preparation of the catalyst and ends with the actual synthesis of the nanotubes. A transition metal is sputtered onto a substrate as the catalyst. Then, either chemical etching or thermal annealing are used to produce catalyst particle nucleation. Thermal annealing causes clusters to form on the substrate, from which the nanotubes will develop. The etchant could be ammonia. The preferred variants of CVD are plasma-enhanced or thermal methods [2].

(II) **Arc discharge**

The most common and easiest way to produce CNTs is by using an arc discharge technique. This produces a complex mixture of components, which further requires purification to separate the CNTs from the soot and the residual catalytic metals present in the crude product [2]. An enclosure, usually filled with inert gas (helium, argon) at low pressure (50–700 mbar), is used with two carbon rods, placed end to end and separated by approximately 1 mm distance. A high-temperature discharge between the two electrodes is created by a direct current of 50–100 A, driven by a potential difference of around 20 V. The discharge vaporises the surface of one of the carbon electrodes, and forms a small rod-shaped deposit on the other electrode, which usually contains MWNTs. The arrangement is illustrated in figure 4.8.

The uniformity of the plasma arc and the temperature of the deposit forming on the carbon electrode determines the yield, type and size of the resultant CNTs. It is possible to selectively grow SWNTs or MWNTs depending on the procedure. If both electrodes are made up of pure carbon,

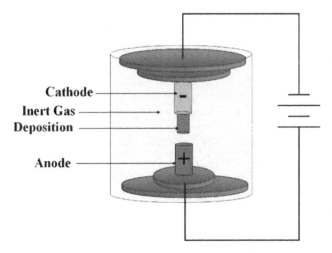

Figure 4.8. Fabrication of CNTs using arc discharge method. (This Arc discharge nanotube.png image has been obtained by the author(s) from the Wikimedia website where it was made available under a CC BY 3.0 licence. It is included within this book on that basis. It is attributed to Somnath2.)

the main product will be MWNTs. In addition, several other forms such as fullerenes, amorphous carbon and some graphite sheets are formed. Purifying the MWNTs results in a loss of structure and disorders the walls. However, much research is under way to attain pure MWNTs in a large-scale process without purification. Typical sizes for MWNTs are an inner diameter of 1–3 nm and an outer diameter of approximately 10 nm. There is no need for an acidic purification step as no catalyst is involved in this process. This leads to MWNTs with a low number of defects. If the arc discharge is carried out in a liquid nitrogen environment or in the presence of a magnetic field or rotating plasma, then MWNTs with more than 95% purity can be synthesised. If SWNTs are preferable, the anode has to be doped with a metal catalyst, such as iron, cobalt, nickel, yttrium or molybdenum [2, 4]. Various parameters such as the metal concentration, inert gas pressure, kind of gas, the current and system geometry determine the quantity and quality of the obtained nanotubes. Usually, the diameter is in the range of 1.2–1.4 nm. This can be controlled slightly by changing thermal transfer and diffusion. By changing the anode to cathode distance, it is possible to enhance anode vaporisation, which improves nanotube formation. Combined with controlling the argon-helium mixture, one can simultaneously control the macroscopic and microscopic parameters of the nanotubes formed.

(III) **Laser ablation**

In this process [6, 7], a block of graphite containing small amounts of cobalt and nickel powder is placed in a tube furnace at 1200 °C. A constant flow of argon is passed through and the graphite block is hit by a series of laser pulses. In a high-temperature reactor, a pulsed laser vaporises graphite. The use of two successive laser pulses minimises the amount of carbon deposited as soot. Successive laser pulses break up the larger particles ablated by the earlier ones and feeds them into the growing nanotube structure. When vaporised carbon reaches the cooler surface, it condenses and develops nanotubes on the cooler surface. The laser ablation apparatus is shown in figure 4.9.

To collect the nanotubes, a water-cooled surface is used. This is followed by heat treatment in a vacuum at 1000 °C to remove the C60 and other fullerenes. This method is used to produce up to 90% pure SWNTs of more than 70 weight percent purity. The material produced by this method appears as a mat of 'ropes', 10–20 nm in diameter and up to 100 μm or more in length. Each rope consists primarily of a bundle of SWNTs, aligned along a common axis. This process is more expensive than the arc discharge and CVD methods [6].

4.3.2 Carbon nanotube FETs

Amongst the many applications of CNTs, CNTFETs deserve special mention. CNT-based transistors have demonstrated excellent properties that are comparable

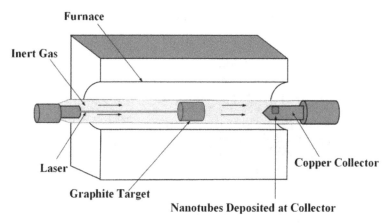

Furnace

Inert Gas

Laser

Graphite Target

Copper Collector

Nanotubes Deposited at Collector

Figure 4.9. Fabrication of CNTs using laser ablation. Adapted from [8]. Copyright (2006), with permission from Springer Nature.

to silicon-based MOSFETs. The high stability and conductivity of SWNTs result in excellent transistor behaviour. High current densities for the ON state (around 2100 mA mm^{-1}) and very small OFF current densities (around 650 nA mm^{-1}) under the same bias conditions can be achieved together with excellent ON/OFF ratios of approximately 10^5. A very high device density may be achieved as the channel length of CNTFETs can be reduced to 5 nm. CNT-based transistors can be divided into two main categories: CNTFETs and CNTSETs. In conventional MOSFET devices the channel is made of silicon. In CNTFETs the channel is formed by a single, semiconducting CNT [8]. A gate separated from the channel by a thin insulator film is placed on top of the nanotube. The process involves purified nanotubes being dispersed in dichloroethane spread over a silicon substrate, which is covered with silicon dioxide (SiO$_2$), an insulating compound. On top of this layer, pre-defined electrode areas are created by conventional lithographic techniques. The nanotubes are randomly spread over the surface in this way. In some cases, a nanotube connects two electrodes. A somewhat distinct approach is to spin-coat the CNTs on a substrate first, after which the electrodes are deposited with electron-beam lithography. The latter methodology can also be used to create top-gated CNTFETs, which is schematically shown in figure 4.10.

As grown, SWNTs show p-type semiconducting behaviour. They can be (partially) doped to an n-type semiconductor in several ways. p-type nanotubes can be formed to n-type semiconductors by exposing them to electron-donating compounds such as alkali metals. An easier way to perform the p-to-n trans-formation is by partially heating the nanotube, which is reversible in nature. CNTFETs' functioning is mainly based on the Schottky barriers present at the metal/semiconducting CNT junctions [4, 8, 9]. This type of transistor is very uncommon in conventional semiconductor physics.

The Schottky barrier appears at the metal/semiconductor interface. This prevents charges from being transported through this interface. Even at low voltages there is

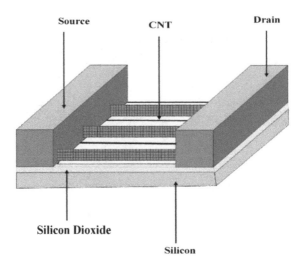

Figure 4.10. CNTFET structure.

considerable charge density in the channel, but these are blocked by the Schottky barriers. By applying a larger gate-source voltage, a stronger electric field develops at the electrodes. As a result, the Schottky barrier becomes smaller. Now the charge carriers can tunnel in a thermally assisted way and the transistor becomes conductive. These Schottky-barrier-based transistors can have very small dimensions, which possibly will allow for the downscaling of silicon devices. The devices function as conventional channel FETs for very small Schottky barriers.

It is possible to make a small quantum dot (QD) on a SWNT, due to the ability to define nanoscale regions of p or n material. The properties of such small doped QDs are influenced by the quantum-mechanical effects of a single electron. The Coulomb force of a single electron or high resistive state can prevent the entrance of a second electron. The so-called tunnelling effect allows the flow of charge. The expression for the energy for one electron to move into the system is [2, 4]

$$E_C = \frac{e^2}{2C}.$$ (4.5)

E_C is called the Coulomb blockade energy. It is the energy by which the previous electron repels the next electron. The electrons move through the system one by one, so the charging and discharging becomes a discontinuous process. If two QDs are joined at a point to form a channel, it is possible for an electron to pass from one dot to the other over the energy barrier by a process called tunnelling. In order to overcome the Coulomb blockade energy E_C, the applied voltage over the QDs should be [4]

$$V_C > \frac{e}{C}.$$ (4.6)

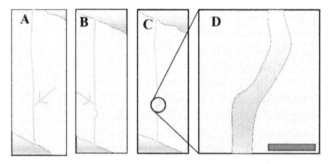

Figure 4.11. Graphical representation of the construction of a SET by bending a SWNT. From [11]. Adapted with permission from AAAS.

In order to observe the Coulomb blockade and tunnelling, the energy of the electron must be higher than the thermal scattering energy [4]:

$$\frac{e^2}{2C} > kT. \tag{4.7}$$

If these QDs are connected to other conductive materials (e.g. metal wires, polymers, CNTs) a SET is formed (see section 4.10 for further details). This consists of a conducting island that is connected by tunnelling barriers to two metallic leads. For temperatures and bias voltages that are low relative to a characteristic energy required to add an electron to the island, electrical transport through the device is blocked. Conduction is possible, however, by tuning a voltage on a close-by gate; this makes the device a three-terminal transistor. It has been demonstrated that a p–n–p nanotube system can be formed by sandwiching a potassium-doped CNT region between polymethyl methacrylate (PMMA)-coated CNTs [10]. Investigations have shown that the QDs are confined in the n region, with the two p–n junctions acting as barriers for the dots. Chemical profiling a single SWNT can lead to on-tube QDs much smaller than the lengths of the nanotubes between metal electrodes.

In order to function properly the system has to be cooled to a temperature less than 100 K. Another technique, found earlier by Postma *et al* [11], showed the fabrication of a room-temperature SET (or RTSET). They were able to produce a region between two tunnelling barriers formed by bending a metallic SWNT with an AFM. The nanotube is placed between gold (Au) electrodes on a Si/SiO$_2$ substrate, as shown in figure 4.11.

4.3.3 Applications of carbon nanotubes

CNTs constitute a material that will be used excessively in the future for many faster, better and cheaper applications. The plethora of potential fields in which CNTs may find a use is unfathomable [2, 4, 7, 12]. These include, but are not limited to, energy storage, lithium-ion batteries, super-capacitors, molecular electronics, composite and nanomaterial science, interconnects, nanosensors and actuators, industrial nanosieves, probing systems, templates, FETs, logic circuits and microprocessors, memory, interconnects, CPUs, field-emitting devices, displays, biomedical devices, solar cells, flexible electronics, and on and on. It should be noted that, in all the mentioned fields, the usage of CNTs has only been proven either conceptually or up to prototype levels. The prospects of all-carbon nano-electronics are equally bright. The challenges are mainly related to the availability of mass-production techniques capable of manufacturing consistent, uniform CNTs [2]. For now, the technology is still too much in its infancy as far as precise device design is concerned, but, speaking optimistically, it can be said that there is huge scope for more research and innovation, too. It will be fascinating to see what this wonderful world of CNTs grows into. CNTFETs boast good scalability, ballistic transport and a GAA structure, but they do face some limitations in terms of purity, placement, contact resistance, density and high switching voltages.

4.4 Tunnel field-effect transistor

The TFET is another experimental type of transistor comprised of p–i–n (p-type region, intrinsic region and n-type region) junctions in a reverse bias state. The concept of quantum tunnelling was developed due to the continuing decrease in the electron barrier with increased transistor scaling. This concept is exploited in a TFET to overcome some of the drawbacks related to subthreshold swing, leakage current, temperature sensitivity and short-channel effects faced by earlier MOSFET device generations [13, 14, 15]. These transistors have more spontaneous current ON–OFF conversions as compared to traditional MOSFETs. The structural and behavioural differences between a conventional MOSFET and TFET are shown in figure 4.12.

The essential structural difference between MOSFETs and TFETs is that, whilst MOSFETs have an identical source and drain, they are of opposite types, with TFETs having an intrinsic channel forming a p–i–n region. MOSFETs use thermal carrier injection for current flow. The gate terminal controls the electrostatic capability of the intrinsic region and builds up the electron concentration in the intrinsic region. As aforementioned, TFETs operate in a reverse bias condition. A higher gate voltage lowers the energy of the intrinsic region. The conduction band of

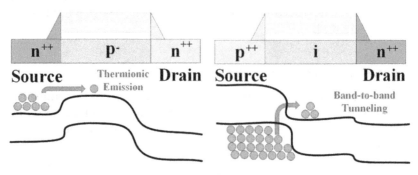

Figure 4.12. MOSFET vs TFET.

the intrinsic region electron overlaps with the valence band of the p-type region electron, causing band-to-band tunnelling (BTBT) and turning the device ON.

Application of negative gate voltage to the gate terminal with a reverse biased p-i-n junction lifts up the band of intrinsic region. This reduces the energy barrier between n-region and i-region allowing the quantum mechanism of BTBT [16]. For lower gate voltage supply, there is a greater difference in the energy of the intrinsic and the p-type regions than the probability for tunnelling, keeping the device OFF. State conversion from ON to OFF, and vice versa, are more spontaneous due to this. This makes TFETs power-efficient devices. Also, they exhibit a very low subthreshold swing at certain low current values. In comparison, MOSFETs have a subthreshold swing of at least 60 mV/dec, due to the thermal insertion of carriers from source to channel. As such, it is difficult to operate them at lower supply voltages.

TFETs can operate at very low voltages if the subthreshold swing is taken below 60 mV/dec. With less leakage current (fA μm^{-1}) and this ability to operate at low supply voltage (around 0.5 V), these devices are promising candidates for research [13]. TFETs may also be used as switches in the near future by using particular structures and varying low-band-difference elements like graphene, germanium and III–V compound semiconductors. In materials for which the band gaps are the same as those of silicon, the process of direct tunnelling is quite negligible. This is because as the height of the barrier is increased, the probability of transmission decreases. For materials that support direct tunnelling, conservation of perpendicular should be kept in mind when direct tunnelling is increased. In other words, the perpendicular momentum must not change when electrons move from the valence band to the conduction band via tunnelling.

4.4.1 Fabrication of tunnel field-effect transistor devices

Due to its sub-100 nm dimensions and radically new physical concepts, TFET technology is not yet in the mass-production stage. However, some laboratories have demonstrated TFET fabrication by using a molecular beam epitaxy technique. One such device, as shown in figure 4.13, was developed by IntelliEPI, Richardson in Texas [17].

The substrate used was p+ indium phopshide (InP) (Zn $\sim 1.7 \times 10^{18}$ cm^{-3}). From the substrate, the structure took a p+ InP (Be $\sim 5 \times 10^{18}$ cm^{-3}) of 300 nm width and a p+ source injector (Be $\sim 1.2 \times 10^{19}$ cm^{-3}) of 12 nm. This was then followed by a n+ $In_xGa_{1-x}As$ of 6 nm. (Here, x was taken to be within a range of 1.0 to 0.53) and a n+ $In_{0.53}Ga_{0.47}As$ of 9 nm with a 5 nm InP layer. The n+ regions had a doping of 1×10^{19} cm^{-3}. Device manufacturing commenced by removing the InP cap coating with diluted hydrochloric acid (1HCL:1H_2O). This was followed by the atomic layer deposition of gate dielectric Al_2O_3 (7 nm in width). Then, in a split wafer experiment, for each sample annealing in nitrogen(N_2) was done at a temperature of 623 K for 3 min. Following this, the deposition of titanium (Ti) or tungsten (W) and masking of SiN_x at the gate was performed. Through optical lithography and reactive ion etching, the FETs were patterned. SiN_x walls were formed by plasma-enhanced CVD. Al_2O_3 dielectric at the gate was removed using a

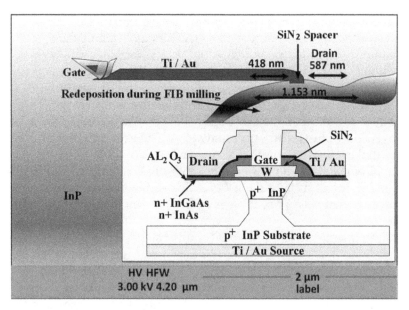

Figure 4.13. Representative fabricated TFET structure. © 2011 IEEE. Adapted, with permission, from [17].

selective wet-etching process. This was followed by metallisation: one application at the back of the wafer with Ti/Au and another at the drain. Selective etching was done on the InGaAs to the InAs coating in a diluted solution of sulphuric acid and hydrogen peroxide ($1H_2SO_4$:$8H_2O_2$: $160H_2O$) to stop at the source. It was followed by a reactive ion etching process that removed SiN_x, exposing the gate. Again, selective etching was done with a solution of hydrochloric acid and phosphoric acid ($1HCl$:$1H_3PO_4$). This removed the InP under the drain and extra SiN_x. At the end, the TFET was passivated with Al_2O_3 (7 nm width) and 3 nm HfO_2. This was done at a temperature of 573 K using atomic layer deposition [17].

4.4.2 Optimisation of tunnel field-effect transistor and scope of device design

In order to outperform CMOS technology, the benchmarks that TFETs should meet are as follows: (i) an ON current in the scope of several milliamperes; (ii) an average subthreshold swing (S(avg)) far below 60 mV/dec; (iii) an ON current/OFF current of at least the order of 10^5; and (iv) a drain voltage less than 0.5 V. To obtain the steep slope and the high tunnelling current, the transmission probability of a source tunnelling barrier should be maximum (near to 1) for a small change in the gate bias. The screening tunnelling length, the band gap and the effective carrier mass, represented as λ, E_g and m^*, respectively, all should be minimised for high barrier transparency, as suggested by the Wentzel–Kramers–Brillouin (or WKB) approximation.

Whereas the band gap (E_g) and the effective carrier mass (m^*) are dependent as a whole on the material system, the screening tunnelling length (λ) is highly influenced by specific parameters such as doping profiles, gate capacitance, dimensions and geometry. Further, the channel should have minimum body thickness, appearing in the best case as 1D electronic-transport behaviour. The unexpectedness of the doping profile at the tunnel junctions is additionally significant. To limit the tunnelling barrier, impurities in the source region should be high and must tumble off to the intrinsic region in as short a width as possible. This requires an adjustment in the doping concentrations of around four to five sets of size inside a separation of just a couple of nanometres. Expanding the impurities in the source region decreases λ and may prompt a marginally smaller energy barrier at the tunnelling junction on account of band gap narrowing [18]. A small screening tunnelling length (λ) can result in a strong modulation of the channel band by the gate, which requires a thin gate dielectric with great permittivity (k).

By choosing an apt gate dielectric, we can improve the ON current and decrease the subthreshold swing. Boucart *et al* [19] compared Si_3N_4 of 3 nm thickness and two high-κ insulators. As the gate coupling gets better with high-κ dielectric, the subthreshold swing improves. For all materials, the OFF current value is less than 1 fA. Increasing the gate capacitance, the ON current of a TFET does not increase proportionally (as happens in the case of MOSFETs). The subthreshold swing of TFETs keeps on improving as the permittivity of gate dielectric increases. In addition, just as with MOSFETs, we can utilise double-gate TFETs. The main advantage of a double-gate TFET is that the ON current in the device is approximately doubled. The OFF current is also doubled, but still remains very low, in the range of fA or pA [15, 18].

TFETs have a major advantage over MOSFETs in that they are less sensitive to changes in channel mobility, because any scattering in the channel can be nullified due to the dominance of transportation through the tunnel junction. As the device is governed by the tunnelling phenomenon, the output current obtained is very small as compared to MOSFETs. For higher currents, higher gate voltages are needed. However, high-κ dielectrics carry some disadvantages in that they have soft and hard dielectric breakdown. This puts a limit on the gate voltage [18]. These devices have very high threshold potential, which is also a drawback. In addition, the devices exhibit ambipolar behaviour; this, however, can be reduced by doping the drain region lightly. Also, using a material with a smaller band gap can be helpful.

4.5 Nanowire field-effect transistors

The conventional MOSFET channels are replaced by a semiconducting nanowire in nanowire field-effect transistors (NWFETs), as shown in figure 4.14. As the name suggests, the diameter of the nanowire is in the nanoscale (approximately 0.5 nm diameter). The nanowire can be made of different types of materials, including silicon, germanium, II–VI materials (ZnS, CdS, ZnSe, CdSe), III–V compound semiconductors (InAs, GaAs, InP, GaP, AlN, GaN) or semiconducting oxides like ZnO [20]. The nanowires provide a larger surface-to-volume ratio along with a smaller-sized channel in comparison to planar devices based on bulk materials. Electrostatic gate control can be obtained on the channel by GAA or gate-surrounding structures of the NWFET. As a result, nanowire transistors, similar

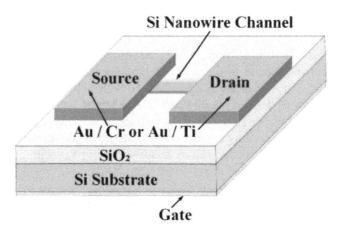

Figure 4.14. Si NWFET. Adapted from [20].

to GAA transistors, provide a solution to various limitations, such as the short-channel effect, that arise due to the scaling of planar MOSFETs.

For the fabrication of NWFETs, a directed self-alignment process is used, which is displayed in figure 4.15 [21]. As a first step of the fabrication, the silicon substrate is deposited with a thin layer of gold catalyst (~1 nm). Lift-off processes and lithography are used to pattern ap thin Au layer on the Si substrate. To grow the silicon nanowires, the catalyst is placed at pre-defined locations in a CVD furnace at low pressure and a temperature of 440 °C for about 2 hr, and SiH_4 streams with a pressure of about 500 mTorr are used. The typical diameter and length of the nanowires are 20 nm and 20 μm, respectively. After the vapour–liquid–solid growth process is complete, the Si nanowires are immediately placed in a furnace for dry oxidation for about 30 min at 750 °C to form a 3 nm-thick SiO_2 layer. As a result, a layer between the nanowires and top-gate dielectrics is generated, which provides a good interface. For the establishment of the source and drain contacts, methods such as the lift-off process and photolithography are utilised. The oxides from the Si nanowires at the source/drain regions are removed by applying a 2% hydrofluoric acid wet etch. After this, electrode metal (Al) deposition happens. The final step is the deposition of the top-gate dielectric (HfO_2) via atomic layer deposition at 250 °C. To improve the interface formed with the Al top gate, a deposition of 5 nm of Al_2O_3 was carried out. The annealing of these final devices in forming gas (5% H_2 in N_2) for 5 min at 325 °C reduces the interface trap density between the dielectric and the nanowire and improves the contact between the gate metal and the nanowire.

Although nanowires show significant improvement over traditional MOSFETs, there are still challenges to be overcome before they can be commercialised. As we have a high surface-to-volume ratios, the surface roughness and surface defects can strongly influence the performance of NWFETs. As seen earlier, NWFETs are generally prepared by CVD technique. However, the development of devices with optimal performance has yet to be achieved due to limitations in the techniques used to fabricate them. The currently used methods deploy fluidic alignment nanoscale probe techniques to harvest, position and align the nanowires. However, such

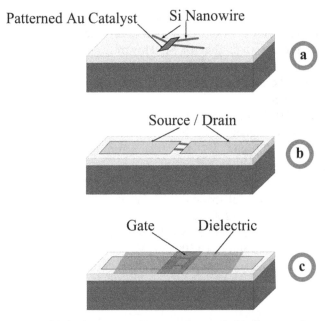

Figure 4.15. Nanowire FET fabrication process. (a) Au catalyst on Si, (b) alignment of source–drain contact, (c) addition of gate dielectric layer. Adapted from [21]. CC BY 3.0.

methods introduce impurities onto the surface of the nanowire that will affect surface roughness and density and, thus, as discussed earlier, the device performance will be adversely affected. In addition to this, even though single devices with a gate length of <40 nm have been made, it is not yet clear whether high-density circuits with a size similar to the size of CMOS can be prepared or not.

In summary, nanowire devices provide yet more possibilities for heterostructure devices and integration with greater yield and further reproducibility. However, they are confronted by several challenges, as the bottom-up approach incurs a decline in integration efficiency, whilst the top-down approach restricts high-speed operation.

4.6 P-type III–V channel-replacement devices

III–V compound semiconductors as a replacement for FET channels have gained considerable attention in recent years due to their high potential in logic applications [1]. III–V semiconductors are seen as a promising component for the development of high-performance CMOS circuits. These materials have already attracted considerable attention as n-type channel-replacement materials. InAs, InSb, InP, InGaSb, InGaAs, etc., are some examples of III–V semiconductor compounds with higher electron mobilities than silicon. InSb as well as GaSb are comprise group III–V materials, in that a number of free electrons are present in the crystal lattice. The lattice constants of InSb and GaSb are 6.479 Å and 6.1 Å, respectively, whereas the lattice constant of Si is 5.43 Å. Therefore, the hole mobilities of InSb and GaSb is 850 cm^2 V^{-1} s^{-1} and 800 cm^2 V^{-1} s^{-1}, respectively, whereas bulk Si has a hole mobility of 500 cm^2 V^{-1} s^{-1} at room temperature (300 K) [5]. The highest electron–hole mobility is observed in InGaAs, which is around 10000–30000 cm^2 V^{-1} s^{-1}. This is 30 times greater than that of Si. However, for high-performance CMOS circuits, p-channel materials that have high mobility are also required. InGaAs does offer the highest bulk electron mobility, but it has a bulk hole mobility almost equivalent to that of Si.

Among all the III–V semiconductor compounds, antimony-based compound semiconductors show higher bulk hole mobility. As aforementioned, the hole mobility of silicon is 500 cm^2 V^{-1} s^{-1}. The antimony-based compounds such as InSb and GaSb show mobilities of nearly 800 cm^2 V^{-1} s^{-1}. There are two ways of further improving these hole mobilities: (i) application of biaxial compressive strain, and (ii) application of uniaxial strain [1]. In this way, the hole mobilities of InSb and GaSb can be increased to 1260 and 1500 cm^2 V^{-1} s^{-1}, respectively. The semiconductor compound InGaAs has superior characteristics as an n-channel material, further showing a hole mobility of 1500 cm^2 V^{-1} s^{-1} when uniaxial strain is applied. The width of the quantum well observed for these compounds is only 5 nm. Further research on the widening of this width can further improve their hole mobilities. Thus, many antimony-based semiconductors are seen as potential III–V materials for the development of CMOS circuits. However, it should be noted that, due to large lattice mismatch, we cannot grow III–V layers on silicon. The lattice constants of various III–V materials are shown in figure 4.16.

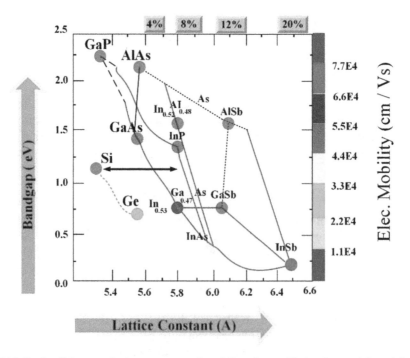

Figure 4.16. Trade-off between lattice constant and mobility along with band gap. Adapted from [1]. Copyright (2018), with permission from Elsevier.

As a rule, the greater the difference in lattice constant, the higher the lattice mismatch. Yet, in the last few years several techniques have been developed to tackle this issue. Some examples of these are confined epitaxial lateral overgrowth (or CELO), strain-relaxed buffers (or SRBs), template-assisted selective epitaxy (or TASE), aspect ratio trapping (or ART), etc. The production of most III–V semiconductors is costly and complex. Thus, typically only the silicon channel is replaced by III–V materials whilst keeping the bulk material the same. One such example is the use of a InGaSb nanoribbon for the realisation of III–V p-FET on a silicon substrate. The peak mobility obtained in this case was 820 cm^2 V^{-1} s^{-1} and the subthreshold slope was 130 mV/dec. However, all the best results for III–V materials have been demonstrated on substrates ranging from 50 to 150 nm. In order to leverage the high-volume manufacturing techniques used for silicon and the standard tools already developed in industry for it, 300 mm wafers of III–V materials are required. But as there are no III–V substrates available for III–V materials, the integration of III–V on silicon becomes necessary.

Two of the most significant obstacles in using III–V materials is their manufacturing cost and the complexity involved in their production process. Another issue can be seen in the case of germanium, in which the quality of the native oxide is very poor as compared to that of SiO_2. This issue is even more complex in the case of III–V materials. The reason for this is that III–V materials are binary, ternary or quaternary materials. This richness in the elements leads to a greater number of possibilities that can be generated at the interface. Due to this, a large number of possible defects are possible. A high-quality gate stack is an critical part of a MOSFET's design. Thus, this challenge must be overcome first before going proceeding with further implementations of III–V CMOS devices. Recent advancements in atomic layer deposition have provided a means by which to address this challenge.

4.7 N-type Ge channel-replacement devices

The high electron and hole mobility in bulk germanium, which is about 2.4 times that of silicon, is considered the main advantage of Ge-based MOSFETs [1]. GeSn semiconductor alloy, which contains a 10% fraction of Sn, has been considered for CMOS and other applications. Theoretically, GeSn nMOSFETs have shown much higher hole mobility than Ge nMOSFETs. Germanium oxides have very poor electrical characteristics due to their many traps. The major advantages of the high-κ/Ge gate stacks are that they offer both high electron mobility and low equivalent oxide thickness (or EOT). Due to the high-κ/Ge gate stacks, a thermally stable passivation layer is formed on the Ge substrate instead of germanium oxide. Luckily, as this layer is thermally unstable it can be removed from the interface by annealing. This helps to overcome the aforementioned problem with germanium oxide. But this is not the case with zinc oxide and silicon interfaces. Thus, for a stable interface, we need to choose both the high-κ material and substrate very carefully. Specifically, for germanium, some dielectrics used are GeYO, GeAlON, GeON, GePO, among others. Here, the positive or negative ions in GeO_2 are

Table 4.2. Comparison of different materials with their mobilities.

Element	Electron mobility (cm^2 V^{-1} s^{-1})	Hole mobility (cm^2 V^{-1} s^{-1})	Mobility ratio	Lattice constant (Å)
Indium antimonide (InSb)	78 000	850	9.1764	6.479
Gallium antimonide (GaSb)	3000	800	3.529	6.1
N-type germanium (Ge)	3900	1900	2.052	5.658
Lead telluride (PbTe)	6000	4000	1.5	6.46
Silicon (Si)	1500	500	3	5.14

replaced by the metal or nitrogen atom. However, a disadvantage of this type of device is that the germanium substrate has a large dielectric constant, which in turn makes the devices prone to short-channel effects. The current drivability of the short-channel effect for Ge-based nMOSFETs is lower than for Si-based nMOSFETs. The current priority for researchers motivated by Ge nMOSFETs is to show the high performance of short-channel devices as compared to that of Si nMOSFETs. A spacer (Y_2O_3) is used to decrease the leakage current and variation in performance by reducing the process variability.

Table 4.2 illustrates the mobilities of some typical materials that can be used in channel-replacement devices. From the table, we can see that InSb has very high electron mobility as well as hole mobility, so it can be used in very-high-speed applications. It has a higher mobility as the lattice constant of the atom is higher than all the other elements listed in the table. As the lattice constant increases, the concentration per unit volume decreases, which results in a lower number of collisions and in turn results in better mobility.

4.8 Potential evaluation—extending MOSFETs to the end of the roadmap

Table 4.3, adopted from the IRDS [14], lists demonstrated and projected parameters. Parameters like the cell size, density, switch speed and energy as well as circuit speed and energy, subthreshold slope, operating temperature are listed, in addition to any challenges associated with the materials. Further, in figure 4.17, the different performance parameters for NWFETs, TFETs, and n-type and p-type alternate channel material FETs are evaluated. For each criterion, a value of 1, 2 or 3 is assigned, where '3' means that the performance of the device is great as compared to traditional CMOS and '1' means that the performance is very poor.

A survey conducted among some of the best minds in the field of technology and logic devices today was used for the evaluation. In this, the so-called Overall Potential Assessment (OPA) = the summation of the potential of all relevance criteria, of which there are eight in total. Thus, if three points are allotted to each relevance criterion, we get a maximum OPA of 24 [14]; if one point is allotted to

Table 4.3. MOSFETS: extending MOSFETs to the end of the roadmap [14].

Device		FET	Carbon-based nanoelectronics	2D material channel FET	Nanowire FETs	Tunnel FET
Typical example devices		Si CMOS	CNT FET	Graphene nanoribbon FET	Ge/Si nanowire FET	Si TFET III-V TFET 2D TFET
Cell size (spatial pitch) [B]	Projected	50 nm	100 nm	100 nm	40 nm [NW1]	CMOS [T1]
	Demonstrated	56 nm	1.4 µm [CNT1]	1.4µm [G1]	300 nm [NW1]	<100 nm [T2]
Density (device/cm^2)	Projected	1.00E+10	1.00E+10	1.00E+10	5.9E+10 [NW1]	CMOS [T1]
	Demonstrated	1.08E+10	5.10E+07	5.10E+07	1.8E+8 [NW2]	Not known
Switch speed	Projected	12 THz	6.3 THz [CNT2]	7 THz [G2]	9.5 THz [NW3]	3 THz [T1]
	Demonstrated	0.4 THz	153 GHz [CNT3]	300 GHz [G3]	103 GHz [NW4]	−10 GHz [T3]
Circuit speed	Projected	61 GHz	100 GHz [CNT4]	Not known	100 GHz [NW5]	500 GHz (FO4 inverter) [T1]
	Demonstrated	~10 GHz	52 MHz [CNT5]	22 MHz [G4]	2 GHz [NW6]	Not known
Switch energy, J	Projected	3.00E-18	Not known	Not known	4E-20 [NW7]	1E-18 [T1]
	Demonstrated	1.10E-16	1E-11 [CNT6]	Not known	2.0E-17 [NW8]	Not known
Circuit energy, J	Projected	1.20E-17	1.5E-18 [CNT7]	6.25E-18 [G5]	2.0E-16 [NW9]	6E-18 [T1]
	Demonstrated	1.00E-15	Not known	Not known	Unknown	Not known
Subthreshold slope, mV/dec	Projected	60 mV/dec	60	60	60	<20 [T4]
	Demonstrated	62 mV/dec	60 [CNT8]	Not known	61 [NW10]	20 [T2]
Operational temperature		RT	RT	RT	RT	RT
Material challenges		Si	CNT density, contacts	Dialectics,substrates, in situ mobility, contacts	Assembly, directed placement	Not known

each relevance criterion, we get a minimum OPA of 8. If OPA $\geqslant$ 20, the potential of the technologies is deemed to be better than silicon CMOS. For OPAs ranging from 16 to 20, the potential of the technologies in this scale range are projected to be slightly better than silicon CMOS. For OPA < 16, the potential of the technologies are projected to be (significantly) worse than silicon CMOS [14].

4.9 Quantum confinement and associated devices

Quantum confinement is the foundation of many 'Beyond CMOS' devices currently under development. In a 3D or a bulk structure, electrons are generally free to move across all three dimensions. Three options are available to scale this bulk structure to the nanoscale, either by choosing to work in one dimension or two or three dimensions, as shown in figure 4.18. If only one dimension is reduced at the nanoscale, a quantum well like a thin film or graphene is obtained, which is a 2D structure. If two dimensions are reduced at the nanoscale, 1D quantum wire-like nanowires, nanorods and nanotubes can be obtained. If all three dimensions are reduced at the nanoscale, a QD (2–10 nm) or a dimensionless structure can be obtained. Quantum confinement is a constraint on the movement of rapidly and randomly moving electrons in a substance to different discrete levels of energy. It is

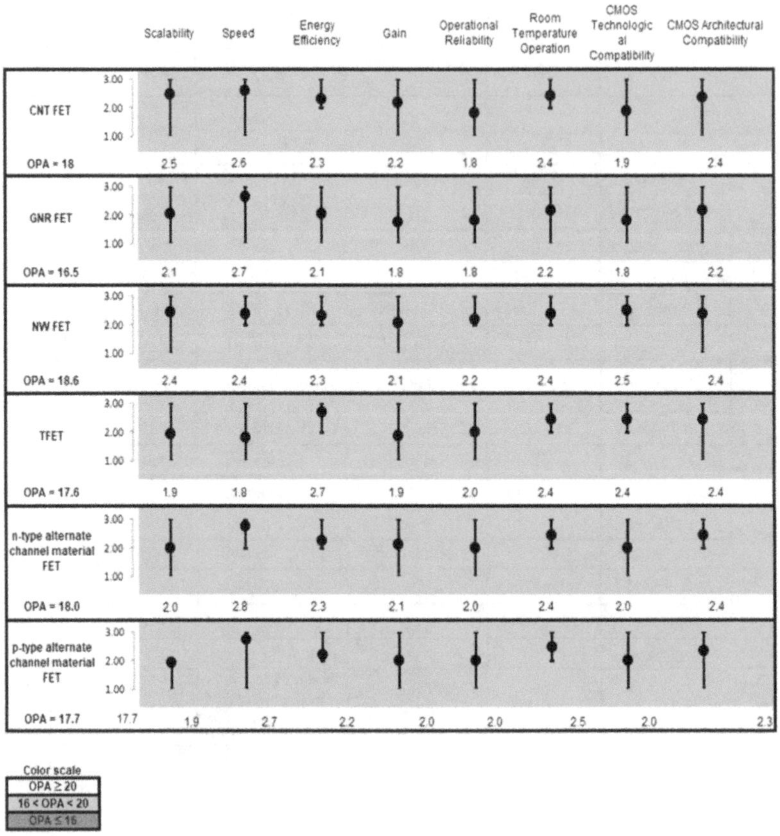

		Scalability	Speed	Energy Efficiency	Gain	Operational Reliability	Room Temperature Operation	CMOS Technological Compatibility	CMOS Architectural Compatibility
CNT FET	3.00 / 2.00 / 1.00								
OPA = 18		2.5	2.6	2.3	2.2	1.8	2.4	1.9	2.4
GNR FET	3.00 / 2.00 / 1.00								
OPA = 16.5		2.1	2.7	2.1	1.8	1.8	2.2	1.8	2.2
NW FET	3.00 / 2.00 / 1.00								
OPA = 18.6		2.4	2.4	2.3	2.1	2.2	2.4	2.5	2.4
TFET	3.00 / 2.00 / 1.00								
OPA = 17.6		1.9	1.8	2.7	1.9	2.0	2.4	2.4	2.4
n-type alternate channel material FET	3.00 / 2.00 / 1.00								
OPA = 18.0		2.0	2.8	2.3	2.1	2.0	2.4	2.0	2.4
p-type alternate channel material FET	3.00 / 2.00 / 1.00								
OPA = 17.7	17.7	1.9	2.7	2.2	2.0	2.0	2.5	2.0	2.3

Color scale
| OPA ≥ 20 |
| 16 < OPA < 20 |
| OPA ≤ 16 |

Figure 4.17. Potential assessment of different materials. Adapted from [14] with permission.

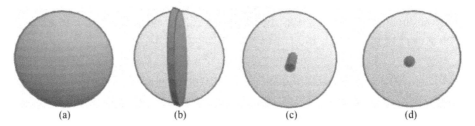

| (a) | (b) | (c) | (d) |

Figure 4.18. Schematic of (a) bulk (3D), (b) quantum well (2D), (c) quantum wire (1D), and (d) quantum dot (0D).

an impact where a typical material's dimensions are comparable to the involved electrons de Broglie wavelength.

In a quantum well only one dimension is confined, so electrons cannot move in that particular dimension. In a quantum wire two dimensions are confined, so that the electrons are unable to move in those two dimensions. In a QD, all three dimensions are confined so the electrons cannot move in any dimension. At the macroscopic level, the interaction between electrons are not taken into consideration because of the negligible effects observed at such a level. However, when downscaling to the nanoscale, the interaction between electrons can no longer be neglected because it gives rise to many effects, such as single-electron effects, that play a major role in quantum physics.

4.10 Quantum-mechanical tunnelling and Coulomb blockade in a single-electron transistor

Electrons behave both as particles and in some instance as waves, as discovered by Louis de Broglie [22] and Erwin Schrodinger in 1920. Study of this behaviour culminated in the formulation of the Schrödinger wave equation, which formed the basis of a new quantum mechanics. According to quantum mechanics, there is a nonzero probability that an electron will tunnel through a potential barrier. If the total amount of energy present in a circuit is lowered, the probability of tunnelling through the barrier becomes large. This phenomenon is known as the quantum tunnelling effect, and the junction through which the electrons tunnel (or not) is known as a quantum tunnel junction, as depicted in figure 4.19. The tunnel junction has a capacitance, C_j, and a resistance, R_t, which depend on the physical size of the tunnel junctions. Tunnel junctions with capacitors form the basic building blocks of SET circuits.

4.10.1 Orthodox theory

The 'orthodox' theory of single-electron tunnelling was developed by K Likharev [23]. It describes charging effects such as Coulomb blockades, Coulomb oscillations and electron transport in a single-electron circuit consisting of tunnel junctions, capacitance and voltage sources. The theory calculates the tunnelling rate across each junction, taking into consideration the change in free energy of the system. The orthodox theory makes the following approximations:

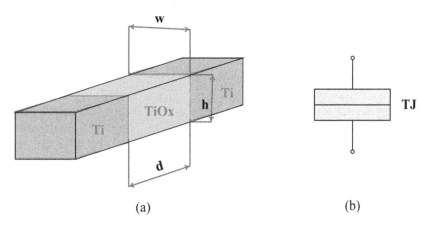

(a)   (b)

Figure 4.19. Tunnel junction. (a) Sketch, (b) diagram.

- The electron energy quantisation inside the conductors is ignored.
- The electron tunnelling time through the barrier is assumed to be negligibly small in comparison with the other timescales.
- Co-tunnelling (explained later in this section) is ignored.

In the absence of tunnelling, the electrical behaviour of a SET tunnel junction corresponds to a capacitor. A capacitor C_j, charged q, contributes $\frac{q^2}{2\,C_j}$ to the energy present in the circuit. Hence, the tunnelling events change the energy of the circuit in steps of

$$E_c = \frac{e^2}{2\,C_j},\tag{4.8}$$

where $e = 1.602 \times 10^{-19}\,\text{C}$ and the energy E_C is referred to as charging energy.

During tunnelling, the electrons move sequentially through the junction [24]. The electron charge can, however, make up any arbitrary portion of the charge on the tunnel junctions. A portion of the electron wave function being on the tunnel junction and a portion of it being scattered throughout the capacitive island in a SET circuit lead to noninteger multiples of the electron charge. Unfocalized charge will arise from this as a consequence. In order for the charging energy, also known as Coulomb energy, to prevail over the quantum charge fluctuations on each individual island, the tunnel junctions need to have a high enough tunnelling resistance. An equation can be used to indicate this:

$$\frac{e^2}{2\,C_j} \cdot R_t \cdot C_j \gg h \Leftrightarrow R_t \gg \frac{h}{e^2} = 26\,\text{k}\Omega.\tag{4.9}$$

Here, $R_t \cdot C_j$ is the characteristic time for charge fluctuations, the time constant for charging capacitor C_j through tunnel resistor R_t, and h is Planck's constant.

- The second effect relevant to the observability of a single electron tunnelling is the thermal energy. If the thermal energy dominates the charging energy E_c, the quantisation effect becomes nonobservable. Hence,

$$E_c \gg k_B T,\tag{4.10}$$

where k_B is Boltzmann's constant and T is the absolute temperature.

The quantity of free energy present in the circuit determines whether a tunnelling event will take place or not.

After a certain amount of time, td, if the amount of free energy drops as a result of a tunnelling event, then the electron will tunnel, bringing the circuit to a lower energy state. Thus, by calculating the free energy before and after the potential tunnelling event, it is possible to predict the tunnelling behaviour of electrons. The sum of the charges stored on each capacitor, including the charges transported by the voltage sources and the charges stored on tunnel junctions, yields the total amount of free energy in a circuit. It is given by

$$F = \sum_{i=0}^{k-1} \frac{q_i^2}{2C_i} - \sum_{j=0}^{l-1} q_j(t)V_j. \tag{4.11}$$

In the above equation, for a circuit consisting of k capacitors C_i each containing a charge q_i, which contributes to the free energy by $\frac{q_i^2}{2C_i}$, and l voltage sources V_j, where each transferred a charge $q_j(t)$ since $t = 0$, contributes to the total amount of free energy as calculated in equation (2.4). By calculating the amount of free energy before (F_{initial}) and after (F_{final}) the tunnelling events, it is possible to calculate the corresponding change in free energy $\Delta E = F_{\text{final}} - F_{\text{initial}}$ and hence the possibility of a tunnelling event.

SET works on the quantum-mechanical principles of the Coulomb blockade and quantum-mechanical tunnelling (QMT). A SET's unit cell is a tunnel junction (barrier) with a size on the nanoscale that is made up of two conductors and a thin oxide layer. In the device seen in figure 4.20(a), there is a small island and one or more gate electrodes that are capacitively connected to the island. The source as well as the drain electrodes are connected to the island via a tunnel barrier. In this case, the tunnel junction capacitances and resistances are C_s, C_d and R_s, R_d, respectively. The operating principle of a SET is the charge balance between the gate electrode and the island coupled via a capacitor C_{gi}.

Electrons tunnel from the source to the island and from the island to the drain terminal to produce the current through the SET. The source and drain terminals are grounded in order to ensure its operating principle. By applying a voltage V_{gi} to the ith gate, a charge of $C_{gi}V_{gi}$ is accumulated at the gate electrode, resulting in a total charge of $\sum C_{gi}V_{gi}$ at the gate. If the total charge of $-ne$ (where n is the number of electrons and e is the elementary charge) on the island is balanced with that in the gate, then $\sum C_{gi}V_{gi} = ne$ is satisfied. As n is a stable number of electrons in this situation, no current passes through the island. This is known as a Coulomb blockade. A further increase in the gate voltage increases the total charge on the gates and the charges become unbalanced. With $\sum C_{gi}V_{gi} = (n+1/2)e$, the electrostatic potential for n electrons and $n+1$ electrons becomes equal, which implies n or

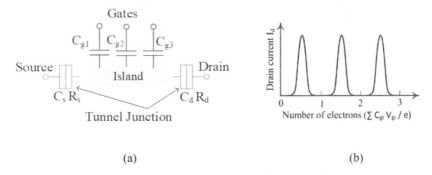

(a) (b)

Figure 4.20. SET. (a) Equivalent circuit of a multi-gate SET, (b) gate voltage vs drain current characteristics of a multi-gate SET.

$n+1$ electrons in the island. Electrons flow from the source to the island when a modest voltage is placed across the source and drain electrodes, leaving $n+1$ electrons in the island. Once an electron has tunnelled from the island to the drain, the number of electrons in the island returns to n. By repeating these procedures, the current consequently flows as a result of single-electron tunnelling. $n+1$ electrons become stable as the gate voltage is increased further. Figure 4.20(b) shows how this results in periodic features of the source–drain conductance as a function of the gate voltage.

4.11 Structure and working principle of single-electron transistors

Figures 4.21(a) and (b) show the 2D and 3D structure of a SET. In comparison with conventional MOSFETs, SETs feature a conductive island and two tunnel barriers at the centre instead of the channel. The gate dielectric and the electrodes (source, drain and gate) are similar to MOSFETs. The voltage sources are connected with the gate and drain electrodes, and the source is grounded. QMT is responsible for current conduction in SETs, and hence it is a desirable event, contrary to MOSFETs. The gate terminal controls the tunnelling from the source to/from the drain through the island [25].

Figure 4.22 shows a SET with two tunnel junctions, J1 and J2, which surround an island with ne initial charges, where n is the initial number of electrons [25]. Electron exchange occurs with the drain and source terminals only through an island. The gate electrode provides electrostatic or capacitive coupling. The electron number in the island is controlled externally through a gate capacitance, C_G. In such a device, the leakage through the gate capacitance is negligible due to the thick gate oxide. However, if one wishes to analyse it accurately, one can consider it as a parallel plate capacitor and perform exact analysis. The contact near the left tunnel junction, J_1, is the source electrode, and the contact near the right tunnel junction, J_2, is the drain electrode. Here, voltages $\frac{V}{2}$ and $-\frac{V}{2}$ appear at the source and drain, respectively. V_G is applied at the gate electrode.

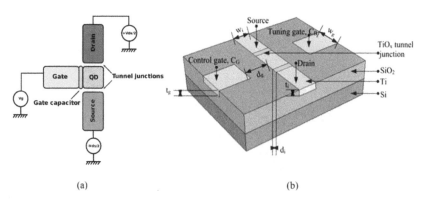

(a)   (b)

Figure 4.21. SET schematic. (a) 2D structure of a single-gate SET, (b) 3D structure of a two-gate SET. (This Set schematic.svg has been obtained by the author(s) from the Wikimedia website, where it is stated to have been released into the public domain. It is included within this book on that basis.)

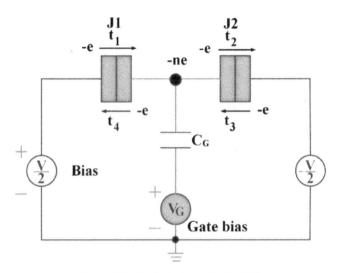

Figure 4.22. Equivalent circuit of a SET.

There are four possible tunnelling processes, t_1, t_2, t_3, and t_4, wherein an electron is added to or removed from n electrons due to the effect of tunnelling [25]. Let the change in the electrostatic energy due to the four tunnelling processes be ΔE_1, ΔE_2, ΔE_3, and ΔE_4, respectively. The change in electrostatic energy ΔE_1 due to the tunnelling process t_1 is given by

$$\Delta E_1 = \frac{e}{C_\Sigma} \left[C_\Sigma \frac{V}{2} - C_G \, V_G - e \left(n + \frac{1}{2} \right) \right], \text{ where } C_\Sigma = 2C_J + C_G, \quad (4.12)$$

where C_Σ is the total electrostatic capacitance seen from the island (junctions J1 and J2 have the same junction capacitance C_J for symmetricity). Similarly, ΔE_2, ΔE_3, and ΔE_4 are obtained by replacing in the above the last term with $+\left(n + \frac{1}{2}\right)$, $-\left(n - \frac{1}{2}\right)$, and $-\left(n + \frac{1}{2}\right)$, respectively. The equation for ΔE_1 is usually depicted via a C_Σ–V against C_G–V plot. The hatched region in figure 4.23 indicates the condition under which the tunnelling t_1 cannot occur when n electrons exist in the island [4, 25]. This structure is called a Coulomb diamond. If source–drain biasing is fixed and the gate voltage V_G is varied, then the SET characteristics pass along a line as shown, which cuts through the 'diamonds' formed by regions of forbidden and allowed tunnelling. This causes the source–drain current to flow periodically, as shown in figure 4.23 [4, 25].

To observe the Coulomb blockade effect in an actual device at 300 K, one must have a $\frac{e}{2C_\Sigma}$ much larger than kT for 300 K. At 300 K, this requires $C_\Sigma < 3.1$ aF, an ultra-small capacitance, requiring the device to be small scale (in a few tens of nanometre range) [4]. To overcome the Coulomb blockade effect and allow QMT to occur in the SET, three conditions must be satisfied:

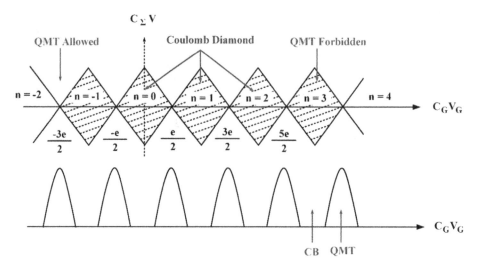

Figure 4.23. Coulomb diamond characteristics of SET.

1. The bias voltage that is applied must be smaller than the ratio of the elementary charge and the capacitance of the junction, $V < \frac{2e}{C}$.

2. The thermal energy, kT, of the system must be less than $\frac{e^2}{2C}$. Otherwise, electrons will gain enough energy from the environment to cross the junction.

3. The tunnelling resistance, R_T, of the tunnel junction must be greater than 25.813 kΩ ($\frac{h}{e^2}$).

It is necessary to determine the total capacitance from the geometry of a SET in order to create SET-based circuits. The circuit's required current determines the tunnel resistance R_T (> 25.8 kΩ). Device designers typically choose the value of R_T intuitively or through trial and error before settling on a suitable figure. Figure 4.24(a) illustrates the output characteristics of a SET. To understand the I–V characteristics, refer to the circuit in figure 4.22. The flat region represents the Coulomb blockade [4]. The effect of the blockade decreases with an increase in gate voltage. Figure 4.24(b) shows Coulomb oscillations whose magnitude increases with increasing drain voltage.

4.11.1 Modelling approaches for SETs

SET modelling must include the principles of Coulomb blockades and QMT with the design parameters of tunnel barriers and the island. The modelling requires one to define tunnel junctions and the island in terms of either standard elements or in I–V form. This helps ensure efficient circuit design and makes connections with other elements straightforward. The following details several different approaches used to model SETs according to the orthodox theory outlined above.

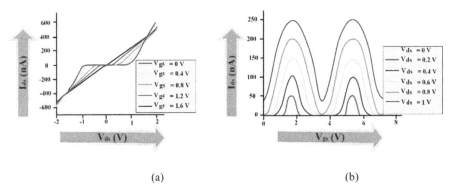

Figure 4.24. SET characteristics. (a) Coulomb blockade region, I_{ds} vs V_{ds}, (b) Coulomb oscillations for various drain bias, I_{ds} vs V_{gs} for a V_{ds} of 0 to 1 V.

Wasshuber *et al* [26] proposed a Monte Carlo (MC) simulation technique for single-electron-based devices and developed a software to this end, SIMON. Their method calculates the probabilities of all possible tunnelling events for one step for bias supply and determines one of the least possible random tunnel times. SIMON can accurately determine tunnelling current, power dissipation and can plot Coulomb diamonds for stability analysis. Other available tools, such as MOSES [27] and SENECA [28], are also MC-based tools for SET, but SIMON is a widely acclaimed and accurate tool, and boasts an easy-to-use graphical user interface. However, it should be noted that these tools can only be used for SET circuits; co-integration with FETs is not possible. In addition, the simulation of large circuits is a very time-intensive task.

Likharev *et al* [29] developed a master equation (ME) approach using the orthodox theory of SET. They also developed a simulator, named SETTRANS, for SET-based simulation. To compute the drain current of the circuit, equations based on statistical mechanics are solved for all possible states occupied by the circuit. The approach has benefits for small circuits, but as the circuit size expands, the number of states occupied by the circuit can be huge, and hence the essential states cannot be found easily. SPICE-based models are simple and flexible to use. In addition, the various features of electronic design automation tools and co-simulation with FETs brings many advantages. Yu *et al* proposed a SPICE macro modelling method [30] for SETs. In particular they presented a SET macro model featuring diodes, resistors and capacitors. They also simulated hybrid SET-MOS circuits. However, the room-temperature operation and realisation of large circuits using a macro model are complex problems.

Uchida *et al* developed an analytical model for SETs based on orthodox theory and a ME [31, 32]. Their model supports hybrid SET-MOS simulation. It is limited to drain bias $\left(\mid V_{DS} \mid < \dfrac{E}{C_{\Sigma}} \right)$, where E is the electrostatic energy and C_{Σ} is the total capacitance of the SET, and background charge is not considered. Further, their model is only applicable to symmetric SETs. Inokawa *et al*, however, developed a modified model that does consider background charge effect and is applicable to asymmetric SETS. The drain bias limitation remains unsolved. Royer *et al* presented a model with $\left(\mid V_{DS}\mid < \frac{E}{C_{\Sigma}}\right)$ based on a ME [33]. However, the model was complex and required a lot of time for large circuits.

Mahapatra *et al* developed an analytical model based on the prior models of Uchida *et al* [32] and Inokawa *et al* [33]. The Mahapatra–Ionescu–Banerjee (MIB) model depends on orthodox theory and a ME and can be applied in both digital and analogue applications. The model is accurate for both symmetric and asymmetric SET devices with single or multi-gate structures [34]. The authors further demonstrated SET-MOS co-simulation using this model. The MIB model is limited to conditions $\left(T < \frac{e^2}{10\,k\,C_{\Sigma}}\right)$ and $\left(\mid V_{DS}\mid < \frac{3e}{C_{\Sigma}}\right)$.

As mentioned, SETs are small in size (few nanometres), consume very low power as the number of electrons transferred are very low, and have a charge sensitivity of 10^{-6} $(e/\text{Hz})^{1/2}$. They have high operating speed and can be used as logic as well as memory devices. However, over the last two decades of their research, room-temperature operation, small feature sizes, fabrication challenges and a lack of available simulation models have been the major drawbacks for SET development [4, 14, 25]. Hence, SET technology is still yet to reach its maturity. However, recent research works on SETs for room-temperature operation [31, 35–38], high current drive [36–38], silicon processing, and hybridisation with CMOS [39–43] still give SETs an edge ahead of many other 'Beyond CMOS' nanodevices. SETs have been investigated by many researchers and explored as a successor to conventional MOS devices in the nanometre regime [14]. They offer the much in-demand potential for scalability and high packaging density [4, 14, 25]. Significant research in fabrication, logic development and modelling of SETs shows that SETs' logic circuits can override those of CMOS devices [31, 40] from a functionality perspective. SETs are also perfect candidates for nonvolatile programmable memory devices [34, 39, 44]. Furthermore, SETs have been taken up by many researchers as potential candidates in very-large-scale integration processes due to their merits over conventional MOSFETs, such as their nanoscale size, ultra-low power dissipation and distinctive Coulomb blockade behaviour. Though SETs exhibit such remarkable features, the prospect of complete replacement of CMOS by SETs remains uncertain. SETs, rather, are helping to overcome limitations pertaining to low current drive, high output impedance, lack of mature room-temperature operable technology, large-scale infrastructure, proven design methodologies, and economic predictability. CMOS technology still has its benefits, such as high speed and voltage gain, that can also compensate for the limitations of SETs. A CMOS-to-SET interface is possible as the range of current is very small and the voltage levels are manageable, usually several millivolts. Therefore, even though a complete replacement of CMOS by SET is very difficult in the near future, hybridising SET with CMOS can meanwhile open a window for new functionalities and better performance. Advanced co-simulation and design environments and compatible fabrication technology platforms with which to achieve successful hybridisation of CMOS-SET are expected [14]. The nano damascene process is the most suitable technique for the fabrication of SETs.

4.12 Other quantum structures and their applications

With a reduction in material size, at the nanometre scale, quantisation effects appear as the movement of electrons is confined. This leads to the formation of discrete energy levels, depending on the size of the structure. Based on this, artificial structures with different properties from bulk materials can be created. Thus, dimension control as well as the composition of the structure make it possible to develop a material's properties as per its intended application. Below are a list of such quantum structures and their applications.

4.12.1 Quantum dots

QDs are often referred to as very small, man-made semiconductor particles, usually no more than 10 nm in size. Their extremely small size makes their optical and electronic properties different from those of bulk materials. It is possible to control the band gap by controlling the size of a QD, i.e. by adding or removing material, the boundaries of a band gap can be adjusted. This obviously changes the frequency of the photon emitted when an excited electron drops down from the conduction band back to the valence band. This means there is also a change in the energy of the photon it absorbs. As $\lambda = h.c./E$, there is a change in the wavelength of the photons that the material absorbs or emits. This shows that QDs made up of semiconductors will exhibit a shift in the optical absorption and emission towards shorter wavelengths as the dots get smaller. This is known as blue shift [2, 4]. CdS and ZnSe dots can be sized so as to emit blue to near-ultraviolet light. By comparison, CdSe dots can emit light throughout the entire visible spectrum. InP and InAs can emit far red and near-infrared light. They are used in tagging and tracking biological species, to create special inks, dyes and paints, in light displays and chemical sensing. CdSe dots can be introduced into cancerous cells in the body. Application of light causes these dots to glow, helping to zero down the scope of erroneous targeting of healthy cells during surgery and treatment. TiO_2 and ZnO particles are used in sunscreen as they absorb harmful ultraviolet rays. In bulk sizes, they also scatter visible light. As such, if used in paints, they help white pigmentation. It is this same property, however, that can form white patches when used on the skin. Yet when they are used in the form of QDs of a size <50 nm, they assume a large band gap. Thus, they become transparent to visible light but also absorb high-energy ultraviolet radiation. The typical energy band gap in solar cells (with a maximum efficiency of approximately 34%) is around 1.4 eV. However, solar rays contain photons from a wide spectrum of energy. Therefore, photons with energy <1.4 eV will not be absorbed. Hence, solar cells are aggregated with semiconductor QDs of varying sizes. Depending on the size, different photons will be absorbed by different QDs. Photon absorption will lead to excitation of electrons from the occupied valence band to the unoccupied conduction band, thus leading to an increase in the solar cell current and hence improving its power output [2, 4, 45]. The overall efficiency of a solar cell can in this way be improved. Based on their size, their highly tunable optical properties are fascinating, resulting in a variety of research and commercial applications including bioimaging, solar cells, light-emitting diodes (LEDs), laser diodes, quantum computing and transistors.

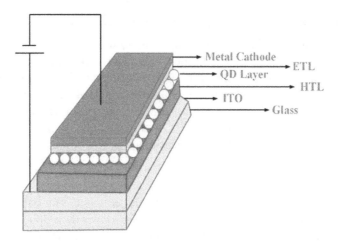

Figure 4.25. QD-LED structure. Adapte from [45]. CC BY 3.0.

In normal LED displays, white LEDs are used; with the help of colour filters, red, blue and green colours are obtained. Here, the colour filters are not 100% efficient: some of the light (energy) is lost. In so-called Q-LEDs, blue LEDs and QDs with different sizes, which can be designed to generate red, blue and green colours, are used. As ascertained, QDs reflect colour according to size. Therefore, pure red, green and blue colours can be achieved with the help of QDs. This has a dual benefit: the making possible of pure/accurate colours and the minimising of light/energy lost due to colour filters [45]. Figure 4.25 shows a multilayer LED structure in which the QDs act as an active layer. These LEDs consist of a hole transport layer (HTL), an electron transport layer (ETL) and the emissive layer of QDs.

QDs have been used in powerful supercomputers, commonly known as quantum computers. Quantum computers use 'qubits' (quantum bits) for processing and storing information that can exist in both an ON and OFF state at the same time. This amazing development allows both the time and memory complexity of information processing to be vastly improved compared to conventional computers.

4.12.2 Quantum wires

Quantum wires are lightweight and have high electrical conductivity, small diameter, low chemical reactivity and high tensile strength. Therefore, they are highly preferable in the manufacturing of electronic devices like double-layer transistors, nanowires, etc. Conventionally, wires are passive components. Yet, if small enough, they acquire new properties and can behave as active components such as sensors, transistors or optical devices. Nanoscale wires help in investigating the electrical transport and mechanical properties of materials. They are crystalline in nature and can be manufactured from metals, semiconductors, superconductors, polymers and insulating materials. They have very high aspect ratios. Their common dimensions are of lengths of tens of micrometers and widths of 1–200 nm. Structures like belts and rods can also be made in this regime. An additional aspect at this scale is that if two wires have different crystal structures, surface conditions or aspect

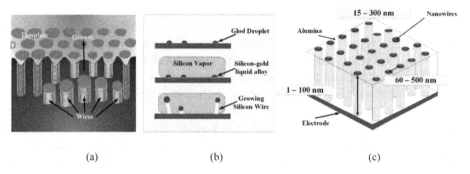

(a)   (b)   (c)

Figure 4.26. Graphical representation of (a) nanowire growth from template, (b) vapour–liquid–solid technique to fabricate silicon nanowires, (c) nanowire formation. [2], adapted by permission of the publisher (Taylor & Francis Ltd, http://www.tandfonline.com).

ratios, then they will behave totally differently even if they are made up of the same material. For example, gallium nitride (GaN) wires with a diameter of less than 100 nm will radiate different wavelengths according to their crystal structure. This is due to their high surface-to-volume ratio [2].

Nanowires can be manufactured by using templates or by using vapour–liquid–solid techniques. In growth from a template, anodic alumina (Al_2O_3) is used as a template. This has tiny pores of diameter 1–500 nm, as shown in figures 4.26(a) and (c) [2]. The wire material fills the holes either in liquid form under high pressure and temperature, in vapour form, or by electrochemical deposition, in which ions are drawn into the template by using a voltage bias. As soon as the material is emplaced, it takes on the shape of the cavity. A wire is frequently formed by adding more wire material, which extends from the template and keeps the templated shape. As illustrated in figure 4.26(b), a vapour–liquid–solid technique can be used to fabricate silicon nanowires. A droplet of a catalyst (in this case, molten gold) is put on the surface. Silicon gas is inserted, which absorbs into the gold until it becomes supersaturated with silicon. Here, a silicon speck is solidified and forms a seed. Further deposition of silicon causes the wire to grow. The molten gold bead remains at the top of the wire.

Nanowires have several interesting features unseen in bulk and 3D materials. These features are seen in nanowires because the electrons are confined at the quantum scale and, thus, they fill energy levels other than the conventional spectrum of energy levels or bands that are not present in bulk materials. In optoelectronic and nanoelectromechanical devices (i.e. NEMS), nanowires can perform an important role as sophisticated components, as metallic interconnects in nanoscale quantum devices, and as leads for bimolecular nanosensors.

4.12.3 Quantum wells

Films, layers and coatings fall into the category of 2D quantum structures. Here, most of the physical interactions are surface interactions. An object's surface type makes it either useful or useless. Appropriate coating onto a product can make it more efficient, less prone to friction and more resilient to wear. Because of their quasi-2D nature, electrons in quantum wells have a density of states as a function of energy with distinct steps. A reduced quantity of active material in the quantum wells leads to an increased performance in optical devices such as laser diodes. Therefore, quantum wells are commonly used in fiber optic transmitters, manufacturing of high-electron-mobility transistors (or HEMTs), solar cells and infrared photodetectors [2]. Below brief descriptions are given of such applications.

The conventional single-junction solar cell's theoretical maximum efficiency is about 34%, primarily due to the inability to absorb several different light wavelengths. Solar cells consisting of multiple n–p junctions of various series-connected band gaps, as shown in figure 4.27, can improve the efficiency, (theoretically) extending the range of absorbed wavelengths, though the complexity of manufacturing and production costs restrict their use to general applications. However, if one or more quantum wells are used in the internal region in the p–n junction of the solar

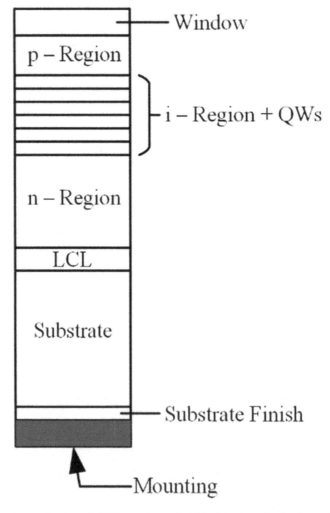

Figure 4.27. Quantum well solar cell. QW: quantum well; LCL: lateral conduction layer. Adapted from [46] with permission.

cells, it will give an increased photocurrent over dark current, which can increase efficiency over traditional p–n solar cells [47].

Quantum wells have shown potential as thermoelectric devices for energy harvesting. It is claimed that they are easier to manufacture and grant the ability to work at normal room temperature. In a thermoelectric device, as shown in figure 4.28, a central cavity is attached to two electrical reservoirs by quantum wells. The central cavity is maintained above the reservoirs at a hotter temperature. The quantum wells perform the role of filters, allowing the passage of electrons over a certain energy. An observational device produces an output power of approximately 0.18 W cm^{-2} for temperature with a difference of 1 K. Quantum wells can transmit electrons of any energy above a certain threshold, whilst QDs can transmit only electrons of some specified energy. One of the possible applications is to convert the unwanted heat back into electricity from electrical circuits, as for example in computer chips, reducing the need for powering and cooling the chip. In this scheme, a central cavity in thermal equilibrium with a hot reservoir of temperature T_h is connected to cold electronic reservoirs with temperature T_C via quantum wells with subband thresholds $E_{L,R}$. The temperature difference drives a current against the applied bias voltage.

A quantum well laser is a type of laser diode wherein the device's active region is so small that there is an occurrence of quantum confinement. Compound semiconductor components that are efficient in emitting light are used to produce laser diodes. The wavelength of light emitted by a quantum well laser is determined not just by the band gap of the material from which it is built, but also by the width of the active region. In a well laser, wavelengths can be altered by changing the width of the quantum well layer, where in standard lasers the layer structure should be changed. This implies that, with the help of specific semiconductor materials, longer wavelengths can be obtained with quantum well lasers than traditional laser diodes. Because of the stepwise structure of its state density function, a quantum well laser has better efficiency than a traditional laser diode.

In stimulated emission of radiation, a photon encounters an already excited electron and causes it to emit another photon. By doing so, the electron drops down to a lower energy state and finally two photons are produced in place of the originally incident single photon. This stimulated emission usually takes place in a material when there is population inversion; that is, where the excited electrons outnumber the electrons in

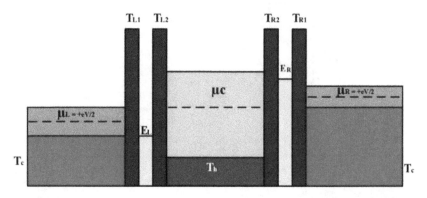

Figure 4.28. Energy diagram of a quantum-well heat engine. Adapted from [47]. CC BY 3.0.

the ground state. In conventional lasers, the energy pumping required for population inversion is achieved by using light, electricity, chemicals or combinations thereof. This can also be achieved by using quantum confined systems [2].

A quantum well is made by constraining electrons into a region of minimal thickness. Typically, the structure uses sandwiched layers of different semiconductors like GaN and indium-doped GaN (InGaN), as shown in figure 4.29(a) [2]. Each layer of doped GaN (3–4 nm thick, high conductivity) behaves like a quantum well separated by pure GaN (low conductivity). A power supply provides electrons.

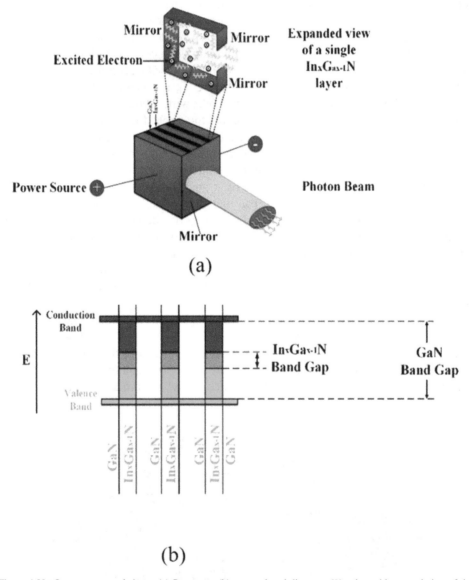

Figure 4.29. Quantum cascade laser. (a) Structure, (b) energy band diagram. [2], adapted by permission of the publisher (Taylor & Francis Ltd, http://www.tandfonline.com).

As shown in figure 4.29(b), $In_xGa_{x-1}N$ has a smaller band gap than that of pure GaN. Any electron arriving in the conduction band of $In_xGa_{x-1}N$ is quickly trapped. However, it lacks the energy to surmount the larger band gap of GaN. Thus, it remains there. Together, these trapped electrons have quantised energy levels and behave like particles in a quantum well. Stimulated emission causes excited electrons to emit photons that reflect between the mirrors on either end of the device, until finally they are emitted out of the aperture. Electrons excited by the power supply usually emit a photon within around 100 ns after excitation by the process of spontaneous emission. However, in this case, due to quantum effects, tunnelling takes place across the GaN barriers. The excited electrons keep on vanishing and reappearing in adjacent wells, and soon there is a build-up of population inversion. This initiates the necessary condition for the laser. However, an excited electron cannot keep on tunnelling forever; eventually, some of the trapped electrons undergo spontaneous emission. The photons emitted encounter the already excited electrons, resulting in stimulated emission with amplification in the number of photons. The dimensions of the quantum well and indium doping decide the wavelength of the laser.

Layers or coatings are usually synthesised by self-assembly methods [2, 8]. Molecules or particles form the individual units of the coating material, and are designed such that they template themselves and form densely packed layers, less than a micron thick. The right conditions will allow the unit to organise and assemble on its own. A coating with a single layer of highly organised molecules is called a monolayer [2]. A self-assembled monolayer (SAM) is shown in figure 4.30. The tail group physically reacts with the environment and defines the surface properties. The backbone chain group is the spacer that defines the thickness of the layer. The molecules are fixed to the substrate by a chemical bond created by the head group. SAMs can change a surface's surface energy or make it chemically inert or receptive to a certain kind of molecule or reaction. They have the ability to make a surface hydrophobic, hydrophilic, adhesive, slick, rough, and electrically positive or negative.

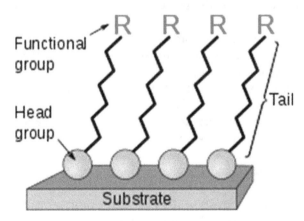

Figure 4.30. A self-assembled monolayer (SAM) [2].

Particles can be dispersed as coatings by using self-assembly, CVD or electro-deposition. Deposition of particles enhances surface area as the interaction sites increase. For instance [2], a spray-on coating containing a photo catalyst and nanometre-sized TiO_2 can create a self-cleaning glass pane. Because the particles are so minute, they chemically bond to the glass and are transparent to visible radiation. A glass that was previously hydrophobic now exhibits hydrophilia due to the coating. As a result, water spreads uniformly and sweeps away the dust that has been broken down by the photo catalyst using energy from ultraviolet radiation, rather than forming dirty steaks. Similarly, a thin layer of TiO_2 particles spread on our skin can absorb cancer-causing ultraviolet radiations.

4.13 Nanoelectromechanical systems

Microelectromechanical systems (MEMS) have been used for a long time. However, recent breakthroughs in nanoelectromechanical systems (NEMS) have conferred even greater benefits such as almost zero current leakage, fast switching and much more. As the name suggests, NEMS are devices with enhanced electrical and mechanical functionality with the special characteristic that they operate on the nanoscale. By integrating transistor-like nanoelectronic components with mechanical actuators, pumps or motors, they can be used to form physical sensors. Moreover, biological and chemical domains are also being exploited for the same. They will soon be used as a technology to make possible combining engineering and the life sciences in ways that are currently impractical with micro-scale tools and technologies. Particularly at the molecular level, NEMS hold the promise of revolutionising measurements of incredibly minute displacements and incredibly weak forces. The following are some merits of NEMS devices [48–50]:

(i) No leakage: in semiconductors, when charge carriers such as holes and electrons tunnel through the gate-insulating region, it constitutes leakage. In MOS transistors, leakage can occur between the source and drain, too, and is known as subthreshold conduction. In NEMS, an air gap is present between the drain and the source in the OFF state. Hence, they have no leakage.

(ii) Fast: due to their extremely small size, NEMS switches are relatively fast, as only a small voltage is enough to change state from OFF to ON. This also reduces power consumption.

(iii) High I_{ON}/I_{OFF}: since in NEMS there is zero leakage and fast switching characteristics, so, I_{ON}/I_{OFF} is very high.

(iv) Tolerance to high temperature: as discussed, graphene and CNTs have the properties of high thermal conductivity, high Young's modulus, etc., which make NEMS devices (which are fabricated from graphene or CNTs) tolerant to high temperature and other conditions.

A NEMS relay with an electrostatically operated cantilever can be utilised by this technology as a storage device. In comparison to a flash memory cell, it offers a lower P/E voltage and reduced latency [51]. A nanomechanical memory cell

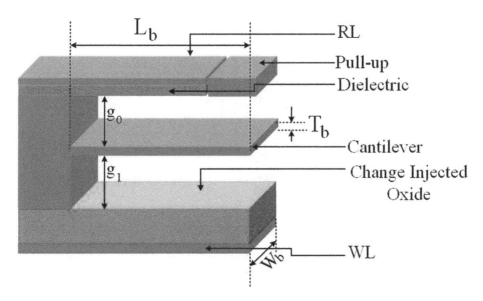

Figure 4.31. Structure of a cantilever switch. © 2022 IEEE. Adapted, with permission, from [52].

with two actuation terminals is depicted in figure 4.31 [52]. The device has four terminals: a pull-up terminal, a cantilever, a write line (WL) and a read line (RL). Starting from the bottom, the WL, a metal terminal made of the same material as the cantilever, is used to pull in the cantilever. An oxide layer sits on top of it and serves three functions: first, it provides insulation between the cantilever and the WL; second, it lowers the effective relative permittivity in the space between the two to strengthen the WL's electric field; and third, it makes the switch bistable when charged during manufacturing.

The cantilever beam, which forms the following layer, works as an actuator. The same material, only with greater dimensions, is used to fix the cantilever at one point to increase path resistance. This can be applied to regulate the current in the electrical path pull-up terminal—cantilever—ground. Another dielectric layer sits above that one and does the same job as the lower layer, but without the charge injection. The RL used to actuate the cantilever during the reading process forms the final layer.

The cantilever experiences an electrostatic attractive force when $-V_{DD}$ is applied to the WL in the absence of charge injection. The cantilever collapses towards the bottom if this force is greater than a specific threshold. The pull-in state is used to describe this. The V_{PI} (pull-in voltage) is the minimal WL potential at which pull-in occurs. The cantilever's spring tension overcomes the electrostatic force of attraction as well as the pressure of the contact adhesion when this voltage is dropped to a certain degree. This voltage, known as V_{PO} (pull-out voltage), causes the cantilever to return to its starting position. The same holds true if the RL is subjected to $-V_{DD}$. However, in this instance, when the cantilever is in the pulled-in position, it will make contact with the pull-up terminal, causing the potential on the beam to

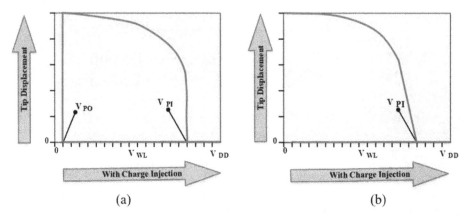

Figure 4.32. Cantilever tip displacement profile (a) with and (b) without charge injection into bottom oxide layer. © 2022 IEEE. Adapted, with permission, from [52].

increase to V_{DD}. Here, the cantilever serves as the switch's output terminal, or the bit line as it is known in relation to memory cells [52].

The ensuing electric field and electrostatic force towards the WL is amplified when carriers (electrons) are injected into the bottom oxide layer. This keeps the cantilever from releasing when the potential at the WL reaches zero. This is how a charge injection transforms a monostable relay into a bistable relay that can be utilised as a nonvolatile memory cell.

Figure 4.32 shows the switch's tip displacement profile both with and without charge injection. Similar to how charge leakage and trapping take place in a floating gate MOSFET (FGMOS), this charge injection is accomplished by applying a high voltage across the oxide–metal interface [53]. The primary distinction is that whilst in FGMOS the charge injection takes place each time data is written to the cell, in the proposed design it is carried out as part of the manufacturing process to make the switch bistable. The cantilever's behaviour is determined by the design criteria, which include material choice, geometry, size and aspect ratio. The electric field equation and the beam deformation equation from solid mechanics are combined in a mathematical model. The displacement of a point on the cantilever in this mdel is given by the below equation [54]:

$$y = -\frac{W_{dist}}{6EI} (2L_b^3 - 3L_b^2 a + a^3),\qquad(4.13)$$

where W_{dist} is the distributed load on to the cantilever, E is the material's modulus of elasticity, I is the cantilever's moment of inertia, and a is the distance of the point at which the distributed load starts the tip of the cantilever. The equation below provides the cantilever's moment of inertia [54]:

$$I = \frac{1}{12} W_b T_b^3.\qquad(4.14)$$

The distributed load on the cantilever is modelled as [54]

$$W_{\text{dist}} = -\frac{V^2 \varepsilon_0 W_b}{2g_0^2}. \tag{4.15}$$

These equations can be used to determine the cantilever's anticipated deflection profile at a given voltage. The tool MEMSLab [55] (on NanoHub.org) uses this mathematical model. It is preferable to use a material for the cantilever that has a lower elastic modulus in order to maximise the amount of deflection and obtain a lower pull-in voltage. Moreover, this will result in a drop in pull-out voltage. The pull-out voltage must be as low as possible for a bistable switch in order to prevent the pull-out with the least amount of charge injection. In the relaxed state, for the read '1' process, the charge distribution at the bottom dielectric layer (−VDD to RL) has a potential of the same polarity as that of the RL, but creates an attractive force in the downward direction. This can cause the cantilever to not pull-in towards R_L. To summarise, it is desirable to have a material with low modulus of elasticity for the cantilever element. Material selection, aspect ratio and air gap all play an important role in the design of NEMS switches, which can be used in designing nonvolatile memory with very small program/erase voltages.

NEMS devices, due to their small size, are extremely sensitive to many factors such as temperature, pressure, mass, etc. Thus, they are extensively used in cars and in mobile phones, as well as in many other applications, as sensors. This property has recently been utilised to demonstrate mass spectrometry of biomolecules using extremely sensitive NEMS devices, as illustrated in figure 4.33(a).

Nanometre-sized oscillating beams have very small mass and hence very large natural oscillating frequencies. For example, a silicon beam 100 μm long, 20 μm wide and 1 μm thick weighs about 5 ng and has a natural frequency of about 140 kHz [2]. The ability to weigh objects as small as a single bacterium is made feasible because of this. New chemical, physical and biological mass-detection applications are further made possible by this sensitivity when detecting extremely small amounts of a given substance. A beam's natural frequency is where the resonance curve of that beam peaks. There is a discernible variation in frequency with increased adsorbed material mass. With this method, chemical and biological compounds can be detected with great sensitivity. NEMS memory elements are composed of nanomechanical beams, which are typically constructed of silicon and are clamped to supports at either end, as illustrated in figure 4.33(b). They have thicknesses and widths of several hundred nanometres and lengths of a few micrometres. Their operation is based on a nonlinear phenomenon that develops in oscillating elements under the influence of very strong driving forces. For a given driving frequency, the amplitude of the oscillations can have two distinct values, which can be depicted as logic 0 and 1 [2]. The device is driven and its deflection is detected by an electromagnetic method. It is possible to measure deflections less than 0.1 nm. It is a volatile memory if the beams only hold a particular state whilst the electricity is ON. If the beam is made up of a material with intrinsic stress, then they do not switch states until a new pulse is applied, even if the power is switched off. Such beams are used in nonvolatile memory [2]. The resonant

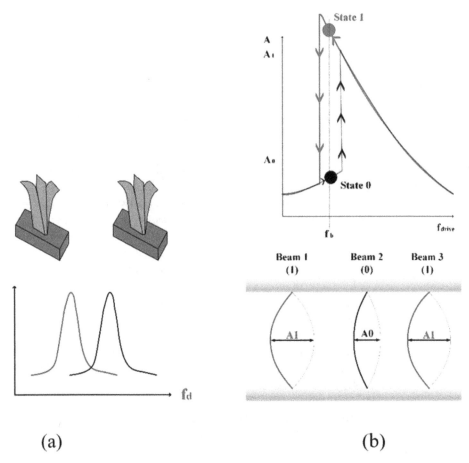

Figure 4.33. NEMS devices. (a) Mass sensor, (b) memory element. [2], adapted by permission of the publisher (Taylor & Francis Ltd, http://www.tandfonline.com).

frequency of these beams is very high due to their lower mass. As such, they can switch between the states very quickly.

NEMS devices are now being printed using 3D printers with nanometre precision. They can be used for many measuring tasks and printed from many types of materials including silicon, polymers, plastics, etc. They provide a positioning accuracy on the substrate of less than 5 nm for sensor dimensions down to 10 nm [2]. With this precision, sensors can be applied accurately at the points where pressure forces actually occur. Yet NEMS switches suffer from a serious limitation: the voltage required to activate the circuit as compared to that required for a typical CMOS switch activation voltage is larger by many orders in magnitude [49, 53]. A charge pump can be employed to create these large voltages. These will, however, perish soon, but the overhead energy and boot-up delay, when seen with reference to commonplace fast Fourier transform processing, is much less. Thus, more research is required to determine better power gating possibilities and their fabrication. Due to the small size of NEMS, fabricating such devices is difficult and research is ongoing in the field of fabrication. The procedures are broadly categorised into two categories [49, 53]:

1. Top-down approaches: these approaches include X-ray fabrication and nano electron-beam lithography. They have proven to be quite successful; the technologies' performance, when analysed, is on par in terms of productivity, process controllability, clump item capacity and similarity with present IC innovation.
2. Bottom-up approaches: these approaches comprise single-atom manipulation and QD self-assembly methods. In the past, MEMS technology was used successfully to construct numerous micro-scale sensors, actuators and micro-devices, the sensitivity being the key issue in the selection of the process.

4.14 Atomic switches

An atomic switch is a nano-ionic device which works on the principle that, during the switching operation, the diffusion of metal ions and their oxidation or reduction process is controlled [56, 57]. These are a class of electrochemical switches, based on their working principle. The significance of atomic switches is that they enable downscaling to less than 11 nm, which is the current size of the technology node.

Some of the novel characteristics of these switches include their lower power consumption, smaller size, lower ON resistance, nonvolatility and the switch possessing learning abilities. In atomic switches, the diffusion of ions and their reduction/oxidation processes can be controlled to form/annihilate a conductive path between the two electrodes of the switch. A schematic of an early atomic switch is shown in figure 4.34(a). As shown in the figure, silver (Ag) and gold (Au) form the electrodes for an Ag-doped As_2Se_3 ionic conductive material. When a bias is applied between the Ag and Au electrodes, it forms Ag filaments, as shown in figure 4.34(b), and hence a conductive path between the two electrodes is established. On reversing the voltage, it annihilates or destroys the Ag filament, which turns the switch off. Using such a setup, it is possible to achieve a switching time shorter than 1 μs.

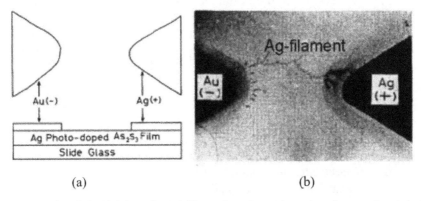

(a)   (b)

Figure 4.34. Atomic switch. (a) Schematic, and (b) top view of an early version of an atomic switch. © 2010 IEEE. Reprinted, with permission, from [56].

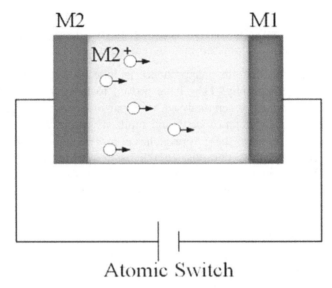

Atomic Switch

Figure 4.35. Schematics of a gap-type atomic switch. M1 is the inert electrode, M2 is the reversible electrode. © 2010 IEEE. Reprinted, with permission, from [56].

In 2001, a gap-type atomic switch, as shown in figure 4.35, was developed. Annihilation and formation of a conductive path in the nanogap between the electrodes is the basic working mechanism of such switches [56]. The two electrodes are different from each other in this switch: one is a reversible electrode on which an ionic conductive material can be formed, whereas the other is an inert metal electrode. During the ON state, a low resistive state is achieved for the ionic conductive material. This can be measured by flowing a current though it. With the application of a positive voltage a bridge is formed between the ionic and the electronic conductive material and the metal electrode, turning the switch ON.

A bias with opposite polarity will turn the switch OFF due to annihilation of the atomic bridge as the cations are redissolved into the conductive material. Such switches can be downscaled to the 11 nm technology node. In addition, such atomic switches can be two or three terminal in design. Two-terminal switches can be used as memory or programmable switches; three-terminal switches can be used as nonvolatile transistors. These designs are briefly described in the following.

4.14.1 Two-terminal switches

Gapless and gap-type are two types of two-terminal atomic switches. The essential difference between these two types is shown below.

The working mechanism of a gap-type two-terminal switch [56] is shown in figure 4.36. Ag is used as a reversible substrate, Ag_2S as the ionic and electronic mixed conductor electrode, and platinum (Pt) as the counter metal electrode. The formation of a bridge in the gap occurs. This represents the ON phase of the switch. Ag and Pt are used as reversible substrates. Ag_2S is the medium between the two terminals. As can be seen, the bridge forms inside the gap between the terminal and substrate.

Figure 4.36. Working mechanism of a gap-type two-terminal atomic switch turning ON. © 2010 IEEE. Reprinted, with permission, from [56].

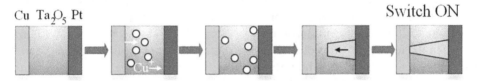

Figure 4.37. Working mechanism of a gapless two-terminal atomic switch turning ON. © 2010 IEEE. Reprinted, with permission, from [56].

The working mechanism of a gapless two-terminal atomic switch [56] is shown in figure 4.37. In this type of switch, copper (Cu) is used as a reversible electrode, Ta_2O_5 is the ionic conductive material and Pt serves as the inert counter electrode. The formation of a bridge in the medium can be seen. This represents the ON phase of the switch. Cu and Pt are used as reversible substrate. Ta_2O_5 is the medium between two terminals. As can be seen, the bridge forms between the two terminals inside the medium substrate. These switches have many potential applications that can help us drastically improve the current technology. They can be used as memory units, can be used as logic gates, and are also programmable, which means we can use them as programmable switches [56, 57].

Atomic switches are bipolar nonvolatile switches. This is why they require positive and negative bias voltages for erasing and writing. Hence, they require novel designs for peripheral circuits, cell structures and architectures. However, atomic switches merit new design development and research, as they have shown considerable potential in being able to achieve a scaling of less than the current 11 nm technology node [55, 56]. One of the important things is that logic circuits (such as logic gates) can be configured using two-terminal atomic switches, in that we can configure all three basic gates using a two-terminal atomic switch (i.e. AND, OR and NOT gates) as opposed to other conventional two-terminal devices which can configure only two (i.e. AND, OR gates). Because of their small size, nonvolatility and low ON resistance, atomic switches are a good fit for field-programmable gate arrays (FPGAs). They may also help to reduce the size of boards by up to 1/30 as compared to current switching circuits that use static RAM (SRAM) and volatile transistors made from semiconductors [56, 57].

4.14.2 Three-terminal switches

Three-terminal atomic switches are another type. As the name suggests, these switches have a third terminal. This third terminal is responsible for monitoring the creation and destruction of the bond between the other two terminals. The metallic bond between the source and drain terminals allows the electron flow from one to the other. The three-terminal atomic switch design has a benefit. There are two lines: the signal and control lines. Both lines are divided into three-terminal atomic switches [56, 57]. Semiconductor transistors show the same behaviour. Because of this, three-terminal switches act like semiconductor transistors. Semiconductor transistors can be used to form logic gates, hence why three-terminal switches are better suited as an option for gates in logic devices rather than as an option for memory devices. A demonstration of a three-terminal switch can be undertaken by using Cu_2S as the conductor filament between the terminals. Pt is chosen as the material for the source terminal and Cu for the gate and drain. A schematic is shown in figure 4.38(a) [56]. The creation and destruction of the ionic path between the source and the drain are monitored by the gate, as shown in figure 4.38(b). This operation is performed repeatedly, and the results are provided in figure 4.38(c).

In addition to logic devices, atomic switches can also be used as memory devices [56, 57]. As aforementioned, atomic switches are bipolar and nonvolatile. Because of this, we need to have a bipolar bias voltage to write or erase something. However, there is a caveat: to do so, we would need to redesign devices such as cell structures and peripheral circuits. These devices, which are commonly used commercially, need their architecture to be modified in order to adapt to the jump from conventional parts to atomic switches. Yet, as mentioned, a strong advantage is that atomic switches may allow the making of nodes smaller than 11 nm technology. This is encouraging and, due to this, it can perhaps be overlooked that an initial redesign is needed. Indeed, using atomic switches, a 1 KB-sized memory chip has been developed. Peripheral circuits, used to read and write on atomic switches, have been developed as well. These circuits use transistors, in which for each transistor there is an atomic switch provided to choose if the switch is to be written or read. Because of the use of one transistor for each atomic switch, it is called 1T-1S technology.

As we have seen before, atomic switches can also be used to formulate logic gates [56, 57]. Two-terminal atomic switches are a good alternative for this. They can be

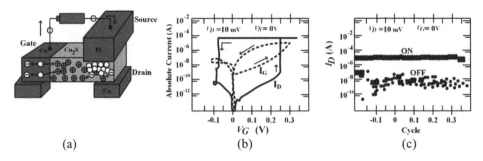

Figure 4.38. Working of three-terminal atomic switches. (a) Three-terminal atomic switch, (b) relation between current and voltage, (c) sequential operating state of the three-terminal atomic switch. From [57]. © National Institute for Materials Science (NIMS), adapted by permission of Taylor & Francis Ltd (http://www.tandfonline.com), on behalf of the National Institute for Materials Science (NIMS).

used to build very sturdy and smaller devices. Let us take a closer look at the mechanics of the logic gates made from atomic switches. First, for an AND gate. This utilises two gap-type atomic switches. Inputs are fed through the Pt terminal of the atomic switches. The output is measured by the Ag_2S terminal. The terminal gives a higher voltage only when the higher voltage has been applied to both the Pt terminals. Thus, if both Pt terminals are at high voltage, only then will the Ag_2S terminals show high voltage, just like how AND gates work. An OR gate also consists of two gap-type atomic switches. However, in this design, two wires are made out of Ag_2S and one wire is made out of Pt. The output of this system becomes high even if only one of the Ag_2S wires is at high voltage. Only when both of the Ag_2S terminals are set to low voltage will the output show the low voltage. It is similar to the workings of an OR gate. A NOT gate, by contrast, is made out of only one atomic switch. It uses extra capacitors and resistors to simulate the working of a NOT gate. Here, the output is registered as low whenever the input feed is high, and for low input feed it registers as high at the output terminal. This behaviour is attained by using discharging and charging currents for the capacitor, which influences the ON and OFF phases of the atomic switch.

Programmable switches are another field where atomic switches have been found to be useful. Their noted characteristic nonvolatility and lower ON resistance, along with their smaller sizes, are the main reasons they are a good alternative for the conventional switches used in FPGAs. Atomic switches can be reconfigured, which is one of their main strengths [56]. Large-scale integrated circuits (or LSIs) that can be reconfigured, i.e. FPGAs, are devices that can be used to achieve many functions whilst using a single board or chip. However, the large area consumed by the switching circuits downgrades the potential performance of FPGAs in comparison to application-specific integrated circuits (ASICs). These switching circuits use semiconductor transistors and SRAMs. If we use atomic switches instead of semiconductor transistors and SRAMs, we can downsize the boards or circuits by up to 30 times their present size [56, 57]. Further, the formation of atomic switches on the layers of CMOS devices would help, too. These layers are made of metal. It will help to reduce the area for the switching circuits. In this way, atomic switches would help to make different types of programmable logic devices, like cell-based programmable ICs (or CBICs). This would help to integrate all the functions and operations on a single board, potentially achieving as high a performance as that of ASICs.

Thus, atomic switches show the significant developments made in the domains of nanotechnology and nanodevices, and have important applications in a variety of fields. Crucially, they have shown the potential to reduce the size of the technology node to less than 11 nm, which is the current size. In summary, we can say that atomic switches are the future for downsizing some current technologies, the miniaturisation of which could help us achieve many goals.

4.15 Mott FETs

To miniaturise a circuit requires radical changes in the material and design, so that additional scale reduction can be achieved. Further reduction of channel length using a SiO_2 gate insulator is near impossible. This is where so-called 'Mott FET' technology comes into play [58, 59]. In the Mott FET approach, the first important task is to discard the drain channel and source-channel p–n junction of MOSFETs. The Mott FET concept replaces the base silicon channel in the circuit by a congenitally conducting material that is nonlinear in nature, dubbed as a Mott insulator.

As evident from figure 4.39, the drain and the source electrodes are segregated from one another by the use of a channel that consists of the Mott insulator material, which happens to be nonlinear [58]. The gate oxide has been fabricated using a material with a very high dielectric constant. The gate oxide channel interface is grown in an epitaxial manner. The base electrode configuration, which involves the source, the drain and the gate electrodes, is similar to the configuration seen in MOSFET. The gate oxide size is comparatively thick, which is not a problem for a material possessing a high-dielectric-constant ferroelectric attribute.

A base comparison of a Mott insulator and semiconducting material at the microscopic level is depicted in figure 4.40 [58]. In the base semiconductor model, a refined pictorial model represents electrons in the form of Si:Si bonds, which have a valence band that consists of four electrons for an individual Si atom. The gap that can be seen in the Si band structure is mildly related to the basic concept of bonding–antibonding excitation energy, which is depicted by a single Si atom. In contrast, electrons are 1:1 bonded to the third-dimensional transition metal atoms in the conventional Mott insulator. This occurs because the interatomic hopping energy t_h is low and the intra-atomic electron-electron Coulomb repulsion U is high in the compact 3D orbitals. It is interesting to note that this involves electron localization via Coulomb blockade, much like a quantum dot array. The requirement that limits the inter-dot conductance to less than one quantum of conductance in order for the Coulomb blockade to occur is the same as the requirement that $U > t_h$ [59].

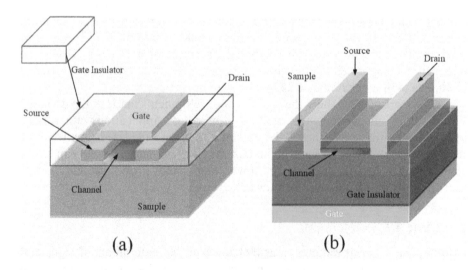

Figure 4.39. Structure of (a) MOSFET, and (b) feasible Mott FET. Adapted from [58] with permission.

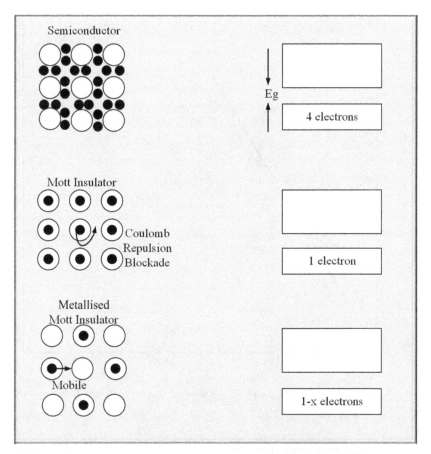

Figure 4.40. Semiconductor, Mott insulator and metallised Mott insulator. Adapted from [59]. Copyright (2000), with permission from Springer Nature.

The lower part of figure 4.40 depicts a metallised Mott insulator as a result of hole doping. Since there are vacant sites available, electrons can move between them without paying an energy penalty. When the system is conceived of as a metal, the Coulomb Blockade gap disappears as the concentration of holes rises due to the improvement of the intersite hopping processes.

At 1 V drain potential, the relatively large Mott insulator gap (1–2 eV) should ensure that there is a potential barrier that spans all or most of the channel width in the "off" state. Therefore, under these circumstances, the MTFET channel ought to be insulating.

Cuprate materials, related to high-temperature superconductors, are well researched examples of Mott insulators that, upon hole doping, become p-type conductors. These materials can be used to frame epitaxial interfaces with high dielectric and high ferroelectric coatings, for example $SrTiO_3$ (STO), and are candidate materials for a p-type Mott FET gadget. Prior work has likewise given an account of superconducting and typical state oxide channel field impacts.

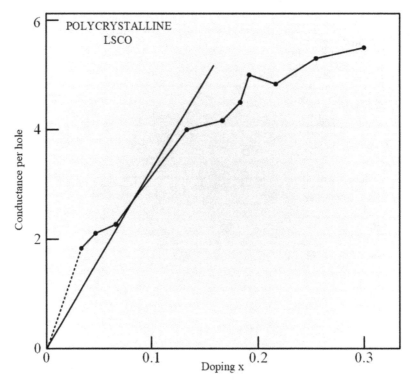

Figure 4.41. Conductance per hole vs concentration x per Cu. Adapted from [59]. Copyright (2000), with permission from Springer Nature.

Figure 4.41 [59] displays some transport information for $La_{2-x}Sr_xCuO_4$ (LSCO) as a component of the composition. The material is acting as a Mott insulator at $x = 0$. The chemical concentration of the hole x gradually metallises the material. With hole concentration, conductivity increases faster than because of the Mott transition. The cuprate material is considered to be metallised at approximately $x = 0.15$, which is also a composition with well-established superconductivity.

The first prototypes of Mott FET devices have been relatively large-scale devices with thick ferroelectric oxide films. Figure 4.42 shows a typical configuration used for such devices. Starting from a 1 atomic % driving STO substrate, which structures the entryway cathode, a layer of STO entryway oxide (with a thickness of approximately 200 nm) is put away by laser examination. Changing targets in the laser analysis chamber, the channel material is collected on the STO layer. In another chamber, the Pt terminals comprising the origin and drain are dissolved through a touch mask. The protection channel in the cuprate layer is at long last quenched by laser treatment. At present, the concept of a Mott FET has only existed at the level of testing and hypothetical support. However, the possibility of Mott FETs as a feasible design in thickly pressed coordinated circuits is as yet a work in progress [59]. Along with this, in modern FET devices, voltage sent through a channel increases the conductivity of the semiconducting material, allowing

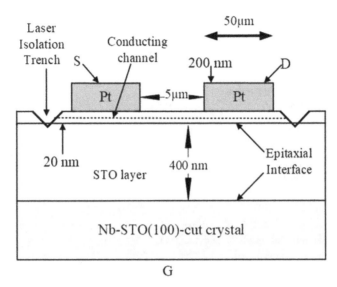

Figure 4.42. Large-scale Mott FET device. Adapted from [59]. Copyright (2000), with permission from Springer Nature.

electricity to move between a source and drain. The net result is a device that can operate as either a gate or a switch.

4.16 Negative-capacitance FETs

By using a negative capacitor in a FET, we can minimise the switching energy and thus power dissipation. As we apply voltage, a normal capacitor stores charge. A higher voltage means more stored energy. A negative-capacitance FET (Neg-CgFET) with a ferroelectric material used as gate oxide can work below the basic Boltzmann 60 mV/dec limit of subthreshold swing [60]. A schematic and equivalent circuit of a Neg-CgFET are shown in figures 4.43(a) and (b). For a negative capacitor, the charge stored across it is inversely proportional to voltage. Now, subthreshold swing is one of the most significant performance parameters of a transistor. Subthreshold swing means the amount of gate voltage (V_G) needed for a one-decade change in drain current (I_{DS}). Lowering the subthreshold swing is advantageous, because then there is less applied voltage and hence less power and energy dissipation overall.

The passive amplification of a gate voltage at the interface between the semi-conductor channel and ferroelectric gate oxide sets a path for the underlying subthreshold swing mechanism for 60 mV/dec operation of an Neg-CgFET. Fundamentally, if we look from a top-down approach, then first the amplification of the surface potential of the semiconductor takes place, due to stabilisation of the field effect at the state of negative capacitance due to the charge balance between semiconductor and field effect, which are two semiconductors. Since the overall performance of the device is largely dependent on the charge balance, several different types of device configuration have been extensively discussed in the

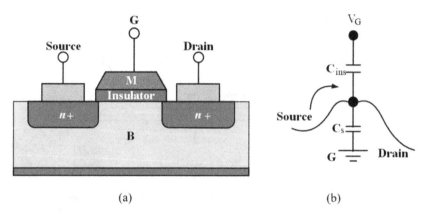

Figure 4.43. Structure (a) of a Neg-CgFET, in which the insulator is replaced by a ferroelectric material; (b) looking from the gate, the insulator and the semiconductor constitute a series combination of two capacitors. © 2018 IEEE. Adapted, with permission, from [60].

literature. Amplification and stabilisation are the two most important characteristics of a Neg-CgFET. The interface between the gate oxide and semiconductor channel leads to differential amplification of the gate voltage, which leads to the sub 60 mV/ dec operation. Meanwhile, usually the negative capacitance is unstable by nature; but if the positive capacitance is properly designed, the ferroelectric in the device structure can be stabilised and lead to stable amplification [61]. There are two main capacitors in a FET: an insulator capacitor and a substrate capacitor. In general, a nonlinear dielectric is an insulator. The capacitance is defined as follows:

1. The inverse curvature (inverse of double derivative) of free energy density (G).
2. The slope of polarisation versus electric field curve:

$$C_{\text{ins}} = \frac{dp}{dE}.$$ (4.16)

There are two types of dielectrics:

1. Paraelectric: no polarisation left when the electric field is removed.
2. Ferroelectric: polarisation is left even when the electric field is removed.

If the insulator capacitance is negative, i.e. $C_{\text{ins}} < 0$, then $\frac{C_s}{C_{\text{ins}}} < 0$, leading to $1 + \frac{C_s}{C_{\text{ins}}} < 1$

Thus, subthreshold swing is given by

$$S = \left(1 + \frac{C_s}{C_{\text{ins}}}\right) * 60 \frac{\text{mV}}{\text{decade}} < 60 \frac{\text{mV}}{\text{decade}}$$ (4.17)

We cannot make bulk FETs with a subthreshold swing of less than 60 mV/dec; but, with this model, we can reduce the subthreshold swing further. So, to the same end, we have to apply low voltage. As such, power dissipation decreases. Some subthreshold swing experiments have reported results with less than 60 mV/dec due to negative capacitance. However, this was only about ~10 pA for very low current (~1 nA) or much higher current (~50 nA) and range [61]. These experiments demonstrate the potential of Neg-CgFETs. Static noise in −ve performance FETs is also improved. More energy-efficient systems can be made using negative capacitance. However, these devices are still in the research phase.

4.17 Alternative information-processing devices

In conventional electronics, electric charge is the basis of all phenomena. It is used for storing and transferring information. With conventional CMOS technology reaching its fundamental limit, the quest for smaller and better devices at the nanoscale arrived at a juncture wherein electron charge became just one of several carrier parameters to be exploited for functionality. At the nanoscale, one can look at various other properties, such as electron spin, magnetic effects, exciton behaviour, etc. The following subsections, then, are devoted to such novel alternative information-processing devices. They present promising future technology, but presently are in need of more in-depth research.

4.17.1 Spintronics and magnetism

Spintronics is a new field that uses the spin property in place of or in addition to the electric charge. Whether we refer to it as spintronics or spin electronics or magneto-electronics, it refers to the same domain: exploiting the spin of an electron for use in information circuits. A major advantage of spintronics is that it does not require special semiconductors but common metals like iron, aluminium, copper, etc., can also be used.

Spin is a quantum-mechanical property. It is a characteristic that makes an electron a tiny magnet, with north and south poles. It has two orientations: up and down, as shown in figure 4.44, similar to positive and negative electric charge. The magnitude of these is $\pm \hbar/2$ (where $\hbar$ is Planck's constant). With these two states of spin, we can draw parallels with how charge is used in electronics to represent 0's and 1's. The usage of the spin property invokes many advantages. All the information can be stored in the form of the orientation of the spin. In addition, the amount of energy needed to change spin is less in comparison to what is required

Figure 4.44. Types of spin.

to generate the current needed to maintain electron charge in a device [2, 14]. Thus, spin devices need less power. Transferring data is faster as spin states can be set swiftly. Furthermore, electron spin not being dependent on energy makes spin nonvolatile—information sent using spin will remain the same even after a power loss. Spin devices have the ability to combine logic functions and storage functionality together, which leads to the elimination of separate components for the two different purposes.

Spin transfer torque is an effect in which we alter the magnetic orientation of a ferromagnetic layer. This alteration of orientation is done in a magnetic tunnel junction with the help of a spin-polarised current. Electrons have two intrinsic properties, associated with their charge and a small amount of angular momentum. Mostly, electric current is unpolarised, meaning 50% up spin and 50% down spin. In contrast, a spin-polarised current contains a majority of electrons with one kind of spin being either up or down. One can generate a spin-polarised current by allowing it to pass through a thick magnetic layer (called the fixed layer). Conversely, one can alter the magnetic orientation in a thin magnetic layer (called the free layer) by passing a spin-polarised current across it. These effects are mostly observed in nanoscale devices [2, 14, 63].

4.17.2 SpinFETs and SpinMOSFETs

One of the most promising alternatives in the nano regime is the idea of using an electron's spin in a new transistor configuration. Such a transistor would be capable of producing low or high output current according to the relative orientation of its ferromagnetic contacts and the spin direction of its electrons. On applying a magnetic field, the magnetism of ferromagnetic materials aligns in the same orientation as the applied magnetic field, and it does not vanish on the removal of the applied magnetic field.

In 1989, Supriyo Datta and Biswajit Das, of Purdue University, performed an experiment that included an electro-optic light modulator. They suggested the device be called a 'SpinFET'. Most of the interest in spintronics in today's world is the product of their work. The electro-optic cell, which has the ability to rotate the angle of polarisation of the light beam by biasing it with the proper voltage, has polarizer and analyser plates located on two sides, as shown in figure 4.45. Both plates are perpendicular to each other. The polarizer polarises the light and, after going through the electro-optic cell, it is received at the analyser. This light can be allowed to pass through the electro-optic cell or it can be blocked. This behaviour is controlled by the voltage applied to the electro-optic cell [62].

Based on this experiment, Datta and Das proposed an electronic device in which instead of using a polarised beam the spin of the electron can be used, as depicted in figure 4.46. In this device, the current flowing through the semiconductor channel is made up of electrons with a certain spin. The ferromagnetic material aligns the spin of the electrons passing through itself in one specific direction, which results in the current so described, and this specific direction is dependent on the direction of magnetism in the ferromagnet.

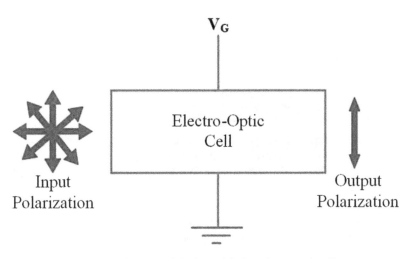

Figure 4.45. Datta and Das's model of an electro-optic cell.

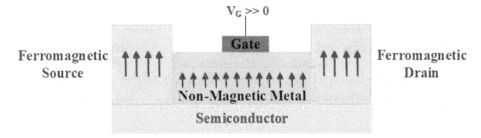

Figure 4.46. SpinFET schematic.

From figure 4.46, we can say that the resemblance of spin-based transistors to existing electronic transistors is uncanny, but the operational principles are completely different, as SpinFETs use the spin injection property at the source contact and the detection property at the drain contact. Spin-selected current is injected in the semiconductor channel. This current has a majority spin polarised to one side because the ferromagnetic source, the contact from which the electrons are injected, does not have an equilibrium among spin states. Ferromagnets inherently have a majority spin down or up, which helps determine their direction of magnetisation. The electrons, most of which are polarised, progress via a 2D electron gas (2-DEG) within the semiconductor channel. If the direction in the bulk spin can be adjusted by the gate bias either parallel or antiparallel to the ferromagnetic drain contact's magnetic orientation, the device's conductance would be high or low, individually. This spin of electrons can be influenced by the gate voltage. The spin precession angle of the electrons in the semiconductor channel is directly correlated to the strength of the applied voltage; this phenomenon is known as the Rashba effect [62]. When the bulk spin positioning of the electrons is parallel to the magnetisation direction of the ferromagnetic drain at the drain end of the channel, current can course through the drain, thus resulting in the ON state of the SpinFET. Changing the voltage of the gate also changes the spin precession angle (it implies an alteration to the axis of rotation of any rotating body) because of the Rashba effect. Using this property, any preferred alignment to electron spin can be induced in the channel. Following the same logic as mentioned earlier, when the spin orientation of electrons at the drain end of the channel is antiparallel to that of the magnetic orientation of the drain, the flow of electrons is blocked and the drain current drops sharply because of the magnetoresistive nature of this phenomenon. This situation is the OFF state of the SpinFET because of the steep drop of the output current.

Figure 4.47 shows how the spin of the electrons can be controlled by the gate electric field. In figure 4.47(a) the gate voltage is positive, thus the precession of the electrons is controlled, allowing the spin to reach the drain with the same polarisation [63]. In figure 4.47(b) the exact opposite is happening, thus inverting the polarity of the electrons. In figure 4.47(c) the gate voltage is zero, thus the injected spin starts processing through the nonmagnetic or 2-DEG channel, thereby reducing the net spin polarisation by the time is reaches the drain. The spin–orbit interaction effect is responsible for the strength of the current flowing through the device. The electric field across the channel can control the spin–orbit interaction strength, and this results in the current modulation. As such, we have a complete transistor working and controlled by the voltage applied at the gate. Here, two parameters determine the high or low output of the transistors: the relative placement of the two ferromagnetic contacts (the source and the drain), which affects the spin direction of the electrons and gate voltage.

In order for this device to work, the spin-polarised electrons must enter from the ferromagnetic source to the semiconductor channel. It may seem that there should not be any problem. But, in reality, it is not straightforward because of big differences in the conductivity of the ferromagnetic contact and the semiconductor channel. This is known as the conductivity mismatch problem. The solution to the problem is to insert a thin layer of an insulator in between them. This layer provides a spin-dependent tunnel resistance. A corresponding schematic is shown in figure 4.48.

There are numerous potential applications for SpinFETs. However, they require specific features for specific applications: for example, more transconductance for greater speed operation, low OFF current for less dissipation of power, and high

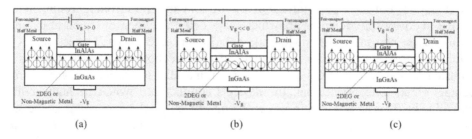

(a) (b) (c)

Figure 4.47. Spin control in SpinFET. (a) Gate voltage > 0, (b) gate voltage < 0, (c) gate voltage = 0. Adapted from [63], with the permission of AIP Publishing.

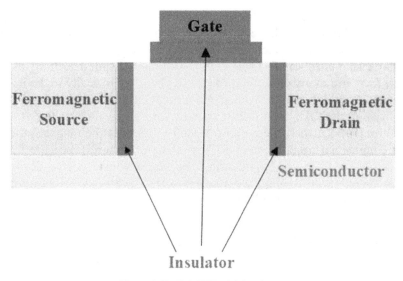

Figure 4.48. SpinFET with insulator.

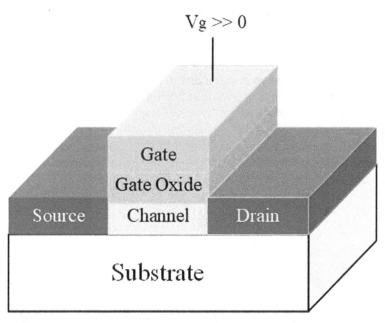

Figure 4.49. SpinMOSFET. Adapted from [64], with the permission of AIP Publishing.

magnetic current ratio for nonvolatile memory. For these, another variation of SpinFETs has been proposed, known as a SpinMOSFET.

The configuration of a SpinMOSFET is shown in figure 4.49, which is almost the same as a SpinFET except for the fact that it has a metal oxide gate. It has the same working principle as a SpinFET for the most part. If realised, SpinFETs can provide many benefits. Researchers have successfully demonstrated tasks like overcoming conductivity mismatch between ferromagnets, reducing spin diffusion time and length, stable spin injection, fast spin transport, accurate spin detection, and others. However, there are many challenges for this technology to really make its place in the semiconductor industry. For one, spin manipulation at room temperature is a mammoth challenge. All experiments to date have been undertaken at very low temperatures [64]. Further, the current needed to reorient the magnetisation is too high for commercial applications and large-scale adoption. Since the making of SpinFETs comes from the combination of semiconductors and ferromagnetic materials, there are also many compatibility issues. Still, the literature reports some notable applications as follows.

4.17.3 Spin torque majority gate

The basic difference between a spin-transfer torque and a spin torque majority gate (STMG) is that the STMG can solve complex logic functions. As shown in figures 4.50(a) and (b), a STMG has three inputs and one output, as in its logic table. The output terminal of the majority gate shows the same logical state (0 or 1) as the majority of the three inputs [65]. The stack of layers and materials are totally compatible with the CMOS process [65]. A three-input STMG consists of four separate ferromagnetic nanopillars and a common free ferromagnetic layer. It operates as a majority gate and is switched via the motion of domain walls due to spin-transfer torque. The magnetisation direction in each layer is represented with a white triangle. The common free layer is at the bottom. The logical state of the input nanopillars is characterised by the voltage applied at them, and that of the output nanopillar is defined by its magnetoresistance. This is shown in figure 4.50(c). The fixed layers with magnetisation are in the middle of the nanopillars. In figure 4.50(d), the squares represented by 'A', 'B', 'C' and 'Out' are the input nanopillars and output nanopillar, respectively. 'a' represents the arms' minimum width, pillars' size

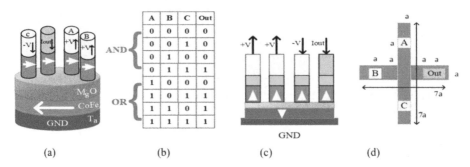

Figure 4.50. A STMG. (a) Structure of three inputs and one output, (b) logic table, (c) cross-sectional view, (d) top view. (b) Adapted from [65]. (d) Adapted, with permission, from [66]. © 2011 IEEE.

and gaps in between them. A STMG can be used as a two-input AND or OR gate by reconfiguring its third input nanopillar. Each nanopillar (both input and output) has discrete fixed layers that are pinned by the anti-ferromagnetic layers above them. From figure 4.50(c), we can see the voltage applied to the inputs, where +V is equivalent to 1 and −V is equivalent to 0. When −V is applied to an input, it drives the current from the bottom to the top, or in an upward direction, meaning the electrons will be moving in the opposite direction, and therefore aligns the orientation of magnetisation parallel to that of magnetisation in the fixed layer. Similarly, +V drives the current in a downward direction and therefore makes the direction of magnetisation in the free layer antiparallel to that of the fixed layer. These changes in magnetisation in the free layer interact and set the direction of magnetisation in the free layer to that of the majority of inputs. Once the direction of magnetisation reaches its final value, the input source can be removed, and the direction of magnetisation can be observed from the magnetoresistance of the stack below the output nanopillar. If the measured value of magnetoresistance between the fixed layer and free layer is high, it represents 1; if the value of magnetoresistance is low, it represents 0. A conventional sense amplifier is used to measure magneto-resistance [66]. The primary advantages over the existing CMOS logic are the smaller area, nonvolatility, radiation hardness, reconfigurability and low power. The disadvantages are lower switching speed and, thus, higher switching energy. Table 4.4 indicates the performance estimates of one bit of a STMG full adder versus that of a high-performance 22 nm CMOS node according to the ITRS [67].

Table 4.4. 22 nm CMOS full adder vs STMG counterpart [67].

S. No.	Parameters	CMOS	STMG	Ratio
1.	Process feature, F, nm	22	22	1
2.	Area factor, F*F	2314	273	0.12
3.	Area per gate, μm^2	1.12	0.13	0.12
4.	Voltage, V	0.18	0.1	0.12
5.	Switching time, ps	0.37	2826	7637
6.	Clocking time, ps	108.9	5651	51.88
7.	Switching energy with external circuits, aJ	10136	257680	25.42
8.	Power per gate, active, μW	93.1	45.6	0.49
9.	Power per gate, standby, μW	0.18	0.0	0
10.	Activity factor	0.005	0.010	1.98
11.	Power per gate, average, μW	1.11	0.46	0.41
12.	Power per unit area average, $W\ mm^{-2}$	1	3.5	3.48
13.	Throughput, mops $ns^{-1}\ cm^{-2}$	8.2	1.3	0.16
14.	Reconfigurable	No	Yes	–
15.	Nonvolatile	No	Yes	–
16.	Radiation hardened	No	Yes	–

4.17.4 Bilayer pseudo-spin field-effect transistors (BisFETs)

BisFETs, as the name suggests, are a type of FET device based on the electrical properties of two layers of semiconductors held in close proximity. The occupation of electrons of the upper or lower layers is treated similarly to spin down or spin up, which is called pseudo-spin. The combined/collective effects observed show similarities to the collective spin found in ferromagnets. Figure 4.51(a) shows the structure of a BisFET [68], and figure 4.51(b) gives a schematic representation [69].

The BisFET structure consists of both types of semiconductors, n-type and p-type, which are made up of graphene and hence referred to as graphene layers. Their carrier densities are large but almost equal in value. Here, both electrons and holes are treated as fermions. Fermions are those particles which have a half-integer spin. However, under specific conditions, the electrons of the n-type layer can pair up with the holes from the p-type layer, which results in the creation of electron–hole pairs. In the case of graphene layers, these excitons condense at room temperature, resulting in bosons [70]. Bosons are particles with an integer spin. They do not follow the general rules of the Pauli's exclusion principle. Their energy distribution is described by Bose–Einstein statistics. According to the Bardeen–Cooper–Schrieffer (BCS) theory of superconductivity, when the electron–hole pairs are created and act like bosons, they condense into a state in which the electrical resistance

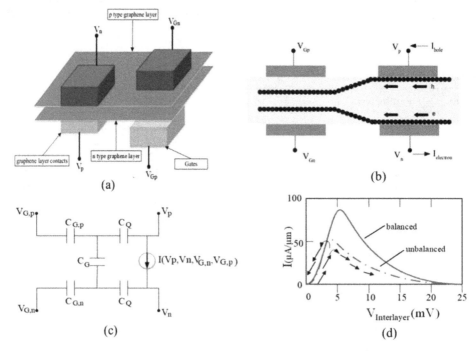

Figure 4.51. BisFET. (a) Structure, (b) schematic representation of circuit, (c) circuit model of BisFET, and (d) interlayer current–voltage (*I–V*) characteristics. (a) Reproduced from [68]. © 2012 ECS - The Electrochemical Society. All rights reserved. (b)–(d) © 2008 IEEE. Adapted, with permission, from [69].

demonstrated is zero. Bose–Einstein condensation alters the quantum wave function in the layers of graphene, which, for a very small interlayer voltage, results in a huge reduction of resistance (tunnel resistance) between the layers of graphene [70]. The equivalent model is shown in figure 4.51(c) [69]. If a large amount of voltage is applied, the condensation is destroyed, which results in large resistance. Hence, BisFETs are only suitable for those devices that will always have low voltage during operation. For energy-efficient information processing there is a need for a steep transition between ON and OFF states over a short range of voltage (nearly 10 mVs). As indicated in figure 4.51(d), BisFETs enable ON and OFF state transitions over ultra-low voltages [69].

The maximum current (in amperes) that can be sustained is approximated as

$$I_{max} = GV_{max} = \left(\frac{2\pi e}{h}\right)\left(\frac{\rho_s}{\lambda_j}\right),$$
(4.18)

where V_{max} is the corresponding interlayer voltage, G is the contact conductance per unit width, ρ_s is the superfluid density and λ_j is the Josephson length. In prototype BisFET inverters, switching energies per device of approximately 10 zeptojoules (10^{-20} J) have been observed [69].

For use in real-life scenarios, with conventional CMOS devices we do not need to convert between the state variables of BisFETs, since a BisFET is a charge-based device, even though it internally depends on pseudo-spin effects [71]. The output characteristics of a BisFET are much different from the output characteristics of MOSFETs. For this reason, logic circuits must work in a different manner. As yet, BisFETs remain only a concept in the literature. Their main challenge involves the fabrication of two perfect layers of graphene separated by a very small distance as well as a perfect gate structure. When fabricating BisFETs, there is a need for a very high degree of control over the graphene; the quality of surface and dielectric also poses many challenges. The practical aspects of the theory of condensation are also not as developed as one would need for further research. Furthermore, these devices have different characteristics compared to MOSFETs, requiring ways of implementing the digital logical circuits that would create a whole new domain of physics [71]. An improved theory of condensation is needed to understand the processes going on and to justify the observations observed, hence helping in reaching viable and useful conclusions [70].

4.17.5 ExcitonicFETs

An exciton is a state where positively charged holes and negatively charged electrons are attracted to each other. Using the properties of excitons, many types of transistors can be developed. The main purpose of an excitonic field-effect transistor (ExFET) is to allow a steep inverse subthreshold slope by creating an energy gap. To create such an energy gap, a gate-controlled formation of an excitonic insulating state is used. The steep inverse subthreshold slope indicates the need for a lower threshold voltage for the same OFF current, which implies that the device can be used at low supply voltages and results in low-power operation. As previously mentioned, conventional CMOS devices have a subthreshold slope of nearly 60 mV/dec. This is a large value when the operations to be undertaken are at the nanoscale, and thus needs to be reduced [71]. During the ON state, as in most conventional FETs, charge is the most important state variable. The excitonic insulator is the state variable that characterises the device during the OFF state. The interaction occurring between two oppositely doped parallel segments results in the formation of an excitonic insulator, as shown in the figure 4.52 [71].

Coulomb interaction exists between the n-type layer's electrons and p-type layer's holes. Under certain gate voltage conditions, the Coulomb interaction results in the condensation of the system to a phase called an excitonic phase. The process results in the opening of an energy gap in the exciton spectrum of a single particle [71]. Using a spatial separation of charges with opposite polarity, recombination of electrons and holes is suppressed. Since we use the gate field to create an energy gap that previously did not exist, the current flow from source to drain is suppressed and the conditions of steep inverse subthreshold slope are satisfied, making the operations power efficient. Once the required conditions for creating an insulating excitonic phase are satisfied, the ExFET device can be switched from a state which is highly conductive into an OFF/nonconducting state [71].

These excitonic devices provide many potential advantages such as interconnection speed, fast operation, small dimensions and scalability, and integration capacity. Further, excitonic switches can operate around a temperature of 100 K [72]. The circuits are composed of an AlAs/GaAs semiconductor structures. The

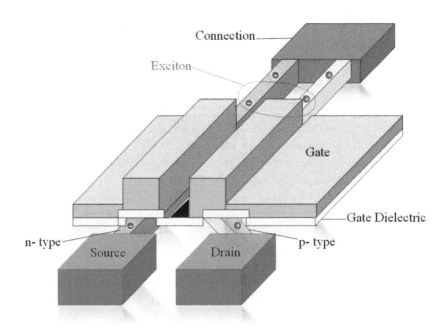

Figure 4.52. Structure of ExFET. Adapted from [71] with permission.

possible device designs also include exciton optoelectronic transistors (EXOTs), excitonic bridge modulators (or EXBMs) and excitonic pinch-off modulators (or EXPOMs). Control of exciton fluxes by electrode voltages are the essential operating principle of excitonic devices. Inputs and outputs can be photonic or excitonic. At the input, photons are converted into excitons, while on the other hand at the output excitons are transformed into photons. The prominent challenges for excitonic devices are the finite binding energy and lifetime of excitons; the lifespan of an exciton is smaller than a nanosecond in the regular semiconductors. Because of the short duration of the lifespan, an exciton can only travel a small distance. After that, the exciton recombines. The problem is solved using the concept of indirect excitons. The holes and electrons in separated layers come together to form the indirect exciton. The lifespan of indirect excitons is much larger than the lifespan of regular excitons. This lifespan exponentially depends on the height of the separating barrier and the separation distance [72]. Due to this lifespan, indirect excitons have the capacity to cover large distances. Due to this, large numbers of excitonic devices can be patterned.

Researchers have proposed many ways to realise an optoelectronic transistor based on excitons. One of the examples is a monolithic optoelectronic transistor (or MOET), which consists of a modulator, FET, detector and resonant tunnelling diode. Another example of an ExFET is the aforementioned EXOT. Researchers at the University of California proposed a model of an EXOT that had a contrast ratio of 30 [73]. If we can control the exciton flux, then scalable EXOTs can become reality.

A comparison of charge-based 'Beyond CMOS' devices and alternative information-processing devices is tabulated in tables 4.5 and 4.6, respectively.

Table 4.5. Comparison of charge-based 'Beyond CMOS' devices [14, 74].

Device		FET	Spin FET and spin MOSFET	NEMS	Mott FET	Neg cap ferroelectric
Typical example devices		Si CMOS	Spin MOSFET	2-terminal [N1] 3-terminal [N2] 4-terminal [N3, N4]	Mott FET	
Cell size (spatial pitch)[B]	Projected	50 nm	100 nm [S1]	100 nm	10 nm [M1]	30 nm [CG1]
	Demonstrated	56 nm	Not known	sub-1000 nm	0.1 μm to 0.5 μm radius [M2]	30 nm [CG1]
Density (device/cm^2)	Projected	1.00E+10	1.00E+10	1.00E+10	1.00E+12	Similar to Si CMOS
	Demonstrated	1.08E+10	Not known	~1E5 [N5]	6.67E+05	Not known
Switch speed	Projected	12 THz	12 THz or less [S2]	~1 GHz [N6]	2 THz (0.5 ps) [M3]	500 GHz [CG2]
	Demonstrated	0.4 THz	Not known	0.18 GHz [N7]	13.3 THz–0.1 GHz (75 fs–9 ns) [M3]	Not known
Circuit speed	Projected	61 GHz	61 GHz or less [S2]	~1 GHz [N8]	Not known	Similar to Si CMOS
	Demonstrated	~10 GHz	Not known	25 KHz [N8]	Not known	Not known
Switch energy, J	Projected	3.00E-18	~3E-18 [S2]	<1E-17 [N9]	0.1 uW [M4]	3.00E-19
	Demonstrated	1.10E-16	Not known	2.3E-17 [N4]	Not known	Not known
Circuit energy, J	Projected	1.20E-17	1.2E-17 [S2]	<1E-17 [N9]	Not known	Not known
	Demonstrated	1.00E-15	Not known	Not known	Not known	Not known
Subthreshold slope, mV/dec	Projected	60 mV/dec	~60 mV/dec [S2]	0.00E+00	≤60 mV/dec [M5]	26 mV/dec [CG3]
	Demonstrated	62 mV/dec	~90 mV/dec [S3]	<0.1 mV/dec [N3]	Not known	40–43 mV/dec [CG4]

Table 4.6. Comparison of alternative information-processing devices [14].

	Spin wave device	Nanomagnetic logic	ExFET	BisFET	All spin logic
State variable	Magnetisation	Magnetisation	Excitonic insulator when OFF, conventional when ON	Presence or absence of superfluid excitonic condensate	Spin/magnetisation
Function	MAJ, NOT	Boolean logic	Gate enabling the transitions	Gate-controlled differential resistance	Nonlinear
Class-example	Adders, counters, special task logic units	Majority gate	Steep subthreshold slope device	Superconducting, pseudo-spin device	Median function
Architecture	Systolic, nonvolatile	Systolic/pipelined	Conventional CMOS	Morphic	Morphic
Application	Boolean and non-Boolean logic	Low power, nonvolatile, radiation hard	Device for low-power applications	Low power, high speed, general purpose logic	General purpose/nonvolatile/reconfigurable logic
Comments	Allows for parallel data processing on multiple frequencies or distinct information channel	Compatible memory technology: MRAM	Excitons have been detected in systems with spatially separated channels, the transition to an excitonic insulator has not been observed	Extremely low power; four-phase clocked power supply	High speed, low power and zero standby power
Status	Room-temperature, GHz-frequency operating prototypes have been demonstrated	Feasibility, CMOS compatible clocking experimentally shown	Experimental work to create this device is ongoing	Simulated	Simulation/some low-temperature experiments
Material issues	Multi-ferroic materials	MRAM/CMOS compatible	Graphene is ideally suited because of the symmetry of the electron and hole dispersion	Low-defect paired graphene layers and compatible thin dielectrics, low-resistance contacts	Spin-coherent channels with reduced electro migration. Magnets with high anisotropy for low-energy operation

In this chapter, we discussed the basic functioning of novel devices, some of their possible applications (such as EXOTs, which is a particular application of excitons exploiting the optical property of boson statistics), as well as some of their drawbacks. These devices are currently in the initial phases of their development, and more research is needed to bring these devices into the semiconductor market. There are numerous challenges in the fabrication of these devices. The research conducted up to the present points the way towards further theoretical and experimental work. Only in the future will we be able to see how much success (research and commercial) these devices get to see.

Questions

1. What are the most used electronic devices other than CMOS?
2. What is difference between a TFET, SpinFET and a MOSFET? Compare at least four performance parameters.
3. Why is graphene a preferred channel-replacement material?
4. State the pros and cons of different CNT synthesis methods.
5. What is the physics behind selecting group III–V materials for channel replacement?
6. Explain the quantum-mechanical effect in a single-electron transistor. What is its significance on the SET performance?
7. Name two quantum structures, their properties and the devices that can be developed using them.
8. What are the advantages of NEMS devices over CMOS? Give examples of such devices.
9. Differentiate between the fundamental principle of a Mott FET and a Neg-CgFET. Compare their performance parameters.
10. Highlight the advantages of atomic switches. What are the implementation challenges?
11. What are the advantages and limitations of spintronics? Are there any prototypical circuits?
12. Do you think ExFET is a potential device that can make for future circuits? If so, why?

References

[1] 2018 *High Mobility Materials for CMOS Applications* ed N Collaert (Cambridge: Woodhead Publishing)
[2] Rogers P and Adams 2008 *Nanotechnology: Understanding Small Systems* (Boca Raton, FL: CRC Press, Tayler and Francis Group)
[3] Li H, Xu C and Banerjee K 2010 Carbon nanomaterials: the ideal interconnect technology for next generation ICs *IEEE Des. Test Comput.* 20–31
[4] Hanson G W 2009 *Fundamentals of Nanoelectronics* (New York: Pearson Education) 1st edn
[5] Hiramatsu M, Deguchi T, Nagao H and Hori M 2007 Aligned growth of single walled and double walled carbon nanotube films by control of catalyst preparation *Jpn. J. Appl. Phys.* 12–6

[6] Osipov A, Arakelian S, Kutrovskaya S and Samyshkin V 2017 The laser induced synthesis of linear carbon chains *Electromagnetics Research Symp.* 548–50

[7] Suzuki S 2013 *Syntheses and Applications of Carbon Nanotubes and their Composites* (London: Intech Open Science) pp 193–222

[8] Mahalik N P 2006 *Micromanufacturing and Nanotechnology* (Berlin: Springer)

[9] Jaiswal M and Singh A 2015 Design and analysis of CNTFET based SRAM *Int. Res. J. Eng. Technol.* **2** 11–15

[10] Daenen M, Fouw R D, Hamers B, Janssen P, Schouteden K and Veld M A J 2003 *The wondrous world of carbon nanotubes* Project Group Report Eindhoven University of Technology The wondrous world of carbon nanotubes

[11] Postma H W C, Teepen T, Yao Z, Grifoni M and Dekker C 2001 Carbon nanotube single-electron transistors at room temperature *Science* **293** 76–9

[12] Hoenlein W, Kreupl F, Duesberg G, Graham A, Liebau M, Seidel R and Unger E 2004 Carbon nanotube applications in microelectronics *IEEE Trans. Compon. Packag. Technol.* 629–34

[13] Verhulst A, Vandenberghe W, Maex K and Groeseneken G 2008 Boosting the on-current of a n-channel nanowire tunnel field-effect transistor by source material optimization *J. Appl. Phys.* **104** 064514–4

[14] ITRS, ERD 2013 https://dropbox.com/sh/qz9gg6uu4kl04vj/AADD7ykFdJ2ZpCR1LAB2XEjIa?dl=0&preview=ERD_2013Tables.xlsx

[15] Vimala P, Sampath N and Priyadarshini D 2021 Boosting on current using various source material for dual gate tunnel field effect transistor *J. Nanotechnol. Nano-Eng.* **7** 1–6

[16] Medhi S 2014 Atlas based simulation study of junctionless double gate (DG) tunnel FET *MSt Thesis* National Institute of Technology Rourkela http://ethesis.nitrkl.ac.in/5613/1/E-62.pdf

[17] Zhou G, Lu Y, Li R, Zhang Q, Hwang W S, Liu Q and Xing H 2011 Vertical InGaAs/InP tunnel FETs with tunneling normal to the gate *IEEE Electron Device Lett.* **32** 1516–8

[18] Esfandyarpour R 2012 Tunneling Field Effect Transistors (PH250 Coursework, Stanford University)

[19] Boucart K 2010 Simulation of double-gate silicon tunnel FETs with a high-κ gate dielectricThesis EPFL

[20] Grieshaber D, MacKenzie R, Vörös J and Reimhult E 2008 Electrochemical biosensors—sensor principles and architectures *Sensors* **8** 1400–58

[21] Zhu H 2017 Semiconductor nanowire MOSFETs and applications *Nanowires—New Insights* (London: Intech Open Science)

[22] Broglie L D 1929 Nobel Lecture: The wave nature of the electron http://nobelprize.org/nobel_prizes/physics/laureates/1929/broglie-lecture.pdf (Accessed: 7 June 2011)

[23] Likharev K K 1999 Single-electron devices and their applications *Proc. IEEE* **87** 606–32

[24] Averin D V and Likharev K K 1991 Single electronics: a correlated transfer of single electrons and Cooper pairs in systems of small tunnel junctions *Dans Mesoscopic Phenomena in Solids* (New York: Elsevier) 173–271 pp

[25] Hamaguchi C 2009 *Basic Semiconductor Physics* (Berlin: Springer Science & Business Media)

[26] Wasshuber C 2001 *Computational Single-Electronics* (Vienna: Springer)

[27] Korotkov A N, Chen R H and Likharev K K 1995 Possible performance of capacitively coupled single-electron transistors in digital circuits *J. Appl. Phys.* **78** 2520–30

[28] Fonseca L R C, Korotkov A N, Likharev K K and Odintsov A A 1995 A numerical study of the dynamics and statistics of single electron systems *J. Appl. Phys.* **78** 3238–51

[29] Likharev K 1987 Single-electron transistors: electrostatic analogs of the dc squids *IEEE Trans. Magn.* **23** 1142–5

[30] Yu Y, Jung Y, Park J, Hwang S and Ahn D 1999 Simulation of single-electron/cmos hybrid circuits using spice macro-modeling *J. Korean Phys. Soc.* **35** 12

[31] Uchida K, Koga J, Ohba R and Toriumi A 2003 Programmable single-electron transistor logic for future low-power intelligent LSI: proposal and room-temperature operation *IEEE Trans. Electron Devices* **50** 1623–30

[32] Uchida K, Matsuzawa K and Toriumi A 1999 A new design scheme for logic circuits with single electron transistors *Jpn. J. Appl. Phys.* **38** 4027–32

[33] Inokawa H and Takahashi Y 2003 A compact analytical model for asymmetric single electron tunneling transistors *IEEE Trans. Electron Devices* **50** 455–61

[34] Mahapatra S, Vaish V, Wasshuber C, Banerjee K and Ionescu A M 2004 Analytical modeling of single electron transistor for hybrid CMOS-SET analog IC design *IEEE Trans. Electron Devices* **51** 1772–82

[35] Dubuc C, Beauvais J and Drouin D 2008 A nanodamascene process for advanced single-electron transistor fabrication *IEEE Trans. Nanotechnol.* **7** 68–73

[36] Beaumont A, Dubuc C, Beauvais J and Drouin D 2009 Room temperature single electron transistor featuring gate-enhanced on-state current *IEEE Electron Device Lett.* **30** 766–8

[37] Wolf C R, Thonke K and Sauer R 2010 Single-electron transistors based on self-assembled silicon-on-insulator quantum dots *Appl. Phys. Lett.* **96** 142108

[38] Sun Y, Rusli and Singh N 2011 Room-temperature operation of silicon single-electron transistor fabricated using optical lithography *IEEE Trans. Nanotechnol.* **10** 96–8

[39] Ecoffey S, Pott V, Mahapatra S, Bouvet D, Fazan P and Ionescu A 2005 A hybrid CMOS-SET co-fabrication platform using nano-grain polysilicon wires *Microelectron. Eng.* **78–79** 239–43

[40] Parekh R, Beaumont A, Beauvais J and Drouin D 2012 Simulation and design methodology for hybrid SET-CMOS integrated logic at 22-nm room-temperature operation *IEEE Trans. Electron Devices* **59** 918–23

[41] Parekh R, Beauvais J and Drouin D 2014 SET logic driving capability and its enhancement in 3-D integrated SET-CMOS circuit *Microelectron. J.* **45** 1087–92

[42] Shah R and Dhavse R 2021 Novel hybrid silicon SETMOS design for power efficient room temperature operation *Silicon* **13** 587–97

[43] Shah R, Parekh R and Dhavse R 2021 Design strategy and simulation of single-gate SET for novel SETMOS hybridization *J. Comput. Electron.* **20** 218–29

[44] Durrani Z A K, Irvine A C, Ahmed H and Nakazato K 1999 A memory cell with single-electron and metal-oxide-semiconductor transistor integration *Appl. Phys. Lett.* **74** 1293–5

[45] Amini P, Dolatyari M, Rostami G and Rostami A 2015 High Throughput Quantum Dot Based LEDs *Energy Efficiency Improvements in Smart Grid Components* (London: IntechOpen)

[46] Ballard I M, Ioannides A, Barnham K W J, Connolly J P, Johnson D C, Mazzer M, Ginige R, Roberts J S, Hill G and Calder C 2007 Quantum well solar cells for concentrator applications *Proc. 22nd European Photovoltaic Solar Energy Conf. (Milan, Italy)*

[47] Science in Society Archive 2009 Quantum Well Solar Cells https://www.i-sis.org.uk/quantumWellSolarCells.php

[48] Massimiliano V, Stephane E and James R 2004 *Introduction to Nanoscale Science and Technology* (New York: Springer)

[49] De Haan S 2006 NEMS—emerging products and applications of nano-electromechanical systems *Nanotechnol. Percep.* **2** 267

[50] Jasulaneca L, Kosmaca J, Meija R, Andzane J and Erts D 2018 Review: electrostatically actuated nanobeam-based nanoelectromechanical switches – materials solutions and operational conditions *Beilstein J. Nanotechnol.* **9** 271–300

[51] Dequesnes M, Rotkin S V and Aluru N R 2002 Calculation of pull-in voltages for carbon-nanotube-based nanoelectromechanical switches *Nanotechnology* **13** 120

[52] Gendepujari K, Rotake D R and Dhavse R N 2022 Design and implementation of NVM Cell using nanomechanical cantilever switch *2022 10th Int. Conf. on Emerging Trends in Engineering and Technology—Signal and Information Processing (ICETET-SIP-22)* (Piscataway, NJ: IEEE) 1–5

[53] Dadgour H F and Banerjee K 2007 Design and analysis of hybrid NEMS-CMOS circuits for ultra low-power applications *Proc. 44th Annual Design Automation Conference (DAC'07)* (New York: ACM) 306–11

[54] Choi W Y, Osabe T and King Liu T 2008 Nano-electro-mechanical nonvolatile memory (NEMory) cell design and scaling *IEEE Trans. Electron Devices* **55** 3482–8

[55] Adeosun O, Palit S, Jain A, Alam M and Jin X 2014 MEMSLab: Simulation suite for electromechanical actuators https://nanohub.org/resources/cvgraph

[56] Aono M and Hasegawa T 2010 The atomic switch *Proc. IEEE* **98** 2228–36

[57] Hino T *et al* 2011 Atomic switches: atomic-movement-controlled nanodevices for new types of computing *Sci. Technol. Adv. Mater.* **12** 013003

[58] Inoue I H 2014 Feasible Mott FET: Concept, Obstacles, and Future http://lptms.u-psud.fr/ecrys/files/2014/10/inoue-sh.pdf

[59] Newns D M *et al* 2000 The Mott transition field effect transistor: a nanodevice *J. Electroceram.* **4** 339–44

[60] Wong J C and Salahuddin S 2018 Negative capacitance transistors *Proc. IEEE* **107** 49–62

[61] Radhakrishna U, Khan A, Salahuddin S and Antoniadis D 2017 *Compact model of negative capacitance MOSFETs (NCFETs)* Technical Manual Nanohub Resource

[62] Modarresi H 2009 The spin field-effect transistor: can it be realized? *Thesis* Universitas Groningen

[63] Thankalekshmi R R and Rastogi A C 2012 Simulation of a spin field effect transistor based on magnetic impurity–doped ZnO *J. Appl. Phys.* **111** 07D104

[64] Sugahara S and Tanaka M 2004 A spin metal–oxide–semiconductor field-effect transistor using half-metallic-ferromagnet contacts for the source and drain *Appl. Phys. Lett.* **84** 2307–9

[65] Nikonov D E, Bourianoff G I, Ghani T and Young I A 2012 Nanomagnetic logic and magnetization switching dynamics in spin torque majority gates arXiv:1212.4547

[66] Nikonov D E, Bourianoff G I and Ghani T 2011 Proposal of a spin torque majority gate logic *IEEE Electron Device Lett.* **32** 1128–30

[67] Nikonov D E, Bourianoff G I and Ghani T 2011 Nanomagnetic circuits with spin torque majority gates *2011 11th IEEE Int. Conf. on Nanotechnology* (Piscataway, NJ: IEEE) 1384–8

[68] Register L F, Mou X, Reddy D, Jung W, Sodemann I, Pesin D and Banerjee S K 2012 Bilayer pseudo-spin field effect transistor (BiSFET): concepts and critical issues for realization *ECS Trans* **45** 3

[69] Banerjee S K, Register L F, Tutuc E, Reddy D and MacDonald A H 2008 Bilayer pseudospin field-effect transistor (BiSFET): a proposed new logic device *IEEE Electron Device Lett.* **30** 158–60

[70] Chen A, Hutchby J, Zhirnov V and Bourianoff G (ed) 2015 *Emerging Nanoelectronic Devices* (New York: Wiley)

[71] Semiconductor Industry Association 2011 2011 International Technology Roadmap for Semiconductors (ITRS) https://semiconductors.org/resources/2011-international-technology-roadmap-for-semiconductors-itrs/

[72] Grosso G, Graves J, Hammack A T, High A A, Butov L V, Hanson M and Gossard A C 2009 Excitonic switches operating at around 100 K *Nat. Photonics* **3** 577–80

[73] High A A, Hammack A T, Butov L V, Hanson M and Gossard A C 2007 Exciton optoelectronic transistor *Opt. Lett.* **32** 2466–8

[74] Bernstein K, Cavin R K, Porod W, Seabaugh A and Welser J 2010 Device and architecture outlook for beyond CMOS switches *Proc. IEEE* **98** 2169–84

[75] Karthikeyan S, Mahalingam P and Karthik M 2009 Large scale synthesis of carbon nanotubes *J. Chem.* **6** 756410

IOP Publishing

Nanoelectronics
Physics, technology and applications
Rutu Parekh and Rasika Dhavse

Chapter 5

Emerging memory devices

Memory represents a significant part of the semiconductor electronic industry. ICs fabricated using CMOS technology primarily use dynamic RAM (DRAM), static RAM (SRAM), electrically programmable read-only memory (EPROM) and flash memory technologies. From a system-design perspective, memory can be treated from two main points of view, that is, RAM, whose contents can be modified very fast and that to theoretically infinite times, and ROM, whose contents cannot be modified. Due to the high jump in customers' demands for sophistication, portability and affordability, the memory market remains one of the fastest growing and most competitive markets in the IC industry [1]. Memory is typically in demand for various applications such as personal computers, consumer electronics, high-performance computation, automotives, sensor-based standalone systems on chip (or SOCs), medical electronics, agro-electronics, robotics, etc.

A detailed study of ITRS [1] and IRDS [2] reports in recent years reveals that at least 30% of chip area is consumed by CMOS-based memory. Further, 80% of the memory market is driven by DRAM and flash technology. The very-large-scale integration (VLSI) design of a memory chip involves the development of a memory core that consists of an actual identical memory storage cell matrix and memory peripheral circuits, which include read–write circuits, decoders, input–output buffers, and so on. Compared to the design and development of high-end logic ICs, memory ICs have seen faster growth from a density/storage capacity perspective. This is due to the number of individual elements to be optimised being much less in a high-capacity memory chip (vis-à-vis the comparable logic development of a particular technology node).

However, the rate of improvement in microprocessor performance outdoes that in DRAM [1, 2]. Thus, memory subsystems have become the most crucial system-level performance bottlenecks, leading to a memory 'bandwidth' wall [3]. Also, growing imbalance in peak compute-to-memory capacity requirements in computer systems has resulted in overprovision of memory size, leading to the emergence of a memory 'capacity' wall and resulting in significant memory underutilisation [3]. Continuous scaling of CMOS technology has revealed many issues in existing MOS-based memory devices like gate leakage, high latency, reliability, scalability, standby power consumption and (re)programming ability. Complex and reconfigurable system architectures, parallelism, high-performance computing and requirements such as power efficiency whilst maintaining fast in-memory computing has further put stringent expectations on the VLSI design of memory structures. Ideally, it is desirable that a memory device should combine the fast writing ability of RAM with the long data retention property of ROM. Ironically, these two attributes are always in conflict with each other. A fast-writing memory cell implies that data stored in the memory cell should get destroyed/altered easily, and long data retention dictates the design of a memory cell that opposes data alteration. As such, there is a huge scope for research in the invention of a so-called 'universal memory' technology.

If benchmark memory technologies are heeded, 3D vertically stacked NAND flash memory will already be produced commercially at the time of writing [1, 2]. However, due to their thinner tunnel oxides, the endurance and data retention are degraded significantly. There is huge latency gap between NAND-based solid-state disks (or SSDs) and DRAM. Additionally, further DRAM scaling faces numerous challenges and is saturated near the 20 nm node. There are many system/circuit-level solutions reported in the literature and adopted by players in the memory market to sustain this scenario, too [1, 2]. Prototypical memory technologies, in particular spin-transfer torque RAM (STT-RAM) and phase-change RAM (PCM), have made improvements in recent years and are now being commercially produced. Emerging memory technologies, especially resistance switching memory (ReRAM) technologies, have been commercialised on a small scale [1, 2]. Other emerging research memory devices include capacitive memory types (ferroelectric field-effect transistors, or FeFETs) and resistive memory types (ferroelectric, nano-electro-mechanical, Mott electronic effect, macromolecular and molecular memories, etc.) [1, 2]. They include carbon nanotubes (CNTs), nanowires, complex metal oxides, transition metal oxides, magnetic materials as well as engineered interfaces between these materials, and so on. However, none of them is yet ready for mass production. Though the underlying physical phenomena responsible for the operation of some of these devices are not fully known, further research may bridge the gaps in the present understanding and lead to deployment of the technology on a commercial scale. In this chapter, we will concentrate on nonvolatile memory devices. The performance of a nonvolatile memory structure is evaluated by the following parameters [4]:

Endurance: the number of SET/RESET cycles a memory can survive before losing the memory window. A flash is expected to have maximum endurance of 10^5–10^6 cycles.

Retention: the amount of time a memory device will retain a logic state after removal of programming bias. It is expected that data should be retained for about 10 years.

Switching voltage, current and power: the SET and RESET switching voltages and currents are the prime factors determining the power. They determine the heat generated during switching of a densely packed memory chip. The voltage decides the compatibility of the 'back end of line' (BEOL), and the current decides the size of the drive transistor.

Switching time: the time taken by the memory to transit from logic state 1 to 0 and vice versa is known as the switching time. It decides the speed of the memory device and is one of the prime parameters of concern for future information-processing systems.

Switching energy: the product of switching time and switching power is the switching energy. It is a key performance metric for applications ranging from high-performance to mobile computing.

Below are detailed descriptions of these emerging memory types.

5.1 Memristors

The three basic components of electronics—resistors, inductors and capacitors—lead to the definition of four cornerstone circuit variables: electric charge, voltage, current and magnetic flux. It is well known that resistors relate voltage with current, capacitors give a relation between voltage and charge, and inductors give a relation between current and magnetic flux. However, none of these passive components could provide a relationship between magnetic flux and charge. To fill in this missing link, L Chua [5], in 1971, came up with the idea of a new element, which he called a memristor. This is depicted in figure 5.1. The term 'memristor' was coined from the idea of a 'memory resistor'. That is, its resistance is not constant as in the case of a resistor but depends on the history of current passed through it. In other words, it depends on the magnitude and direction of charge flow through the device in the past. This property of a memory of its own history is also called a 'nonvolatile property'. For instance, if we power OFF a device and turn it back ON, its resistance would be equal to where it was left off. There are various types of memristors, however in this chapter we will focus on one specific type, titanium dioxide (TiO_2) memristors, in order to understand the basic concepts.

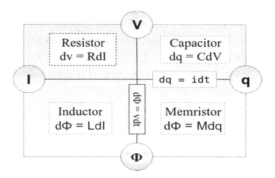

Figure 5.1. Relationship between passive components and cornerstone circuit variables. Adapted from [5]. CC BY 3.0.

Chua's theory Every memristor is identified by its memristance. This is a charge-dependent quantity describing the rate of change of flux with charge [6]. It is denoted by $M(q)$:

$$M(q) = \frac{d\phi}{dq}. \tag{5.1}$$

Now, using Faraday's law, the flux can be written as an integral of voltage over time. Also, by definition, electric charge is integral of current over time:

$$M(q) = \frac{d\phi/dt}{dq/dt}, \tag{5.2}$$

$$M(q) = \frac{V(t)}{I(t)}. \tag{5.3}$$

This expression in equation (5.1) is very similar to $R = V/I$. Hence, the memristance of a memristor is analogous to resistance, and it is expressed in the same unit as that of resistance (ohms). However, they differ from resistors in that memristance is not a constant but depends on the charge flow through the device. In special cases where $M(q)$ indeed is a constant, however, then the expression reduces to Ohm's law, and the relationship between $V(t)$ and $I(t)$ becomes linear. Such behaviour is observed in the case of alternating current where there is no net charge movement and fluctuating change in q may not cause much change in M. In general, the relationship is quite complex, as nonzero current implies time-varying charge, which in turn varies memristance, causing the V–I relationship to vary with time.

Semiconductors are often doped into p-type or n-type to increase their conductance. For the fabrication of a memristor, one can use titanium dioxide (TiO_2), a highly resistive semiconducting material. If an oxygen atom is missing from its usual location in TiO_2, it creates a vacancy with a positive charge, which increases the conductivity of the material. Two layers of TiO_2 are connected with an ohmic contact, one doped with lots of positive vacancies and the other an undoped layer with high resistance. Upon application of a positive voltage to the doped side, oxygen vacancies are repelled; if the voltage is sufficiently positive, they move into the undoped side, making the length of the undoped region lower. It thus reduces the potential barrier for conductivity, and we say that the device is switched ON. By applying a voltage in the reverse direction, the positive vacancies are pushed back in place, increasing the length of the undoped region and hence reducing conductivity. This is how the resistance of the device can be varied—an important property of a memristor. The phenomenon is shown in figure 5.2.

As previously noted, another property of a memristor is the ability to remember its previous state. This property is achieved using the proposed fabrication, as well. When the applied bias is removed, the positively charged titanium (Ti) ions do not move, keeping the length of the undoped region the same, hence equivalent to storing its resistance. It commences in the same state when voltage/bias is next applied.

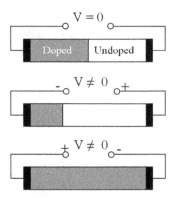

Figure 5.2. Conduction mechanism in memristor. Top: No voltage is applied. Middle: Negative potential is applied to the doped side. Bottom: Positive potential is applied to the doped side. Adapted from [6].

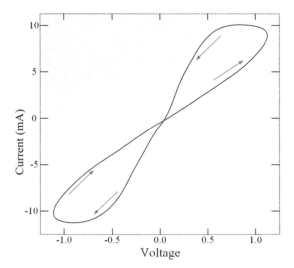

Figure 5.3. *I–V* characteristics of a memristor. Adapted from [6].

In the case of a resistor, current and voltage are linearly related, hence I-V curve is a straight line. However, as discussed above, the case of the memristor is different as memristance depends on the history of charge flow through the device. In fact, it is observed that *I–V* characteristics of a memristor are very similar to hysteresis curves found in magnetic materials. This phenomenon is shown in figure 5.3.

Hysteresis is a concept in physical science in which the state of a system depends on its history. In other words, the output of a system depends not only on its input but also on its history of past outputs. This historical dependence is the basis of memory in memristors and other memory devices such as the hard disk drives (HDDs) used in computers. The resistance of a memristor depends on the past history of currents flowed through it [6]. This behaviour is possible only because of hysteresis. In a memory system, we require two states (0 and 1 in binary) separated by an energy barrier to store information. In order to switch a bit, we need to overcome such a barrier by supplying energy to the system, which we typically do by increasing the magnitude (usually of current or voltage) above a certain threshold. This is precisely what we do to operate a memristor as a memory system. Figure 5.3 shows the shape of a hysteresis loop, which looks like two straight lines joined by curves at the two extremes. These two straight lines can be considered as the two states of the memristor, with different resistances [6]. Since $R = V/I$, the slope of the line in the figure denotes the inverse of resistance. Hence, a steeper line denotes a low-resistance state, and a gradual line denotes a high-resistance state. When voltages are kept within a certain threshold, the memristor stays in its high- or low-resistance state. If the applied voltage is outside the threshold range, it is sufficient to overcome the barrier and change the resistance state; that is, we switch from one curve of the hysteresis loop to another.

Memristors can be used as a nonvolatile type of memory in solid-state storage in computers [7]. They would permit a higher density of data than a HDD while maintaining access times close to that of DRAM. It can also potentially replace flash memory as used in USB flash drives and cameras, which requires the ability to write over and over quickly to the memory. Moreover, it uses less energy than flash memory. Memristors can also provide for faster boot times in computers. Apart from these, they have potential applications in various fields such as programmable logic, neural networks and brain–computer interfaces. However, memristors are commercially unavailable in the market due to their cost and technical difficulties in fabricating nanodevices. The speed of current iterations of memristors is about one tenth that of DRAM.

5.2 Magnetoresistive effect for memory applications

Memory types based on the magnetoresistive effect are regarded as volatile RAM. Their operation depends on a change in the resistance induced by a magnetic field. Two phenomena due to this effect, giant magnetoresistance (GMR) and tunnel magnetoresistance (TMR), are explained in the following.

5.2.1 Giant magnetoresistance

The introduction of GMR led to the inception of spin-based transistors as potential replacements of (then) current electronics and use of spin in the field [8]. GMR is a way to control electrical resistance at the nanoscale using magnetic resistance. The quantum magnetisation effect is observed in thin-film structures with alternating nonmagnetic and ferromagnetic layers. The term 'giant' in GMR would seem to be an oxymoron, considering that the topics under discussion are related to nano-technology. However, this giant refers to the case wherein multiple layers of ferromagnetic or nonmagnetic materials are piled up together in the presence of a magnetic field: a significant (and hence giant) change in the resistance occurs as compared to the maximum sensitivity of similar materials present in magnetic sensors. The primary reason behind this significant change is due to the noticeable orientation of ferromagnetic layers and nonmagnetic layers, which act as partition layers in the GMR structure [9]. These sensors are used for sensing elements in hard drives whose temperatures increase due to certain external agents. The mathematical model for GMR is as follows:

- MR = $(R{\uparrow}{\downarrow}-R{\uparrow}{\uparrow})/R{\uparrow}{\uparrow}$, where MR is the so-called magnetoresistance level.
- $R{\uparrow}{\downarrow}$ represents the resistance when the magnetic orientation is antiparallel (maximum resistance).
- $R{\uparrow}{\uparrow}$ represents the resistance when the magnetic orientation is parallel (minimum resistance).

The resistance of a device is determined by the number of electrons that are scattered in it. In nonmagnetic conductors, the scattering of electrons takes place without any consideration of the spin of the electrons, whereas in a ferromagnetic conductor, electrons scatter differently depending on the type of its spin.

The GMR structure is analogous to a sandwich wherein the ferromagnetic layers act as bread slices and the nonmagnetic layer as the middle portion of the sandwich, as shown in figure 5.4(a). However, this nonmagnetic layer is conducting, which

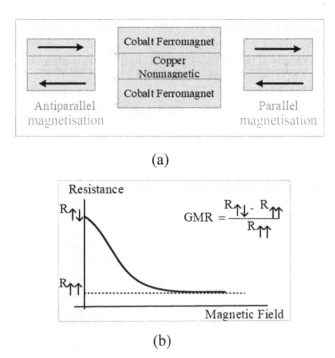

(a)

(b)

Figure 5.4. Concept of GMR. (a) Sandwich-like structure, (b) magnetic field vs resistance. (a) Adapted from [8], (b) adapted from [9].

creates an antiferromagnetic symmetry there, leading to opposite magnetisation in those ferromagnetic layers. This induces a high resistance in the current. Figure 5.4(b) shows a graph of the magnetic field versus resistance. The magnitude of GMR can be higher than 100% at low temperatures for Fe/Cr and Co/Cu multilayers. Copper is extensively used as the nonmagnetic layer due to its excellent conductivity and high resistance change. This scattering and spin of electrons around the central layer causes the overall change in the resistance.

The above described phenomena occur when no external magnetic field is applied. However, when an external magnetic field is applied, it affects the magnetic moments. The magnetic moments in the conducting layers get ordered (in the same direction), overcoming the antiferromagnetic effect. This change in the magnetic moment results in a change in the resistance of the structure. Consider that data is stored in binary form in a HDD, encoded magnetically on rotating magnetic disks within the drive. An electromagnetic field is produced when electrons travel via coils in the drive's write heads, which move radially through the face of the plates, adjusting the position of the magneto-sensitive entities on the plate surface. If the flow of electrons is inverted the field is also inverted. The two directions represent 1 and 0, that is, binary bits. Reading from the disk requires the reversal of this process. As discussed earlier, a GMR drive head comprises two ferromagnetic layers, the first with a fixed magnetic field direction and the second free to align with the magnetic field coded on the disk, with a nonmagnetic layer in between. Aligning electrons in a set direction and pattern can reduce the resistance [8, 9]. Currently, there is an ongoing effort to introduce this technology to computer memory, to replace electronics-based DRAM with spintronics-based magneto-resistive RAM (MRAM) [10].

5.2.2 Tunnel magnetoresistance

The TMR effect occurs in magnetic tunnel junctions (MTJs) [2]. A MTJ is a component consisting of two ferromagnets separated by a thin insulator, as shown in figure 5.5. The magnetic design of a TMR element is quite similar to that of a GMR element. However, in GMR, the current flow is horizontal to the film surface, whereas it is perpendicular in TMR elements. The TMR element is a thin-film element and the barrier layer is made of a thin insulator, 1–2 nm in thickness, which is kept in the middle of the two ferromagnetic layers. In TMR if we see the directions of the magnetisation then the orientation of the pin layer is fixed, whereas the orientation of the free layer changes accordingly due to the external magnetic field [11].

The electrical resistance of the TMR changes according to changes in the orientation of the free layer. If the magnetisation direction of the pin layer and free layer are parallel, then the electrical resistance is low, causing a large current to flow into the barrier layer. When the magnetisation direction of the pin layer and free layer are antiparallel, then the electrical resistance is extremely high, causing a very small current to flow into the barrier layer. The TMR effect is the result of a combination of the two spin channels in ferromagnetic layers and the quantum

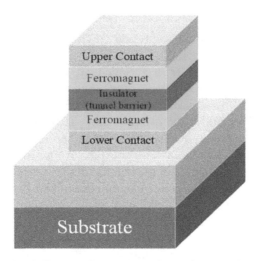

Figure 5.5. Structure of a MTJ. (This Magnetic Tunnel Junction.svg image has been obtained by the author(s) from the Wikimedia website where it was made available under a CC BY-SA 4.0 licence. It is included within this book on that basis. It is attributed to Fred the Oyster.)

tunnelling effect. TMR junctions have a resistance ratio of about 70%. The main application of TMR is in the read heads of modern HDDs. Also, TMR is the basis of MRAM, a new and promising type of nonvolatile memory [11].

5.3 Magnetoresistive RAM

MRAM is a type of RAM where the data is stored in the magnetic domain, unlike the orthodox electric field or current flows. As illustrated in figure 5.6(a), a MRAM cell has a ferromagnetic plate/insulator/ferromagnetic plate structure, in which both plates are capable of holding magnetisation. One plate has a variable magnetisation, which is set to match the external magnetic field, for storing memory; the other plate has a permanent magnetisation. This is called known as a MTJ. In a typical scenario, if two plates have similar magnetic alignment, i.e. a low resistive state, then it will be considered as a 1; if they have opposite alignment, the bit is considered as a 0. An array of MRAM cells can be utilised, as illustrated in figure 5.6(b).

A cell can be read by computing its electric resistance. The cell is powered by powering its access transistor. This allows the current from the supply to pass through the cell to the ground. Subsequently, there will be a change in the relative magnetisation orientation between the plates due to the MTJ. This will change the resistance, and hence the current flowing through the cell. By calculating this resulting current, the electrical resistance within a specific cell can be found, the result of which is the writable plate's magnetic polarity [12].

5.3.1 Writing in MRAM

In between the write lines that are oriented orthogonally all the cells lie, and the plates are parallel to the cells. The creation of an induced magnetic field that the

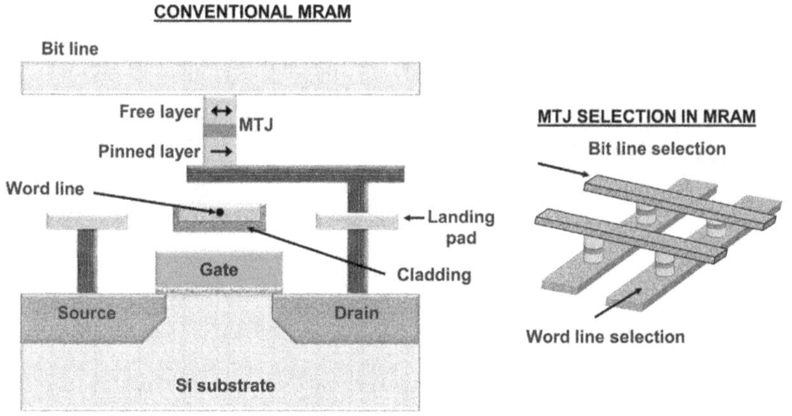

Figure 5.6. MRAM. (a) Structure, (b) array formation. Adapted from [12], Copyright (2017), with permission from Elsevier.

writable plate can receive takes place upon the passing through of a current. To write a 1, the current is run left to right through the bit line and perpendicular to the write word line. Here the magnets would be aligned parallel to each other, showing a low-resistance state, i.e. 1. To write a 0, the current is run from right to left through the bit line and perpendicularly into the page through the write word line. Here the magnets would be aligned antiparallel to each other, showing a high-resistance state, i.e. 0. Once the writing is done, the current is turned off, and then the magnet's state will be maintained until some external field causes it to change [12]. The main drawback of writing in MRAM is that when the device is scaled down, there comes a point at which the induced field overlaps with the adjacent cells over a small area, leading to potential false writes. Also, MRAM designs require a relatively high current to write each single bit [12].

5.4 Spin-transfer torque magnetic random access memory

Spin-transfer torque magnetic random access memory (STT-RAM) can provide faster and more cost-effective solutions for future applications. It is a nonvolatile memory type and utilises magnetic materials for data storage [1, 13, 14]. The endurance of STT-RAM is predicted to be over 10^{15} cycles, with write times of less than 1 ns. Since the data is not mainly stored in the transistors, its OFF state power dissipation is also much less. Thus, STT-RAM is a strong potential candidate for the highly touted 'universal memory'. As explained in the above section and depicted in figure 5.7(a), MRAM operates using magnetic fields which are generated from a word line (write line 1) and a bit line (write line 2). The MRAM element to be written is placed at the intersection of a word line and bit line. The total magnetic field generated from the two lines is larger than the switching threshold required to change the magnetisation of the concerned MRAM element. Other elements experience the field produced from a single

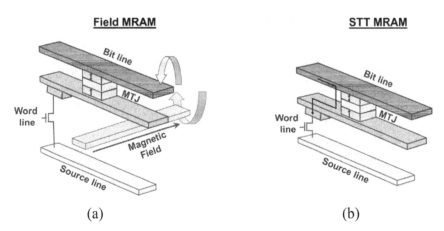

Figure 5.7. Magnetic memory. (a) Field MRAM, (b) spin-transfer torque MRAM design. Adapted from [12], Copyright (2017), with permission from Elsevier.

line, which is not sufficient enough to bring about switching. The window of operation for such MRAM designs is slim and degrades with scaling, which is known as the half-field selection problem. Therefore, it is difficult to achieve high-density MRAM using this write scheme.

This issue can be overcome by using STT-RAM. Figure 5.7(b) shows a STT-RAM design operated by spin-transfer torque [1, 14]. Such a design does not require a magnetic field. The writing current directly goes through the memory element under the word line (transistor) control, thus reducing write errors. The current through the element is given by equation 5.4:

$$I_c = \frac{1}{\eta} \left(\frac{2e}{\hbar} \right) \frac{\alpha}{\cos\varphi} \ (a^2 l_m M) \ (H_k + 2\pi M + H), \tag{5.4}$$

where η is the spin polarisation ratio, α is the damping constant, H is the applied field and H_k is the anisotropy field. The switching current can be reduced by increasing the magnetic spin polarisation ratios or by using a thin layer with smaller damping ratio. It can be seen that as the writing current is proportional to the device area, it has better scalability. Thus, STT-RAM demonstrates advantages with regards to its scalability, speed and power consumption. A typical STT-RAM cell consists of one transistor to supply the current for writing and reading and one MTJ element for the actual storage of data. The generated voltage, which is proportional to $I_{sense} \times R_{low} \times$ TMR, where I_{sense}, R_{low} and TMR are the sense current, low (parallel) resistance state and TMR value, respectively, is then compared to a reference MTJ element to determine the state of the memory bit.

5.5 All-spin logic

Magnets are used in all-spin logic (ASL) to represent nonvolatile data in binary form, and the communication between them is established through spin-coherent channels by the energy that is derived from the supplied input power [15].

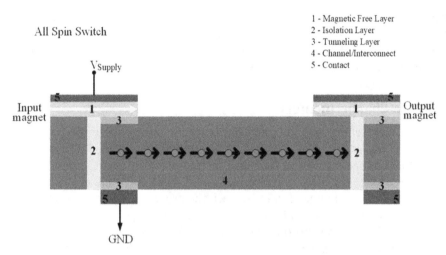

Figure 5.8. All-spin switch. Adapted from [15]. Copyright (2010), with permission from Springer Nature.

Figure 5.8 shows the basic structure of an ASL switch. The two nanomagnets can alter their states through the input torque in a bistable form that can be represented in binary form. Spin current can be generated with the help of stored spin information. This stored information is in the form of the direction of magnetisation. This spin current can be flowed to the desired location through spin-coherent channels. At the desired location, the final magnetisation state can be found with the help of the spin-torque phenomenon. Finally, the resulting output is a change in the magnetisation, according to the magnetic information provided using the supplied energy. The energy and spin current are passed through the channel spin current. When a negative voltage is applied, it injects majority spins, i.e. spins that are parallel to the input magnetisation state, and results in the parallel magnetisation of the output and the input spins. When a positive voltage is applied, it extracts majority spins, and on removing the majority spins the output spin magnetisation is antiparallel to the input spin magnetisation. The energy required to switch between the magnetisation is very low, which is a major possible advantage of ASL devices over conventional CMOS devices. The switching energy can be reduced to kT_{room} per magnet in place of per spin. This can be done by the self-correcting feature of the magnets.

The concept of ASL has reduced the difference between spintronics and magnetics, which has in turn helped increase the chances of creating a device that can provide a low-power alternative to charge-based information. Recently, there have been two advances in this field. First, spin is injected into metals and semiconductors with the help of a magnetic contact of two materials. Second, the orientation of a second magnet is altered with the help of the injection of spins. These two approaches have aided all-spin methods for information processing. Magnets induce spins and these spins turn the magnet, forming a complete system with the benefit of spin currents and bistable magnetic properties. It has been proven that ASL helps in minimising the energy delay required in switching. However, in doing so there are major challenges to overcome. One is room-temperature operation, and the other is the introduction of anisotropic magnetic materials. Some other factors that need to be addressed are the current density and proper selection of the material for the channels.

5.6 Phase-change memory

PCM is another technology that reached maturation during the writing of this book. It can offer DRAM-like features such as bit alteration, fast read and write speeds and good endurance, as well as flash-like features such as nonvolatility, all using a simple device structure. Thus, the introduction of PCM in the memory ladder would allow the seamless exchange of data between the processor and storage [16]. PCM is also expected to extend beyond the scaling limit of existing memory devices [16].

A PCM cell has a mushroom-like structure, known as a T-cell, made up of a highly resistive heater element, as shown in figure 5.9(a). The name 'mushroom' is inspired by the shape of the programming region. PCM uses the large resistivity contrast between the crystalline (low-resistivity) and amorphous (high-resistivity) states of a material to store information. As shown in figure 5.9(b), to reset the cell, a short electrical pulse (typically <50 ns) is applied to the bottom electrode contact. The heat generated in the phase-change material–heater interface causes a region of the phase-change material to melt. The thermal pulse is quenched rapidly to cause the molten region to cool to its amorphous state. To set the PCM cell, an electrical pulse sufficient to increase the temperature of the programming region above the crystallisation temperature and a time period sufficiently long to crystallise the phase-change material is applied. To read the state of the cell, an electric pulse is applied to measure the cell resistance. It is kept small enough so that it does not destruct the state of the cell.

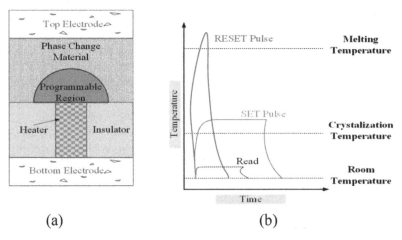

(a)   (b)

Figure 5.9. Phase-change memory. (a) Structure, (b) SET–RESET and read pulses of the cell. Adapted from [16] with permission from the Association for Computing Machinery.

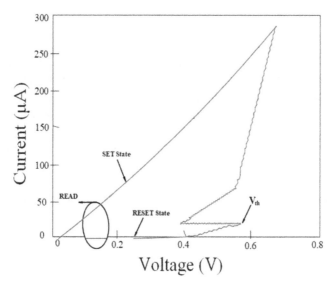

Figure 5.10. Current–voltage characteristics of a PCM cell during SET–RESET–READ operations.

The SET and RESET states' current–voltage characteristics are shown in figure 5.10. A large resistance contrast between the SET and the RESET state for voltages below the threshold switching voltage (V_{th}) can be easily observed. The cell in the RESET state has a high resistance below V_{th} and shows a negative differential resistance. In the subthreshold region, the PCM returns to its original amorphous state after removal of the electrical stimulus. For the voltages above V_{th}, when the electrical stimulus is retained for a sufficient amount of time, the phase-change material undergoes memory switching into a low-resistivity, crystalline state. The physics of the electronic threshold switching process is yet to be fully understood. The conduction of high current through the amorphous region is enabled only by this switching process in order for the crystallisation process to occur. RESET programming consumes the largest power since the cell needs to reach a melting temperature of about 600°C.

5.7 Resistive random access memory

Resistance-based memory devices have been proposed by many researchers in the past, and a significant amount of research is currently being carried out to develop resistive memory devices. This memory type works by altering the strength of the unique dielectric material of a memristor, whose resistance differs based on the voltage applied. ReRAM implementation methods are distinguished based on their distinct switching materials. In memristor technology a passive two-terminal electronic system is designed to convey only the property of an electronic component, which allows it to remember the last resistance it had before it was shut down (i.e. memresistance).

ReRAM is based on a basic three-layer framework of a top electrode, a switching medium and a bottom electrode, as shown in figure 5.11 [17]. When a voltage is applied between the two electrodes, the resistance switching mechanism occurs based on the creation of a filament in the switching material. As in figure 5.12, the two stable states of resistance are the high-resistance state (HRS) and low-resistance state (LRS). The procedure that switches the state from a high to low resistive state is referred to as SET, while RESET denotes the opposite phase switch from low to high. The read process is performed by applying a low constant voltage that does not change the state and senses the current state. The write operation can be carried out using SET and RESET.

ReRAM stores data that uses ions (charged atoms) instead of electrons as changes in electrical resistance. Ions act like a battery on the nanometre scale. For example, the cells have two electrodes made from silver (Ag) and platinum (Pt),

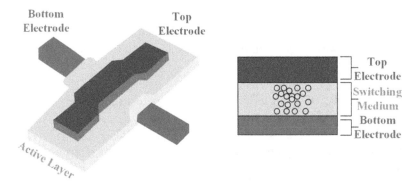

Figure 5.11. Basic three-layer framework of ReRAM. Adapted from [17]. Copyright (2017), with permission from Elsevier.

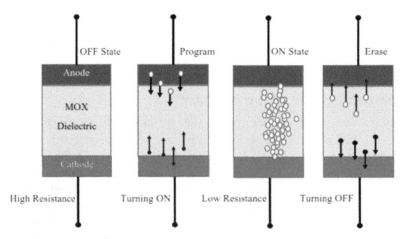

Figure 5.12. Program and erase mechanisms of ReRAM. Adapted from [18]. Copyright (2012), with permission from Springer Nature.

where the ions dissolve and then precipitate again. This increases the electrical resistance that can be used to store data.

ReRAM technology is low power, scalable, fast and completely compatible with CMOS processes. As per the ITRS, the current DRAM scaling limit is 20 nm, which is roughly 4 times that of ReRAM's scaling limit (~5–10 nm). ReRAM has demonstrated a lower read latency and a faster write performance over flash memory. In addition, ReRAM has a higher speed and longer endurance than other traditional devices. Examples of ReRAM with below 2 mA and 3 V specification with 10^6 SET/RESET operations and a read cycle of 10^{12} have been developed [17, 18].

5.8 Ferroelectric RAM

Three types of ferroelectric memory are shown in figure 5.13. Ferroelectric RAM (FRAM) is a type of RAM that combines the quick read and write access attribute of DRAM with nonvolatile capability. FRAM has a transistor-accessed ferroelectric capacitor that stores information in polarisation directions. FRAM was conceptualised more than half a century ago, with the first commercial version appearing in 1988. But, in the light of widely adopted CMOS technology and lack of the right fabrication technology, it lagged behind [19]. With advancements in processes that could integrate ferromagnetic material with CMOS technology, FRAM is presently being considered as an experimental memory technology with several commercial applications. Two major emerging ferroelectric memory technologies available are (i) ferroelectric field-effect transistors (FeFETs), and (ii) ferroelectric tunnel junctions (FTJs). In FeFETs, ferroelectric dielectrics can be integrated in the gate stack of a field-effect transistor (FET) whose channel conductance can be switched by modulation of the ferroelectric polarisation. A FTJ has a memory element based on a very thin layer of ferroelectric tunnel barrier between two metal electrodes. The tunnelling current can be switched by ferroelectric reversal of the thin tunnel barrier.

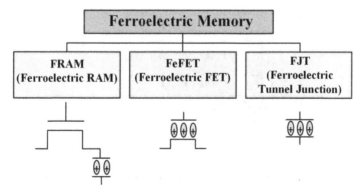

Figure 5.13. Three types of ferroelectric memory. [19] John Wiley & Sons. Copyright © 2015, John Wiley and Sons.

Both FeFETs and FTJs as memory devices combine an access transistor and a storage node in the form of 1-transistor-1-capacitor (or 1T1C) for FeFET and 1-transistor-1-resistor (or 1T1R) for FTJ. However, FeFET is a 1T memory, similar to flash memory devices. Both have indicated extraordinary prospects in emerging memory technologies because of ongoing advances in ferroelectric substances. Memory types based on ferroelectrics depend on an equivalent physical switching component. Switching is brought about by an inversion of thin ferroelectric film by a quick electrical polarisation (Pr switch) [19] under an electrical field.

5.8.1 Ferroelectric field-effect transistors

Several variations of FeFETs have been demonstrated since their inception in 1974, based on combinations of buffer layer and ferroelectric materials. As illustrated in figure 5.14, a FeFET is designed by inserting a ferroelectric capacitor into a MOSFET gate stack [1]. The nonvolatile memory applications of FeFETs are based on the residual polarisation in absence of an external field and the reversibility of polarisation direction under an applied field. By coupling the ferroelectric polarisation directly to the channel of a FET, a FeFET enables capacitor-less ferroelectric memory with simplified device design and nondestructive readout. However, FeFET retention can be degraded by the presence of a depolarisation field and gate leakage [1]. The performance of a FeFET is strongly affected by the interface between the ferroelectric gate and semiconductor channel. Insertion of a buffer layer (e.g. high-k), that is, a metal/ferroelectric/insulator/semiconductor, may improve the properties of the interface.

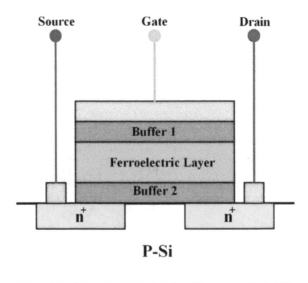

Figure 5.14. Structural schematic of a FeFET. [19] John Wiley & Sons. Copyright © 2015, John Wiley and Sons.

In a FeFET, the process of switching can occur due to the Pr switch in the ferroelectric gate. The polarity change is achieved when a positive or negative voltage pulse is applied at the gate. When a positive pulse is applied, the ferroelectric retains an electric field due to the residual polarisation in the ferroelectric. The direction of this electric field is from the channel to the gate electrode. This increases the threshold voltage (V_{TH}) of the transistor. This electrical field can be considered as the negative portion of the gate field, which makes switching on the system more difficult. In contrast, a negative gate pulse produces a residual electrical field pointing in the direction from gate to channel, which increases the gate voltage [19]. Usually, this electrical field absorbs the negative charges, and thus an inversion layer is induced in the nFET, hence resulting in a normal system.

Retention in FeFETs is still considered the primary issue. A 10-year retention time has been predicted. But these readings are extrapolations of the real measurements that lasted 12 days or less. The retention may be weak because of two main problems that arise while inducing a ferroelectric condenser into the gate stack of a MOSFET [19]: the first is the depolarisation field, and the other is the trapping of charges at the interface of the ferroelectric semiconductor or buffer. Therefore, the occurrence of the process of depolarisation is due to the interactions of a semiconductor with a ferroelectric capacitor.

The charge compensated due to polarisation on the side of the semiconductor cannot be possessed by the ferroelectric due to the presence of semiconductor capacitance. The other issue, the charge trapping at the interface between the semiconductor and the buffer of the ferroelectric, is induced due to the residual field resulting from the polarisation that takes place on this interface. This charge partially regenerates the field generated by the ferroelectric's remaining polarisation and negates the V_{TH} change productively [19]. The injection of charge on highly scaled devices appears to be particularly harmful. Also, it is difficult to coordinate between CMOS and FeFET technology. Furthermore, FeFETs also require a FET structure, and therefore it is unfeasible to integrate them as BEOL memory, as is the case with FTJs and ReRAM. Nonetheless, FeFETs can combine amazingly high durability in a design with low-energy switching and high speed, offering great potential.

5.8.2 Ferroelectric tunnel junctions

FTJs, also known as ferroelectric polarisation reversal memory, is an uneven, metal–insulator–semiconductor or metal–insulator–metal resistive switching memory device. Through the electrodes the estimated resistance of a thin ferroelectric film varies notably because of the giant tunnel electroresistance (TER) effect. Perovskites can be used as regular ferroelectric insulators, for example $Pb[Zr_xTi_{1-x}]O_3$ (also known as PZT) and $BaTiO_3$ (also known as BTO). One terminal (electrode) is usually a metal, for example Au or Pt. The other (for example, base) anode is a regularly conducting perovskite, for example $SrRuO_3$ or La_xSr_yMnO [14]. As of late, $Nb:SrTiO_3$ (niobium-doped strontium titanate crystal) is used to improve TER as a semiconducting bottom electrode [19]. The resistance switching mechanism of

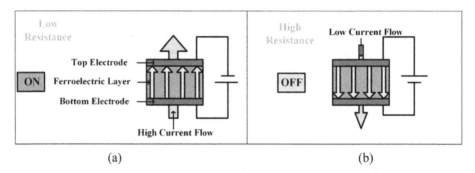

Figure 5.15. Physical representation of an FTJ device in (a) an ON state, showing low resistance and high current flow, and (b) an OFF state, showing high resistance and low current flow. [19] John Wiley & Sons. Copyright © 2015, John Wiley and Sons.

FTJ devices is clarified with the help of figures 5.15(a) and (b). A LRS of the device can be switched to a HRS and vice versa by applying a pulse of appropriate amplitude and polarity to the electrodes. In the ferroelectric insulated layer, the electric field polarises in the direction of the applied field.

As a scalable, ultra-low switching energy technology, FTJ memory offers promise, but it is still relatively new and faces a number of research hurdles that must be resolved before it can be seriously considered as a substitute for NAND flash, SCM, or DRAM [19]. One of the important areas of research is to demonstrate reasonable endurance with significant data retention. The potential scalability of FTJs has also not yet been demonstrated.

5.9 Mott memory

Mott memory is a type of capacitive RAM that uses Mott Insulators. The phase of the cell of Mott memory is toggled using a Mott transition. The switching time of Mott memory is on the scale of nanoseconds and the state of a cell can be accessed electrically. It utilises MOSFETs created from Mott insulators.

Materials which are supposed to be conductors under band theory but are insulators at normal temperature are known as Mott insulators. Mott insulators can be turned into conductors by changing certain parameters such as pressure, voltage, magnetic field, etc. They can be used to make switches, transistors, memory devices, and so on. The so-called Mott transition is used in creating memory from Mott insulators. Electronic phase transitions often cause a change in the dielectric properties of a material. Because of this, Mott memory which utilises the change of capacitance is also possible. The structure of a single cell is shown in figure 5.16(a). A Mott insulator is inserted between two layers of oxide. This forms a MOSFET array. The maximum capacitance of the gate differs for the Mott insulator's isolating and metallic stages. If we imagine an ideal case, when the Mott insulator layer is in its insulating phase, the total capacitance of the gate is the capacitance of the two oxide layers in series and the Mott insulator layer. However,

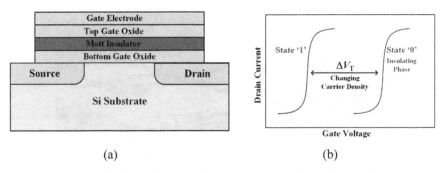

Figure 5.16. Mott memory device. (a) Structure of capacitive memory using a Mott insulator, and (b) shift in threshold voltage due to changes in gate capacitance. © 2015 IEEE. Adapted, with permission, from [20].

when the Mott insulator layer changes its phase only the top and bottom gate layers will contribute to the capacitance. This change in capacitance will result in a change in the MOSFET's threshold voltage, which is shown in figure 5.16(b). As a result, we can read the state of the device from the conductance between the source and drain at a specific gate voltage. The switching mechanism for Mott memory can be via electric field or thermal depending upon the materials used in Mott memory. The switching mechanism helps in the resistive transition of the Mott memory cell. An experiment on NbO_2 with a Mott transition induced thermally with a memory of 110×110 nm^2 allowed a switching speed of <2.3 ns and a switching energy of 100 fJ. From this example, we can deduce that Mott memory has very fast switching speeds and low power consumption. Because of these attributes, it may be possible to use Mott memory in smart card IDs, automobile equipment and other electronic appliances [20].

5.10 Carbon-based emerging memory devices

CNTs, because of their nanostructure and the strength of the bonds between carbon atoms, have very high tensile strength and thermal conductivity. In addition, they can be chemically modified. CNTs exhibit rapid switching speed, ballistic con-duction, high carrier mobility, giant mean free path [21, 22], low heat dissipation and suitable contact resistance [23, 24]. These semiconducting properties of CNTs demonstrate outstanding performance over MOSFET. Transistors based on CNTs can fill in like a molecular electronic gadget in the THz range in correlation with orthodox transistors based on silicon (Si) material. For memory applications, a crossbar arrangement of rows and columns of nanotubes separated by supporting blocks can be used, as in figure 5.17. Application of an appropriate voltage between the desired column and row bends the top tube into contact with the bottom nanotube. The consequent resistance drop is used for bit storage. Van der Waals forces maintain the contact even if the voltage is removed. Separation of the CNTs is achieved by application of a voltage pulse of equal polarity at the crossing point.

Figure 5.17. CNT crossbar memory structure. From [25]. Reprinted with permission from AAAS.

5.11 Molecular memory

Molecular memory uses molecular species (molecules) as the data storage element. It has several advantages over the use of circuits, magnetics, inorganic materials or physical shapes as a data storage element. The storage function of molecular memory can be described as a molecular switch (with more than one stable state), which can be induced by a change in either the resistance or the capacitance of the cell. In an ideal molecular memory device, every bit of data would be represented by an individual molecule, meaning the device would be capable of massive data capacity. Although, in practical applications, a large number of molecules is used for each bit [26]. The two types of molecular memory are capacitive molecular memory and resistive molecular memory.

The storage system of capacitive molecular memory is based on capturing and releasing a charge inside a cell. We can see from figure 5.18 that capacitive molecular memory is generally made up of a redox-active (i.e. reduction–oxidation active) single layer that is attached to a Si substrate. This monolayer can capture or release electrons. By applying different operations on the electrons, the monolayer can be either reduced (i.e. gain an electron) or oxidised (i.e. lose an electron). Redox-active molecules can lose either one or two electrons to gain a neutral state. Thus, the molecule assumes either a mono-positive or di-positive charge state, which can be assigned as a charge storage centre to the molecule. The active single layer is surrounded by a silicon dioxide (SiO_2) insulator to prevent charge loss, and an electron-bounded and ion-conductive electrolyte solution behaves as a further barrier. The structure of resistive molecular memory, by comparison, generally consists of an organic layer between two metal electrodes. Different resistance states can be obtained by applying an external voltage. A readout can be measured by the final resistance difference of the molecular cell [26].

As molecular memory uses small-sized molecules, it can lead to dense circuits. Thus, it possesses extraordinary scalability compared to other nonvolatile emerging memory types. In addition, molecule switches utilise molecules of the same material, so they possess identical characteristics; the issue of compound variability is prevented.

Several examples of molecular memory are detailed as follows.

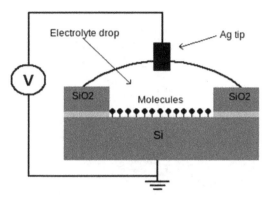

Figure 5.18. Schematic of capacitive molecular memory. Reproduced from [27]. CC BY 3.0.

5.11.1 Single-molecule magnets

Molecules have the ability to remember the direction of the magnetic field applied to them for a relatively long time period after the magnetic field has been switched off, so it is possible to write or save information to the molecule. However, this potential application can only work at very low temperatures; if the molecule is heated by more than just a few degrees above its absolute temperature, the intrinsic property or memory often suffers losses or vanishes [26].

5.11.2 Porphyrin-based polymers

Certain compounds are capable of storing strong electric charges, such as porphyrin-based polymers. These materials behave like material electric capacitors: as the material oxidises, it releases electric charge once a certain voltage threshold is achieved; further, this is a reversible process. In comparison to DRAM memory, they provide much greater capacitance per unit area in ICs, whilst also being cheaper and smaller [26].

5.11.3 Nonvolatile molecular memory

Nonvolatile molecular memory is a capacitive-type molecular memory. Approaches such as synthesis and self-assembly are used to construct molecules and molecule–semiconductor interfaces that can be integrated into existing semiconductor platforms to achieve new device functions. One of the chief aims is to translate the chemical and electrochemical properties of various classes of molecules into desired device characteristics through surface chemistry engineering.

For example, a class of molecules were prepared that has a core structure that readily exchanges electrons, a peripheral structure (linker) that modifies the electron tunnelling barrier, and an end group that anchors the molecule onto a semiconductor/metal surface to create a robust contact. This structure is shown in figure 5.19. With proper molecular engineering, a data retention of over 600 hrs has been achieved so far. This extraordinary nature has also enabled preparation of the

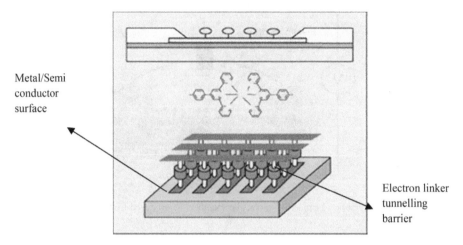

Metal/Semi
conductor
surface

Electron linker
tunnelling
barrier

Figure 5.19. Nonvolatile molecular memory. Reproduced from [28].

first multilevel molecular memory. The operation of molecules accepting or giving electrons adjusts the electrical conductance of nanowires in eight discrete levels. These levels represent the eight possible combinations of three bits: 000, 001, 010, 011, 100, 101, 110 and 111. The ability to store multiple bits of information on the same cell represents a significant step towards high-density memory storage devices. The current focus is to push this molecular memory technology forward to create more robust and reliable devices that are suitable for space applications [28].

In summary, molecules have the ability to remember the direction of the magnetic field, device characteristics can be tailored through surface chemistry engineering and some types can exhibit strong electric charge. Furthermore, molecular memory is very fast and has a very long lifetime. This shows as electronically addressed solid-state data storage in terms of computer memory. Molecular memory can be seen as an area suitable for long-term research, which is currently in progress.

5.12 Macromolecular memory

Before addressing the topic of macromolecular memory, let us define some key terminology.

5.12.1 Macromolecules

Macromolecules are very large molecules, such as proteins. They are typically composed of thousands of atoms or more. They have large diameters ranging from 100 to 1000 Å [29].

5.12.2 Polymers

Polymers are a type of macromolecule composed of a large number of small repeating units, known as monomers. Polymers are commonly created by a process known as polymerisation.

Note: all polymers are macromolecules but not all macromolecules are polymers, as macromolecules may or may not contain repeating monomers.

5.12.3 Dipole moments

Molecules consist of several atoms and various types of bonds. When a bond consists of different types of atoms, the electrons of the bond shift slightly towards the atoms with higher electronegativity; hence an atom with a lower electro-negativity value will acquire a partial positive charge, while an atom with a higher electronegativity value will acquire a partial negative charge. Electric dipole moment is used to measure the polarity of a chemical bond inside a molecule. The larger the difference in electronegativity, the larger the dipole moment (figure 5.20).

5.12.4 Bistable dipole moments

When molecules are placed under an electric field, the dipole moments in the molecules are aligned in the same direction as the field. In bistable dipole moments, dipole moments remain aligned in the same direction even if the electric field is turned off; to reverse the direction of the dipole moments, an external electric field must be applied.

In macromolecular memory, a chip consists of many cells. In each cell, polymers along with a rectifying diode are sandwiched between electrodes, as shown in figures 5.21(a)–(c). Usually, the width of the polymer is about 50 nm. One of the electrodes is connected to a positive voltage, while the other is connected to a negative voltage. An electric field is maintained across each cell. The electric fields across the cells are independent of each other, meaning the direction of the electric field of a particular cell can be changed without affecting the directions of the fields of other cells.

Making digital memory means inventing a way to represent the 0s and 1s of computer logic. In macromolecular memory, the resistance of polymers is used to represent these 1s and 0s. An electric field is applied across the polymers, due to which the dipole moments align in the same direction as the electric field [29]. If the direction of the electric field is in the direction of the current flow, then the resistance

(a)   (b)

Figure 5.20. Dipole moment. (a) HCl molecule, (b) magnitude and direction.

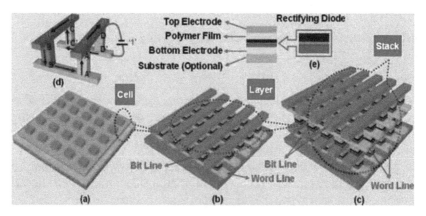

Figure 5.21. Macromolecular memory. Reproduced from http://homepage.ntu.edu.tw/~ntuipse/File/polymer
%20memory%2020090602.pdf.

of the polymers drops, becoming almost negligible; this represents a 1. If the direction of the electric field is in the opposite direction to the flow of current, then the dipole moments align in the opposite direction and hence oppose the flow of current; the resistance of the polymers increases; such high resistance logically represents a 0. A rectifying diode is necessary in each cell; without a diode, faulty readings are possible.

As shown in figure 5.21(d), there is high resistance between D′–D as the dipole moments oppose the electric current. However, current still flows along the path D′–C′–C–B–B′–A′–A–D, and a reading will show less resistance between D′–D. In the presence of a diode, the flow of current would not be possible between B–B′, and hence there will be no current flow and a reading will show high resistance, as expected (figure 5.21(e)). As dipole moments are bistable an electric field is only necessary to change the direction of the dipole moment; once aligned, the electric field is not required to maintain the direction, and hence macromolecular memory is nonvolatile [29]. The advantages of macromolecular memory include the following:

(i) Macromolecular memory is nonvolatile: as dipole moments are bistable, an electric field is not needed to maintain the direction of the dipole moments.
(ii) It requires less energy due to the dipole moments being bistable, with an electric field required only to change state.
(iii) It demonstrates faster read and write speeds than normal disks and drives (e.g. flash memory).
(iv) No cell standby power or refresh required.
(v) It has an operational temperature of −40°C to 110°C.
(vi) The absence of moving parts offers a substantial speed advantage as compared to mechanical storage systems such as magnetic hard disks and optical storage (figure 5.22).

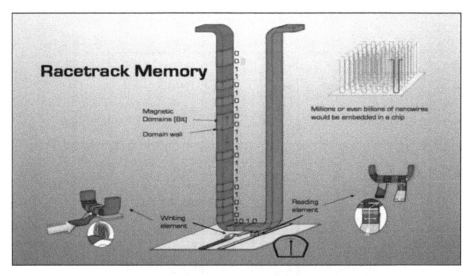

Figure 5.22. Racetrack memory. Reproduced from [30].

5.13 Racetrack memory

Racetrack memory, also known as domain-wall memory (or DWM), is a non-volatile type of experimental memory being developed by IBM's Almaden Research Centre. Some of the advantages of racetrack memory over conventional memory elements are as follows [29]:

1. Higher storage density than comparable solid-state memory devices (like flash memory) and comparable to conventional disk drives.
2. Offers much faster disk read/write speeds.

Racetrack memory is built using spin-coherent electric current to move magnetic domains across a nanoscopic permalloy (a permalloy is an alloy made using 80% nickel and 20% iron). Magnetic domains are nothing but regions in a magnetic material for which the magnetisation is in a uniform direction. These regions are composed of electrons with similar spin, causing the magnetic moment arising because of this spin to align itself and form a domain. The permalloy wire is roughly 200 nm wide and 100 nm thick. As current passes through the wire, the domains pass through magnetic read/write heads positioned close to the wire. The magnetic effect of these read/write heads is capable of altering the domains to consequently alter the pattern of stored bits. We can visualise this as two different sets of domains along the permalloy, one of them representing 1s and the other representing 0s. Figure 5.22 gives an overview of the design of racetrack memory. IBM researchers have attained an unprecedented understanding and control over the magnetic movements inside racetrack memory devices.

5.14 Comparison of memory types

A brief comparison of the existing and potential types of memory is presented in table 5.1 [31].

Table 5.1. Comparison of memory types. Adapted from [31]. Copyright (2016), with permission from Springer Nature.

	DRAM	MRAM	FeRAM	PCRAM	ReRAM	NAND FLASH	STT-MRAM
Storage Cell	Capacitor	Magnetic Tunnel Junction	Ferroelectric Capacitor	Chalcogenide Glass	Metal/Oxide/Metal	Floating Gate	Magnetic Tunnel Junction
Writing Mechanism	Relocation of electrons	Relocation of electrons	Relocation of atoms	Relocation of atoms	Relocation of atoms	Relocation of electrons	Relocation of electrons
Advantages	• Fast • Endurance • Economical	• Fast • Endurance • Non-volatile	• Fast • Non-volatile • Power efficient	• Non-volatile • Economical • Scalable • 3D-integration compatibility	• Non-volatile • Economical • Scalable • 3D-integration compatibility	• Non-volatile • Economical	• Fast • Good Endurance • Power efficient
Disadvantages	• Leakage Current • Volatile	• Slow write for high density architectures	• Limit to scalability • Medium Endurance • Destructive read operation • Difficulty in 3D-integration	• Slow write for high density architectures • Less Endurance • Power hungry	• Slow write for high density architectures • Less Endurance	• Slow write for high density architectures • Less Endurance	• Limit to scalability • Presently 3D-integration is difficult • Costly fabrication
Typical Application	Work Memory	Work Memory	Storage Memory	Storage Memory	Storage Memory	Storage Memory	Storage Memory

In summary, emerging memory devices like STT-RAM, PCM, FeFETs, FTJs, MRAM, Mott memory and CNT-based memory, among others, can be viewed as viable options for cutting-edge memory design. Although these devices have either exhibited or promise better performance, much work is required to establish their reliability, ease of use and scaling limits. Nonetheless, there are trade-offs in the ever-increasing demand for information storage device technology; and these devices are expected to drive it further. We can say there are endless possibilities first with material science, as every element in a periodic table has n dimensions depending on the nanoscale size and shapes, their hybrid compounds and properties. A great deal more research is yet to be done. Then, depending on the physical structure, different electrical properties like resistance, and whether capacitor or transistor type, information can be stored in binary or multi-value logic. Future memory devices have a potential to achieve a so-called universal memory, which will further simplify the complex architecture of computer systems with different types of memory. This will bring many advantages like greater speed, power and design.

Questions

1. State the limitations of present memory and its impact on computer architecture. How can they be resolved?
2. What parameters define memory performance? Compare such parameters of at least two memory types used presently with any three emerging memory types.
3. How can high density in memory be achieved?
4. Name three emerging memory types you think have the potential for future applications and why. Explain the underlying physics.
5. Differentiate between GMR and TMR. How can it be used to build dense memory?
6. State some applications of emerging memory that are not attainable with today's technology.

References

[1] Semiconductor Industry Association 2005 *International Technology Roadmap for Semiconductors (ITRS)* (https://semiconductors.org/resources/2005-international-technology-roadmap-for-semiconductors-itrs)
[2] Institute of Electrical and Electronics Engineers 2021 *Beyond CMOS: International Roadmap for Devices and Systems (IRDS)* (irds.ieee.org)
[3] McKee S A and Wisniewski R W 2011 Memory wall ed *Encyclopedia of Parallel Computing* (Boston, MA: Springer)
[4] Rabey J M, Chandrakasan A and Nikolic B 2004 *Digital Integrated Circuits: A Design Perspective* (New York: Pearson Education) 2nd edn (3rd Indian reprint)
[5] Chua L 1971 Memristor—the missing circuit element *IEEE Trans. Circuit Theory* **18** 507–19

[6] Das M 2015 Seminar report on memristor *BTech Thesis* Maharana Pratap Engineering College

[7] Palmer J 2012 Memristors in silicon promising for dense, fast memory (https://bbc.com/news/science-environment-18103772)

[8] Ahmad J 2017 Giant magnetoresistance (https://slideshare.net/jamilahmedawan/giant-magnetoresistance-75972220)

[9] Pitcher G 2017 Will MRAM replace flash in leading edge processes? (http://newelectronics.co.uk/electronics-technology/will-mram-replace-flash-in-leading-edge-processes/150041/)

[10] Apalkov D, Dieny B and Slaughter J 2016 Magnetoresistive random access memory *Proc. IEEE, Inst. Electr. Electron. Eng.* **104** 1796–830

[11] Dmitri E and Ian A 2013 Overview of beyond-CMOS devices and a uniform methodology for their benchmarking *Proc. IEEE* **101** 2498–533

[12] Bhatti S, Sbiaa R, Hirohata A, Ohno H, Fukami S and Piramanayagam S N 2017 Spintronics based random access memory: a review *Mater. Today* **20** 530–48

[13] Takeuchi K and Chen A 2014 Ferroelectric FET memory *Emerging Nanoelectronic Devices* ed A Chen, J Hutchby, V Zhirnov and G Bourianoff (New York: Wiley) ch 6 pp 110–22

[14] Chanthbouala A, Crassous A and Garcia V *et al* 2012 Solid-state memories based on ferroelectric tunnel junctions *Nat. Nanotechnol.* **7** 101–4

[15] Behin-Aein B, Datta D, Salahuddin S and Datta D 2010 Proposal for an all-spin logic device with built-in memory *Nat. Nanotechnol.* **5** 266–70

[16] Song S, Das A, Mutlu O and Kandasamy N 2020 Improving phase change memory performance with data content aware access *Proc. 2020 ACM SIGPLAN International Symposium on Memory Management (London, UK)* (New York: ACM) pp 30–47

[17] Kumar D, Rakesh A I, Umesh C and Tseung T 2017 Metal oxide resistive switching memory: materials, properties, and switching mechanisms *Ceram. Int.* **43** S547–56

[18] Lu W, Jeong D S, Kozicki M and Waser R 2012 Electrochemical metallization cells-blending nanoionics into nanoelectronics? *MRS Bull.* **37** 124–30

[19] Chen A, Hutchby J, Zhirnov V and Bourianoff G (ed) 2015 Emerging nanoelectronic devices *Emerging Memory Devices: Assessment and Benchmarking* 1st edn (New York: Wiley) ch 13 pp 246–75

[20] Zhou Y and Ramanathan S 2015 Mott memory and neuromorphic devices *Proc. IEEE* **103** 1289–310

[21] Pop E 2010 Energy dissipation and transport in nanoscale devices *Nano Res.* **3** 147–69

[22] Claus M, Blawid S, Sakalas P and Schröter M 2012 Analysis of the frequency dependent gate capacitance in CNTFETs *Proc. SISPAD (Denver, CO, USA, September 5–7)* pp 336–9

[23] Geim A K and Novoselov K S 2007 The rise of graphene *Nat. Mater.* **6** 183–91

[24] Buldum A and Lu J P 2001 Contact resistance between carbon nanotubes *Phys. Rev. B* **63** 161403

[25] Rueckes T, Kim K and Joselevich E *et al* 2000 *Science* **289** 94

[26] Ling D Q, Liaw J D and Zhu C *et al* 2008 Polymer electronic memories: materials, devices and mechanisms *J. Prog. Polym. Sci.* **33** 917–78

[27] Wang L, Yang C-H, Wen J and Gai S 2014 *J. Nanomaterials* **2014** 927696

[28] Dunbar B 2008 Ames technology capabilities and facilities (https://nasa.gov/centers/ames/research/technology-onepagers/nonvolatile_memory.html)

[29] Meena J S, Sze S M and Chand U *et al* 2014 Overview of emerging nonvolatile memory technologies *Nanoscale Res. Lett.* **9** 526

[30] Quick D 2010 IBM researchers bring Racetrack memory another step closer to reality (https://newatlas.com/ibm-researchers-bring-racetrack-memory-another-step-closer-to-reality/17414/)

[31] Yoda H 2016 MRAM fundamentals and devices *Handbook of Spintronics* ed Y Xu, D D Awschalom and J Nitta (Dordrecht: Springer) ch 26 pp 1031–64

IOP Publishing

Nanoelectronics
Physics, technology and applications
Rutu Parekh and Rasika Dhavse

Chapter 6

Modelling and simulation

In the semiconductor industry, modelling and simulation is required at four distinct levels corresponding to various entities, i.e. process, device, circuit and system. A model describes the mathematical relationship between the inputs and outputs of the concerned entity, whereas simulation refers to the process of using the mathematical model to determine the response of the entity to various sets of stimuli, which are intended to represent different real-world operating conditions. In the field of very-large-scale integration (VLSI), simulation is broadly divided into two categories, namely, device/process simulation and design/circuit simulation [1]. Technology computer aided design (TCAD) tools are used for design automation of technological processes and conventional and novel devices. TCAD simulators usually use material properties, process models and device models to mimic the actual fabrication process and to verify the device characteristics at the software level. Electronic design automation (EDA) tools are used to deal with circuit/system-level design. Technological design can be classed into two categories: front-end design, which deals with the verification of the functionality of the circuit, and back-end or physical design, where one generates the complete chip, ready to tape-out.

It is well-known that fabrication is a very cost-intensive process. Fabricating an inaccurate circuit can incur undesired financial overheads, ultimately increasing the cost of manufacturing the chip. To avoid this scenario, simulators allow the designer to run the circuit through various batteries of tests to identify all the critical cases before fabrication. This lowers the probability of failure, thus saving the cost of refabricating the circuit, thereby ensuring it is an affordable process [1]. Not only is fabrication an expensive process, it is also a time-intensive process. There are not one but several steps associated with the actual fabrication of a circuit. Thus, repeating fabrication iteratively until the circuit functionality is practically verified would make for a tedious process. This could result in delaying bringing the circuit to market, and it might end up suffering commercially as a result. Thus, simulators speed up operations and consequently reduce the time to market.

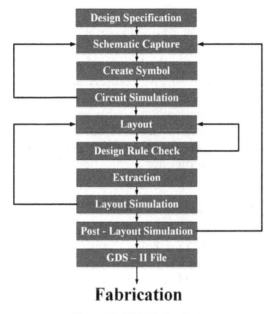

Fabrication

Figure 6.1. VLSI design flow.

Figure 6.1 shows a generic VLSI design flow [1]. For any chip design and fabrication process, initially a design entry is done either textually or graphically on the basis of design specifications. This schematic then undergoes circuit simulation for evaluation of the response of the circuit. Device models present in a simulator help designers to check the performance and characteristics of their design. The primary step in checking the working of a design is to verify its response to a set of predefined test stimuli. If the results obtained deviate from the expected behaviour, the design can be modified until the desired results are obtained. The simulator also provides the flexibility to iteratively check the correctness of the design so that the verification process can be broken down into smaller repetitive steps. Furthermore, it aids the study of the circuit in different working conditions by varying its device parameters. This helps in determining any unexpected behaviour and corner cases, thus indicating to the designer any possible causes of failure that may occur in future. Moreover, replicating the functionality of a design in a software environment provides an added advantage of simplifying the actual design process that would be performed in later steps. Once the simulation results are verified, the physical layout of the circuit is developed and verified. Design specifications are also verified at this level. Utilities like timing analysis, power report generation, register transfer level (RTL) compliers, etc., help to determine the scope of improvement within the circuit design. Following that, a post-layout simulation is performed that generates a GDSII file. A GDSII is a binary file that contains design data such as texts, geometrical shapes and other information regarding its layout. Finally, this file is sent out to the fabrication labs for the circuit to be fabricated on chip. Some of the major advantage of performing simulation are (i) error-free complex design, (ii) easy

and fast design optimisation, (iii) accurate estimation of power, area and performance, (iv) ease of scaling, (v) reduced design cost and (vi) reduced time to market.

In this chapter, we give an overview of modelling and simulation for engineers
who want to embark on their journey in the field of semiconductor devices, processes
and designs. We do not intend to give comprehensive details of modelling and
simulation at the conceptual level. Rather, we want to make our readers aware of the
various techniques and tools presently available in the market for modelling and
simulation of various entities. For instance, for undergraduate students who wish to
explore the various possibilities of the VLSI CAD tools market, this chapter would
equip them with prerequisite knowledge for their detailed survey.

6.1 Technology modelling and simulation

The silicon-based transistor has experienced rapid growth, following Moore's law,
since its first development more than half a century ago. Coming up with accurate
device models is crucial for producing necessary projections and evaluations of the
overall device performance. TCAD modelling and simulations play a vital role in
boosting prototype timing and also help to reduce costs during the development
cycle for nanoscale devices. Typically, different modelling schemes are employed at
the device level and a hardware description language (HDL) is used to build custom
models. As per the ITRS reports, TCAD covers the following [2, 3]:

(i) Device modelling: there is a ranking of models for the operational
description of devices based on fundamental physics. This includes modelling of semiconductor devices such as p–n junctions, bipolar junction
transistors (BJTs), metal-oxide-semiconductor field-effect transistors
(MOSFETs), fin-shaped field-effect transistors (FinFETs), and many more.

(ii) Equipment/feature scale modelling: in this type, there is a pecking order of
models to simulate the effect of the fabrication equipment at every point of
silicon (Si) wafer fabrication, including deposition, the chemical mechanical polishing process and etching, that commences with the equipment
setting and its geometry.

(iii) Lithography modelling: this covers the modelling of display of the mask that
is done by lithographic equipment, the process and its photoresist properties.

(iv) Front-end process modelling: this is used to simulate the physical results
obtained in the manufacturing stages that are used to make transistors up
to metallisation. Lithography is excluded.

(v) Interconnect and integrated passive modelling: this includes the operational
feedback properties such as the mechanical properties, thermal properties
and electromagnetic properties of the back-end architecture.

(vi) Circuit element modelling: this involves compact models for the various
circuit components that includes active components, passive components
and parasitic components; in addition, the circuit elements that are taken
from new device architectures are included in this category.

(vii) Package simulation: this includes physical effect modelling such as mechanical, thermal and electrical modelling of chip bundles.
(viii) Materials modelling: this is used to predict the physical characteristics of materials using simulation tools, and also the subsequent electrical characteristics in some cases.
 (ix) Reliability modelling: modelling of accuracy and relevant effects that occur on the device, process and circuit levels are covered under this category.
 (x) Modelling for manufacturing yield and design robustness: this suggests the development of simulation models for the use of TCAD. It also helps in analysing the influence of process variations and dopant concentration fluctuations on the performance of ICs. Further, it helps in studying the parameters and fabrication viability of the turn design [2, 3].

6.1.1 Device modelling

When a device's behaviour under certain operating conditions is represented in the form of mathematical expressions, an equivalent electrical circuit or a pictorial format with appropriate justifications, it is known as device modelling [2]. Device models are used in device analysis. Device analysis leads to better and accurate device models. Thus, they both depend on each other recursively. A device model gives the direct relation between terminal quantities [2–7]. A device model consisting of carrier transport models and other physical models is useful for simulation of a general device. Device models are helpful in analysis, simulation and design of devices and circuits. They help in the assessment of new device concepts, and can yield accurate distributions of charge, current density, field and potential within the device. Device models are broadly categorised into two categories: qualitative models and quantitative models [2–7]. Qualitative modelling help to visualise physical phenomena without involving as much detail as equations. It is achieved by logical reasoning (i.e. why or how), and assumes the form of equivalent circuits or equations/boundary conditions or diagrams/tables. Device modelling starts first with this qualitative phase as it is based on approximations. Quantitative models help in estimating device terminal characteristics (i.e. how much or how many).
 (i) Types of device modelling strategies:
    (a) Numerical modelling: numerical simulation tools enable a user to describe electronic device structures in two- or a three-dimensional (3D) frames. Each frame is composed of many tiny finite elements. The numerical model also finds solutions to differential equations at the grid points of the mesh created by these tiny finite elements [2–7]. Usually, the tool takes complete details about the device architecture, including material parameters, geometry details, doping concentrations, etc., as input [2–7]. Numerical solvers are then prompted about the appropriate models of physics pertaining to the device under consideration within the frame of its architecture. The solution provides invaluable information about the device's state, including complete

electrostatics in various operating conditions. The method is quite elaborate and has very high computational complexity. Thus, it demands minutes or sometimes even hours to simulate a single device, based on its architecture and the type of stimuli involved. However, this technique is unsuitable for circuit-level analysis.

(b) Lookup table modelling (LUT): these use a group of tables where the external operating conditions of a device are mapped with its static and dynamic behaviour [2]. The information stored in the LUTs is either derived from actual experimentation or from TCAD numerical simulators. The LUT model covers typical biasing environmental conditions in the form of a table. If the actual operating condition being simulated differs from the available ones, then mapping is done using an interpolation with the nearest available points. This approach is very simple and fast, but not very accurate. Further, the size of the LUTs needs to be limited to keep them practical. So, usually, there are many missing biasing conditions [2]. Thus, this approach needs various interpolation schemes for the missing bias conditions. The accuracy of the simulation depends on differential aspects of the performance of the device, which is usually nonlinear, and more particularly on higher-order effects. For example, intermodulation heavily depends on the third and the fifth derivative of the current.

(c) Compact modelling: modern electronics essentially uses compact modelling, which offers the advantages of both numerical and LUT modelling approaches. Compact model development is intensely rooted in the device physics. Modifying physical mechanisms within the limits of some practically acceptable approximations leads to certain formulae that connect the internal state of the device and its electrical behaviour with the outer operating conditions [2–7]. However, approximations affect the accuracy of the model with respect to the exact physical model. Nonetheless, a set of compact formulae is sufficient enough to justify the inaccuracy of the model. To increase the model accuracy, approximation should be done quite smartly. During the development of a compact model for a device, certain physical properties of the device are related with some definite model parameters. These parameters are in turn dependent on the material properties, fabrication nonidealities and the device geometry. Parameters like substrate doping, which do not have single value in a real device, are approximated in the model by a single value that is extracted under an optimal behaviour criterion. Thus, compact models are flexible yet adaptable and suitable for wide range of technologies and devices. Berkeley short-channel insulated-gate field-effect transistor models (BSIM) are some of the most popular compact models used [2–7]. Some types of BSIM compact models used as industrial standards are BSIM compact models, BSIM common multi-gate (or BSIM-CMG) FET models, BSIM independent multi-gate (or

BSIM-IMG) FET models, BSIM silicon-on-insulator (SOI) FET models, etc. In the semiclassical approach to device modelling, techniques like compact modelling for DC and AC behaviour, drift-diffusion models using Poisson's equation and continuity equations, the hydrodynamic equation and Boltzmann transport equation, i.e. Monte Carlo (MC) methods, are popularly used.

(ii) Typical steps for deriving a device model [2]:

(a) Structures and characteristics: the process flow of a device must be either known beforehand or should be developed. In the case of established or mature devices, the process flow will be standardised by a certain foundry. If a new device is to be fabricated, then researchers usually propose a process flow either based on collaboration with the foundry or after exploring compatibility with various processes reported in the literature. In either case, process simulation must result in a stable and accurate target device structure with appropriate dimensions, geometries, doping profiles and interfaces. A cross-sectional view, top view, 3D view, doping profile and material characteristics should be visualised and analysed. This information helps in developing a qualitative understanding or making approximations.

(b) Qualitative model: a qualitative model is a visualisation of a phenomenon by logical reasoning without involving equations and too much detail; for example, visualisation of the direction of magnetic/electric field in a component, carrier concentration and potential distribution of a semiconductor device.

(c) Determine equations: once a qualitative model is obtained, a designer can start working on the equations that govern the device operation.

(d) Determine boundary conditions: the boundary case behaviour of the devices should be incorporated in the model equations.

(e) Summarise all approximations: approximations based on specific needs and ease of calculation should be made such that they do not lead to large variations in the solutions when compared with the actual physical device. If such approximations are not incorporated, the model may be accurate, but computationally it will be highly complex.

(f) Solutions derived from approximations: once proper approximations are made, the appropriate differential equations are solved and the model is ready for testing.

(g) Testing the solution: testing of the device model is required to ensure that it behaves the same as the device. The following criteria can be useful in verification:
  • Generality: model equation should cover various devices of the same class.
  • Continuity: model equations should be continuous over the area of interest.
  • Accuracy: model equations should match the behaviour of the actual device.

(h) Improvement of approximate solutions: based on results from testing, the previous solutions can be revised by modifying equations or changing some approximations. This is an iterative process. Changes are made and testing is repeated until a satisfactory model is obtained.

(iii) Challenges for nanoscale device modelling:

Novel device structures and extreme CMOS scaling need increasingly accurate modelling. Engineered devices are turning out to be standard and novel gate stacks are of important significance. It is necessary that the calculations are computationally proficient and powerful enough to fathom the conditions numerically. In scaled engineered gate stacks, self-warming will be a significant factor, particularly for devices manufactured on SOI wafers or utilising layers containing germanium (Ge) or a Ge substrate [2]. Material properties like aggregate turn, ionic transport or electromechanical impacts should not be left out while demonstrating novel device models. Device models should also complement augmentations to the regular CMOS FET channel. To this end, completely drained ultra-thin-body SOI wafers, multi-gate FETs, nanowires, and FinFETs have been explored. For these structures, quantum transport and ballistic models are imperative as far-reaching, versatile models for discretionary channel bearings [2]. Silicon strips or channel lengths of a couple of nanometres cannot be precisely represented without use of models with ballistic transport, which addition-ally incorporate quantum impacts identified with the phonon spectra and electronic band structure. Reliable approaches need thorough legitimations for their approximations or otherwise will be restrictively computationally intense; less complex plans may depend on self-steady Poisson–Schrödinger equations. Recent techniques use Wigner's or Green's methods, Wigner's transport equation or the many-molecule quantum Liouville equation. Reliable portability models for the modified local density approximation (or MLDA) and density gradient models are also of great significance [2]. More specific structured devices are being investigated, and some are as of now being utilised by numerous makers with the latest Si advances, incorporating nonplanar or raised source/drain structures, and strained with transport-built devices like Si, SiGe, Si:C or Ge, or even hybrid substrates employing III/V materials on silicon, for which a correct and far-reaching representation of stress and strain impacts, morphology, band structure and imperfections/traps is a fundamental necessity [2]. Among the most encour-aging possibilities for emerging channel materials with upgraded mobility are gallium arsenide (or GaAs) or lanthanide gallium arsenide layers for nFETs and germanium layers for pFETs. Emerging memory device advancements are utilising materials with paramagnetic and ferroelectric characteristics [2]. They require modelling of the turn, attractive collaboration and polarisation phenomena that are caused by electrical field, along with transport in amorphous materials and stage changes. The transport displaying of non-crystalline materials requires a predictable treatment of the impact of limited and broadened electronic states [2].

Commercially accessible device models will be required so as to empower the development of novel devices. Their materials and technological processes should also be modelled [2]. These models should not only incorporate all the physical impacts of previously referenced devices, materials and procedures but should further concentrate on demonstrating both the operations and the effects of their evolving geometry and interfaces. They must empower the planning and assessment of structured devices and designs past conventional planar CMOS in an effective manner [1]. It is becoming increasingly important for each new device to demonstrate unwavering performance. Commercially available devices will be required to enable the production of exploration devices to be enhanced, as will the related materials and procedures. Such devices should incorporate all the abovementioned physical impacts and should also focus on both the operational display of these devices as well as their geometry and interface effects. Further, it is crucial to improve the framework and evaluation of standardised devices and designs from the past.

(iv) Primary challenges faced by the device modelling fraternity:

(a) Gate stack: the thin size of the gate dielectrics in newer technology nodes often results in tunnelling gate current. Thus, comprehensive quantum modelling of the entire gate stack needs to be considered [2]. Owing to gate stack engineering, the models must include effective dielectric constants of constant dielectric stacks, charge and trap concentration distributions in high-κ dielectric, interface states and dipoles, and many other parameters. The main focus of the new model should be channel mobility, but some other effects like built in charges caused by the metal and dielectric interface effect, hysteresis effect and flat-band effects by Fermi-level pinning are also important [2]. In recent years, the mechanisms of hot carrier injection and substrate current have been modelled exhaustively. Development of infinitesimal test systems has permitted the generation and understanding of the dynamics of the hot carrier domain in detail. As a result of the ultra-thin dielectric layers used in engineered scaled devices, further improvement is required, particularly concerning the catching and de-catching components of transport in dielectrics [2]. Device degradation and reliability investigation depends on proper modelling of the alterations which occur during device operation like the movement of hydrogen or metallic particles, trap formation, stress-dependent void creation, etc. Capacitance, threshold voltage and reliability characteristics are of major importance. Models should include breakdown of many gate dielectric materials, trap evolution and dielectric degradation phenomena.

(b) Stress and strain: the characteristics of small and specially engineered devices are determined by stress and strain fields due to multiple material layers in the channel, source and drain regions and the uneven heat flow during processing [5–7]. A full tensorial stress field decides all possible channel orientation currents. To model them correctly, effects on band structure, band edges, effective state density and masses, nonlinearities of the stress fields, the effects induced on silicon and the effective mobility of new channel material should be taken into account. For evolving memory devices, piezoresistivity, which is caused by relaxing energy and pressure depending on the rate of saturation, should also be taken into account.

(c) Contact resistance: with the shrinking of device dimensions, the contribution of contact to the total device resistance will increase. This should be considered in the predictive simulation of transconductance and the current–voltage characteristics. Accurate modelling of the state-of-the-art silicon–silicide contact and Si sheet resistance is very important for accurate demonstration of the product and contact engineering.

(d) 3D modelling and fluctuations in dopants, geometry and structure: as aforementioned, the sizes of semiconductor devices have been constantly shrinking, which has led to geometric complications. Dopants are responsible for charge carriers in semiconductors. At atomic scales, where the channel length is of the order of only a few atoms, even a change of a few atoms of dopants in the channel can have a tremendous impact on the overall device characteristics. Since similar devices should have a consistent behaviour in similar environments, maintaining a constant concentration of dopants plays a huge role in device manufacturing. Minor fluctuations in gate dielectrics, ultra-thin-body SOI Si layer thickness, trap concentrations and poly grain sizes also have a large impact on device performance [5–7]. Complete 3D device modelling should take into account doping distributions, 3D geometries for very narrow devices, strain enhancement and coupling among spatial directions [5–7]. These fluctuations increase the number of device parameters required to optimise the device manufacturing and behaviour. As such, each single device type must be represented by a completely random distribution of devices whose doping concentration will also be random, which can be done using MC methods [2–7]. This might also require 3D simulations. Accurate, suitable and efficient descriptions of dopant concentration distribution results are mandatory for evaluating key performance measures (e.g. static RAM noise margins). This further increases the computation cost. Device optimisation studies should also be focused on. During said device optimisation study, the effects that impact the device's output characteristics should be accurately considered. Since effects such as width dependence and edge roughness impact the output characteristics of the device, they should also be looked into. The importance of 3D simulations is increasing and such simulations are no longer optional [2]. However, they are computationally expensive. Hence simulators, emulators and other important algorithms such as meshing algorithms and solver algorithms need to be optimised to the extent that the computational requirements and complexity requirements are on par with two-dimensional (2D) simulations.

(e) Radio frequency (RF) requirements: the development of bipolar-specific models is lagging behind that of CMOS models by quite a high margin.

If numerical treatment of analysing small-signal AC and large-signal transient behaviour is properly enhanced, then support of radio frequency (RF), bipolar CMOS (or BiCMOS), bipolar circuit design and analogue and mixed-signal CMOS can be made possible [5–7]. Analysing device performance, minimising simulation time and costly RF measurement requirements, characterising static non-quasi-effects and providing predictive accurate data requires efficient tools. Device simulations coupled with RF circuits and mixed-mode simulation is required for complete optimisation, but this demands highly efficient algorithms to make it possible. Moreover, hardware and software support are absolutely critical for the parallel computations required during coupling of device and circuit simulations because calculations for different devices will run parallel to each other. DC models like surface quantisation, impact ionisation, direct gate tunnelling, stress effect, etc. should be taken into consideration for the employed model. From a range of sub-kHz to (at minimum) 100 GHz, important noise sources must be covered by comprehensive internal noise modelling, and should also include $1/f$ and random telegraph signal noise. Effects like the self-heating of devices, physical parameters' dependencies and accurate gate resistance should be described in the model. To couple the comprehensive descriptions of external noise sources and transport equations more flexibly, efficient modelling schemes for coupling of substrate noise have to be provided.

6.1.2 Device modelling tools and technologies

Day by day, technology is becoming more complex. Hence, in the semiconductor industry, dependency on TCAD is increasing. TCAD speeds up the process of development and research of novel semiconductor devices and reduces expense. TCAD uses computer simulations to develop and optimise the performance of semiconductor devices and technologies [8]. To accurately analyse and predict the structural, electrical, thermal and optical properties of semiconductor devices, TCAD deals with partial differential equations (like transport and diffusion equations) solvers. As such, all leading companies in the industry utilise TCAD simulations to reduce the number of expensive and time-consuming wafer runs, while developing novel semiconductor technologies or devices. TCAD offers a suite of tools that includes simulation of industrial fabrication processes, device physics models, a powerful GUI-based simulation environment and plotters [8]. Thus, TCAD software is used to develop processes as well as devices. In addition to this, TCAD can display and extract interconnect information and generate basic parasitic data to increase chip efficiency. A typical TCAD suite is shown in figure 6.2.

TCAD simulators support a broad range of uses such as in the design of novel semiconductor devices, memory devices, solar cells, image sensors, power devices, CMOS devices, RF devices and analogue devices. Previously, engineers have used TCAD in exploring alternatives for product design, such as in developing substrates

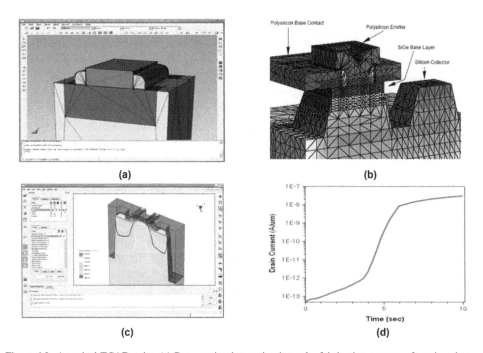

Figure 6.2. A typical TCAD suite. (a) Process simulator: simulates the fabrication process of semiconductor devices; (b) device simulator: simulates the electrical, thermal and optical behaviour of devices; (c) GUI: provides user-friendly design entry; (d) plotter: easy visualisation of process/device simulation results. Reproduced from [8, 9] with permission.

to improve channel mobility and meet performance goals when faced with a scarcity of experimental data. During process integration, TCAD tools help in carrying out split runs of a simulation, such as in design of experiment (DOE), to identify or improve the simulation. This ultimately provides an optimised path of action to conduct a reduced number of actual experiments on real wafers. TCAD tools also yield optimum process recipes to be used in mass production, thereby improving the parametric yield [8]. TCAD software bundles variously use drift-diffusion models, thermodynamic models, hydrodynamic models, energy transport models and Boltzmann equations, etc., to demonstrate carrier transport, energy transport and electromagnetic fields in a device. They utilise additive and subtractive process models to demonstrate oxidation, diffusion, deposition, etching, doping, etc. To generate a new device, TCAD couples material and process parameters. In these approaches, deterministic numerical methods such as box integration, finite difference and finite element are included. The MC method, which is a stochastic approach, is also included. In employing TCAD, a device or process developer has to select an appropriate mix of models and strategies to obtain accurate device/ process results with the least possible computational load. Means must be given to encourage switching between similar models to encourage robust design [5–8]. Device reproduction incorporated with circuit test systems or mixed-mode simulation, where device models are integrated with Simulation Program with Integrated Circuit Emphasis (commonly known as SPICE), requires productive calculations [5–8]. As aforementioned, self-warming of chips must also be considered. With contracting gadget sizes, contact obstruction will assume an increasingly significant role in predictive simulation of the current–voltage qualities and trans-conductance. For bigger bulk gadgets, consistency of reproduction should be investigated. So far, TCAD tools in the industry have concentrated on mass silicon. The possibility of various substrate technologies and 3D integration techniques should be explored. Power enhancers or optical gadgets are typically obtained from numerous transistor cells connected together through a massive interconnect framework. As conventional MOS-based gadgets continue scaling, potential scaling milestones should be considered where a huge variety of new device models can be proposed.

Two kinds of TCAD software tools are available, commercial and open source. Many commercial TCAD tools are widely used in the industry, such as Sentaurus TCAD, Silvaco TCAD, Visual TCAD, etc. for simulation of processes and nano-devices. The three most useful models used in these simulators are the drift-diffusion model, the thermodynamic model and the hydrodynamic model. These models are necessary for understanding the structure and components of semiconductor devices, and as such it is useful to experiment with various scenarios and different carrier transport models for these devices [8].

The drift-diffusion model solves the Poisson's and carrier continuity equations with specified boundary conditions in designated device regions in a self-consistent manner. The thermodynamic model extends this approach, i.e. drift diffusion, for the thermal effect on electrons (also known as the electrothermal effect) under the condition that the carriers and their charges are in thermal equilibrium as a whole lattice. In real life, it can help to solve the lattice thermal condition (heat) equation and Poisson's equation and continuity of carrier equations. The Poisson's equation and the carrier stability equations in the hydrodynamic model address the carrier temperature and the heat distribution equations. Usually, in all TCAD tools, the drift-diffusion model is set as the default model for simulation and analysis [8–11]. As such, there is no need to add any extra keywords while analysing the drift-diffusion transport model. It can be applied to device simulations such as for MOSFETs, BJTs, etc. when a diffusion current exists due to holes/electrons in the semiconductor device. On the other end, when an electric field is exerted and the motion of charge carriers creates a drift current, the set of template equations to be solved for drift diffusion contains Poisson's formula and the formulas for electrons and holes for carrier continuity. In case of substantial self-heating, the thermodynamic transport model is needed. This covers high-current system operations, such as ON state MOSFET operations and active bipolar device operations. In these regimes, substantial Joule heating is emitted inside high-current areas, resulting in a significant increase in the temperature of the lattice. Some coefficients of the transport method, such as the flexibility of the carrier and generation–recombination, are lattice temperature functions. Therefore, to predict the conduct of the electrical device accurately, it is important to have the right lattice temperature distribution.

Recent studies have shown that effective simulation of bipolar VLSI systems should include self-consistent accounting of lattice heating. The ability to simulate high-power structures like bipolar transistor, MOS, insulated-gate bipolar transistor (IGBT) and thyristor machines is a significant feature. Another important use is the design of safety systems for electrostatic discharge. Because of the high thermal conductivity of buried oxide structures and because of the relatively low thermal conductivity of these components, thermal effects are also significant in the application of SOI devices and in devices produced in III–V product systems. The hydrodynamic transport model, along with the Poisson and carrier continuity equations, addresses the carrier temperature and the lattice power equations. The model is particularly useful for simulating systems with deep-submicron and heterostructure designs. This requires the use of nonlocal effects in the simulation, such as velocity overshoot, which can drastically change the output of the system. Because of the local field assumed in the drift-diffusion model, the use of the hydrodynamic transport model avoids the onset of premature breakdowns in a device breakdown simulation. Its carrier transport is usually used for models of heterostructure systems where an important role is performed by energy transfer through heterointerfaces. DEVSIM is a popular open source TCAD software that uses the finite-volume method. It solves partial differential equations on a mesh. Its Python interface allows users to specify their own equations [12]. nanohub.org offers a plethora of tools for aiding research and development of emerging devices and related applications, with more than 2000 reputed publications at the time of writing this book.

6.2 Circuit simulators

As modelling of nanoelectronic devices moves towards circuit implementation and device sizes shrink, modelling is acquiring a growing significance. In this, it is important to adopt a modelling technique that is close to the needs of experimental activities. To provide various levels of abstractions, HDLs were developed for designers.VLSI design requires simulation tools that can cope with the full spectrum of top-down design procedures from register transfer down to circuit level [13]. It is much more complex for an engineer to create an IC from scratch, i.e. by specifying the transistors and wires individually. HDLs allow the description of a program at a high level, which is then converted by simulation synthesis programs to generate a gate-level description. VHDL (IEEE standard 1164) and Verilog (IEEE standard 1364) are the two main languages of interest. In academia, VHDL is preferred because of its higher level of program constructs in field-programmable gate array (FPGA)-based designs. Verilog models the actual hardware in a better fashion and is compact, hence it is more popular amongst application-specific integrated circuits (or ASIC) designers. Systems-on-chip (SOCs), sensor technologies, portable product developments, etc. demand the design of mixed-signal or hybrid ICs. Hence, with the goal of managing interactions between digital and analogue signals, a new language, Verilog-AMS [14], was created. A subset of Verilog, Verilog-A, was created with the express purpose of describing analogue behaviour and an ability to interface with digital behaviours (up to certain extent). In order to study the flow descriptions for electrical, mechanical, thermal and other systems, as well as to understand their potentials, Verilog-A turns out to be a handy language [14]. Verilog-A is a procedural language. It uses constructs similar to C and other computer languages. Using this language, model behaviour can be portrayed to a simulator. A description of the models is easily obtained from the simulations. The relationships between inputs and outputs and the names and ranges of parameters are obtained from the model creator, whereas the interactions between model and simulator are taken care of by the Verilog-A compiler, which handles the simulator interface [14]. Verilog-A can provide high-level behavioural modelling with an emphasis on the physics of the model. These Verilog-A files can be used in simulation platforms like Cadence, Synopsis, Mentor Graphics, etc. to simulate any custom device with its own custom set of parameters and custom set of physics equations. Cadence Design Systems is one of the multinational companies that develops EDA tools, and which provides an environment for development, testing and verification of ICs and SOCs.

These tools are involved in developing various types of products used for design and verification tasks, such as their Custom IC Technology, System and Verification Technology, Digital and Signoff technologies, and many more. These platforms, which fall under custom IC technologies, provide tools for analogue, RF, mixed-signal, standard cell design, FPGA and memory design. Currently, these tools are capable of simulating device models for CMOS technology for channel lengths down to 7 nm [14, 15]. They can also simulate multi-gated devices. Efforts are being made to incorporate nontraditional technologies to explore other possibilities on the basis of emerging technologies; even though the model files for CMOS technologies are commercially available, it is not the case for every emerging technology. However, EDA tools like Cadence [16] provide a way to model these emerging technologies using so-called 'compact modelling'. A compact model of a device is a group of mathematical equations that govern the physics behind the working of the device. It is capable of replicating the behaviour of the device through a range of operating conditions in a fast, precise and robust manner. It uses mathematically proven parameters of IC design applications to model the electrical behaviour of a component present in the circuit design. Compact modelling makes it possible to simulate advanced technologies such as single-electron transistors (SETs), FinFETs, carbon nanotube FETs (or CNTFETs), gate-all-around FETs (or GAA-FETs), negative capacitance FETs (or Neg-CgFETs), and many more. Initially, compact models were developed in FORTRAN or C using lines of complex code. To simplify, condense and standardise the code across various simulation platforms, Verilog-A came to be a necessity of modern-day design. It provided an added advantage over (then) traditional languages by automatically generating derivatives, thus avoiding the need to explicitly hand-code them. It does not require any simulator interface code since Verilog-A itself is compatible with design simulators. Furthermore, it is independent of simulator-specific algorithms and data structures. These features make Verilog-A easy to implement, portable and widely compatible with various simulators, which delivers a comparatively enhanced module quality. It should be noted that for any emerging device, the circuit is usually simulated as a

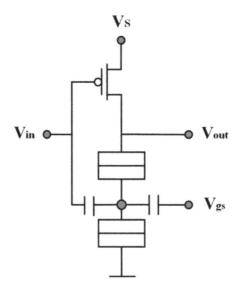

Figure 6.3. Hybrid model for a SET-MOSFET inverter.

hybrid model with CMOS technology. Since the current market is flooded by CMOS-based technology, it is important that a device works in synchrony with CMOS technology. Thus, hybrid models for these devices are preferred over homogeneous circuits. An example of such a hybrid circuit is shown in figure 6.3.

Let us understand the steps for simulating the circuit pictured in figure 6.3 in Cadence EDA tools [17]:

STEP 1: Model the novel device using Verilog-A.
STEP 2: Create a new file and select the device model from the library.
STEP 3: Use the device to create a schematic of the desired circuit, combining it with the standard cell library models. (Note: Make sure to connect the ground and the power supply pins.)
STEP 4: Add the CMOS Spectre Technology file to the schematic design.
STEP 5: Check the schematic for any connection errors and save it.
STEP 6: Open the circuit simulator and set the required working parameters.
STEP 7: Simulate the circuit for the desired simulation time to generate the required response.

Figure 6.4 shows the design flow for the simulation of any schematic design in Cadence Virtuoso [17]. As per the flow chart, it can be seen that, for a hybrid model, Virtuoso takes the Verilog-A model of the device along with the standard cell library and the CMOS technology file to generate the response of the design.

In order to perform circuit simulation beforehand, one of the most prominent software in current use is SPICE [18–20]. There are various versions of SPICE, such as HSPICE, PSPICE, LT-SPICE, and others. These versions are differentiated on the basis of their utilisation and the company selling them. Some are compatible

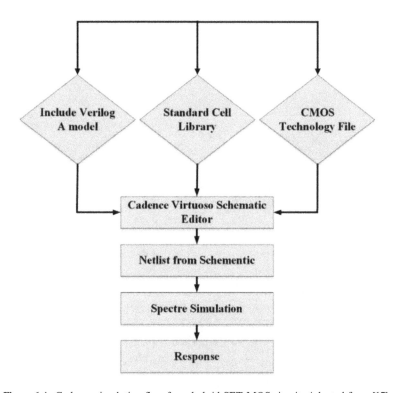

Figure 6.4. Cadence simulation flow for a hybrid SET-MOS circuit. Adapted from [17].

with Windows operating systems, and some run on Linux/Unix. PSPICE is one of the versions used for commercial purposes, sold by Cadence, but HSPICE is preferred by fabrication laboratories that deal with huge datasets. HSPICE is sold by Synopsis. Since these versions are very expensive to obtain, one of the more accessible versions is LT-SPICE, since it is available open source and free of cost. Therefore, LT-SPICE is commonly used in educational fields or by individual users to understand the process of circuit design and simulation. The workflow of SPICE can be observed in figure 6.5 [18–20].

As can be seen, the software requires the topology, technology and test benches as inputs from the user, and further analysis can be done in terms of performance estimation or designing. The topology consists of the connected electrical elements. Further, values for the input parameters can be given and the appropriate technology file attached. Test benches are written (as in the case of HSPICE) or they can be generated (as in the case of LT-SPICE) for output generation. While creating a test bench, it is necessary to specify the type of simulation such as sweep, AC analysis, DC analysis, and so on. Further, the input and output nodes need to be specified and probes applied to them. A netlist can be generated through creation of the schematic. Figure 6.6 shows a circuit diagram for a CMOS inverter and the corresponding HSPICE code to simulate the DC and transient characteristics of the CMOS inverter through .dc and .transient commands.

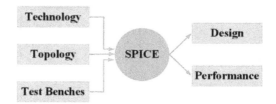

Figure 6.5. Workflow of SPICE simulation.

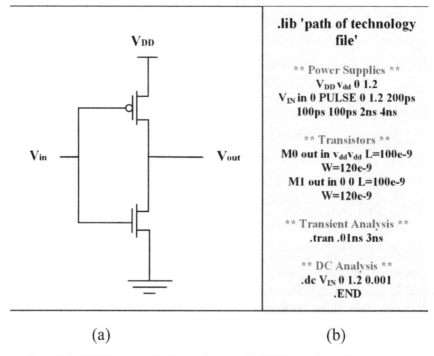

(a) (b)

Figure 6.6. CMOS inverter. (a) Circuit diagram, (b) SPICE code for netlist generation.

6.3 Monte Carlo simulation

Monte Carlo (MC) simulation fundamentally refers to the generation of results due to varying input objects or elements by utilising the relationship between dependent and independent variables. This is in contrast to parametric analysis [2, 3, 9, 21]. A parametric analysis takes into consideration the variation of the parameters one at a time, whereas in MC analysis the parameters are varied simultaneously, thereby giving a cumulative effect. Thus, randomness or variation can emerge as a major aspect while demonstrating the framework of a design. MC is a mathematical method based on the principle of probability. Using MC simulation, the effect on the performance of the design as a whole with respect to the variation of different

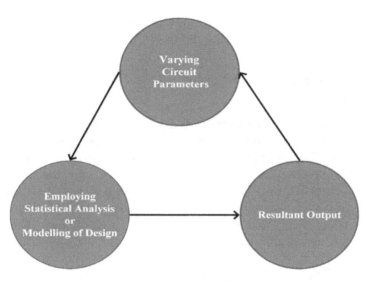

Figure 6.7. Monte Carlo methodology.

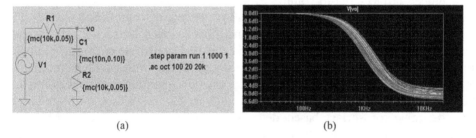

(a)   (b)

Figure 6.8. Monte Carlo simulation. (a) Circuit design in LT-SPICE, (b) simulation results.

parameters can be determined. This grants an insight into randomness and how it can be used to validate the performance of a particular model.

The workflow of MC methodology is depicted in figure 6.7. MC analysis starts by defining the varying input parameters simultaneously. Usually, the minimum and maximum values are fixed and appropriate sampling is done according the number of samples required. Then, the varying inputs are submitted to the design, or alternatively a set of equations defining the relationship between the input and output parameters. After simulation, the output is obtained. The resulting values give an overview of the impact of each independent variable. Both Cadence and SPICE simulators can be used to perform MC simulation.

An example of MC analysis in SPICE is given in figure 6.8. If a circuit has varying parasitic components, then MC simulation can be performed to find the corresponding output in SPICE. In figure 6.8(a), a circuit has been designed by varying resistance and capacitance simultaneously with a tolerance of 5% and 10%, respectively. The obtained result gives an overview of the overall impact on the output due to this variation, as seen in figure 6.8(b).

6.4 Microelectromechanical/nanoelectromechanical device simulators

Microelectromechanical systems (MEMS) are a technology used for miniaturising mechanical and electromechanical elements, which are made using the techniques of microfabrication. The basic physical components of MEMS devices can extend from 1 μm on the lower end of the dimensional range to a few millimetres. MEMS devices can assume a range of forms from moderately simple structures with no moving components to complicated electromechanical frameworks with different moving components under the influence of coordinated microelectronics. They generally comprise a focal unit that processes the information (i.e. the chip) and a few parts that interface with its surroundings, for example as in microsensors. Nanoelectromechanical systems (NEMS), by contrast, are a category of devices incorporating electrical and mechanical utility on the nanoscale. This distinguishes them from MEMS frameworks, where the basic auxiliary components are on the micrometre length scale [22]. In contrast to MEMS, NEMS connect smaller masses with higher surface-area-to-volume ratio together, and are subsequently of great interest for applications with regard to high-recurrence resonators and ultrasensitive sensors. Usually, NEMS incorporate transistor-like nanoelectronics with mechanical pumps, or actuators or motors, and in this way constitute chemical, biological and physical sensors. They offer advantages of low mass, high mechanical reverberation frequencies and possibly enormous quantum-mechanical impacts. For example, zero-point movement and a high surface-to-volume ratio are considered desirable for surface-based detecting components. Applications of NEMS usually incorporate accelerometers or indicators of synthetic substances noticeable all around. A generalised classification scheme for electromechanical systems [23] is given in figure 6.9.

MEMS also find typical use in pressure sensors for measuring fluid/gas/air pressure, and in manipulating mechanical and electrical systems in optical designs. NEMS, in comparison, are more popular in the design of accelerometers, nanomedicine and thermal/mechanical actuators. The challenges faced by MEMS/NEMS devices include lack of mass-production techniques, use of nonbiodegradable materials, difficulty in material handling, etc.

MEMS/NEMS devices need to be modelled differently than conventional electronic circuits. Due to the large surface-area-to-volume ratio of MEMS, incorporation of electromagnetism (e.g., electrostatic charges and magnetic moments) and liquid elements (e.g., surface pressure and consistency), they manifest significant behavioural variations than larger-scale mechanical devices. As there are mechanical components present in both MEMS and NEMS devices, there is a need to pay attention to important mechanical properties of the devices such as their elasticity, thermal, electrostatic and magnetic properties, as well as how the devices respond differently to different environmental conditions. This cannot be done using

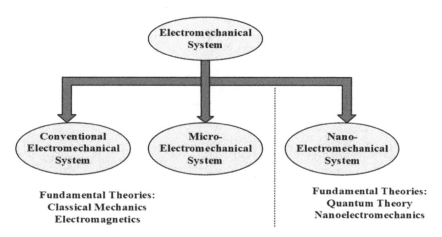

Figure 6.9. Electromechanical systems.

conventional modelling and simulation techniques, and so there is need to use specialised modelling techniques and MEMS/NEMS device simulators such as COMSOL Multiphysics® [24, 25], CoventorWare tools [26, 27], ANSYS [28], etc. All the currently available commercial simulators are intended for MEMS. These simulators cannot be used for the simulation of NEMS devices. Simulators adapted for NEMS designs are still in the research phase.

6.4.1 COMSOL Multiphysics® simulator

For modelling and simulating scientific and complex engineering problems, COMSOL Multiphysics® software, which provides an interactive environment for users, is widely used. It can manage a project in its entirety—from characterising geometries, material properties and the physical science that depicts specific phenomena, to understanding and post-processing models to deliver exact and reliable outcomes. Several modules are accessible in COMSOL, sorted by application area, such as electrical, mechanical, fluid mechanics, chemical, multipurpose physics, and interfacing devices with electronics. An application programming interface (or API) for Java and LiveLink for MATLAB can also be utilised to control the product remotely [24].

An app builder can be utilised to create autonomous, custom, domain-specific simulation applications. Clients can utilise simplified instruments (Form Editor) or programming (Method Editor). COMSOL Server is unmistakable programming for the administration of COMSOL simulation applications in organisations. While simulating thermal actuators in COMSOL, the MEMS module incorporates exclusive physics interfaces for thermal stress calculations with abundant post-processing and representation abilities, including principal stress, stress fields, displacement fields, equivalent stress, and more.

In figure 6.10, the temperature inside a thermal actuator appears in the middle section of the figure, and the current density distribution is shown at the top [24].

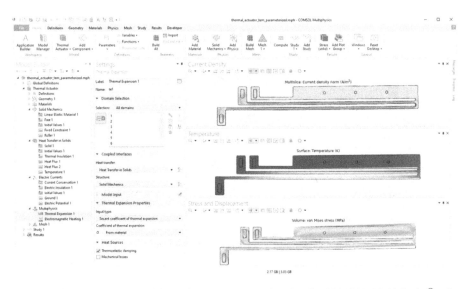

Figure 6.10. Thermal actuators and thermal stress. Image made using the COMSOL Multiphysics® software and provided courtesy of COMSOL.

Figure 6.11. Simulation of an electrostatically actuated MEMS resonator using COMSOL. Image made using the COMSOL Multiphysics® software and provided courtesy of COMSOL.

Similarly, the COMSOL MEMS module presents an exclusive physics interface for electromechanics, which is utilised for MEMS resonators to calculate the change of the resonant frequency with applied DC bias. As indicated in figure 6.11, the frequency diminishes with applied potential because of the conditioning of the coupled electromechanical framework [25].

To model piezoresistive sensors, the COMSOL MEMS module provides many exclusive physics interfaces for piezoresistivity in shells or solids. To pull or tune the resonance of quartz oscillators, a series capacitance is often used, and the MEMS module allows combining 3D and 2D models with SPICE circuits, as used for combined simulations. The stress field in a piezoresistive sensor can be computed with the already defined physics interfaces for piezoresistive materials, as shown in figure 6.12 [24]. Thus, the COMSOL Multiphysics® environment is immensely useful for MEMS developers.

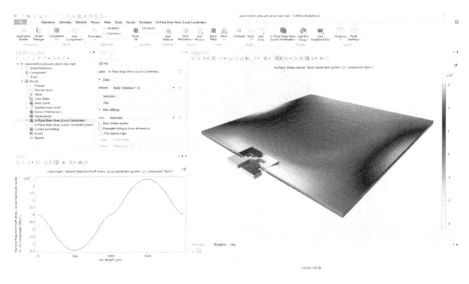

Figure 6.12. The stress field in a piezoresistive sensor. Image made using the COMSOL Multiphysics®
software and provided courtesy of COMSOL.

6.4.2 CoventorWare tools

Another very popular, robust platform for MEMS planning, simulation, checking
and procedure demonstration is CoventorWare. This is an integrated suite of design
and simulation software that has the accuracy, capacity and speed to address real-
world MEMS designs. It caters to MEMS-specific engineering difficulties, such as
multi-physics science collaborations, process varieties, MEMS+IC integration,
MEMS+package interaction, etc [26]. In a matter of hours or days,
CoventorWare tools can demonstrate devices' parametric impacts that would
have taken a great deal of time to physically build and test in a laboratory.
CoventorWare tools can be used to develop an extensive range of MEMS devices,
from motion sensors to types of microphones and optical displays. In
CoventorWare, the workflow is divided into two sectons: the designer and an
analyser. The designer section is preconfigured with a database of more than 40
commonly used MEMS materials. In addition, custom materials and their parameters such as Young's modulus and Poisson ratio, residual stress distribution and
magnitude, electrical conductivity, piezoresistive properties, etc. can be defined by
the user. The inputs to the designer stage consists of materials data, process data,
layout (GDSII or manual). This then gives a 3D model as its output, which is in turn
fed as an input to the analyser section. The analyser provides the capability to
examine and simulate the conduct of MEMS gadgets when exposed to different
physical forces. Figure 6.13 gives an overview of the CoventorWare designer and
analyser stages.

To better understand the CoventorWare simulator, let us look at an example
simulation of an accelerometer. From conceptualisation to improvement and

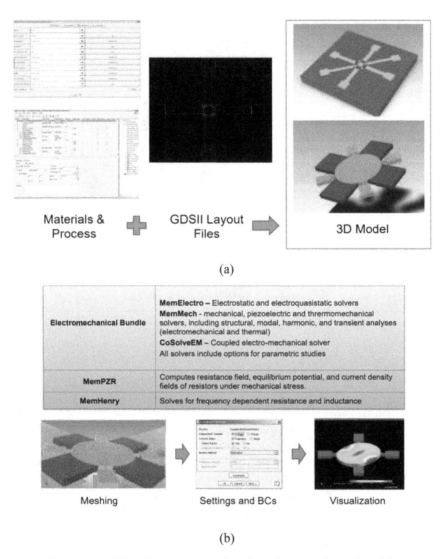

Figure 6.13. CoventorWare. (a) Designer overview, (b) analyser overview. Adapted from [26].

performance upgrading, CoventorMP® can simulate a wide scope of basic specifications identified within a MEMS accelerometer structure. These include various
factors like noise, sensitivity, linearity, bandwidth, nonlinear physics, including
damping, thermal performance, shock resistance, etc. An accelerometer model built
using the multi-physics science components in the MEMS+ module of
CoventorMP® is shown in figure 6.14. It has sensing material, capacitors, and force
feedback control. The model is completely parametric so that the MEMS design
parameters can be shifted to optimise linearity, sensitivity and transmission capacity.
The lower part of the figure shows the effect of pressure on bandwidth [27].

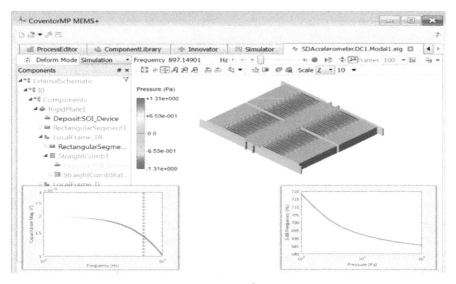

Figure 6.14. Accelerometer simulation using CoventorMP® MEMS+ module. Reproduced from [27] with permission. © Lam Research.

Thus, multi-physics environment-based tools can assist in the development of MEMS devices from the choice of initial material through to manufacturing process optimisation, building structure and modelling device performance.

6.5 System-level design

System-level design is where the designer takes care of each and every part of a system while planning the end solution. It is crucial to take into account both hardware and software as a single unit when modelling and analysing a complete system. Further, at the system level, the number of devices and their interconnections may become complex and unmanageable if analysed using circuit simulators. Troubleshooting will also become very cumbersome. System-level design is thus a multifunction heterogeneous fabric (HUB) on rigid and/or flexible substrates [29] and needs to be simulated in a radically different manner. An example of such a structure showing a cut-through of a graphics card that uses high-bandwidth memory (HBM) based on through-silicon via (TSV) 3D IC technology is illustrated in figure 6.15. The system-level design consists of three phases:

- Architecture exploration: deals with developing a suitable system architecture and distribution of the framework tasks in the specification onto those components of MEMS [9].
- Software scheduling/real-time operating systems (RTOSs): RTOSs are needed since each and every part has a single control, and it is important to serialise the tasks in each and every part [29].
- Communication synthesis: to design and refine the information transfer and interfaces in order to utilise the communication design [29].

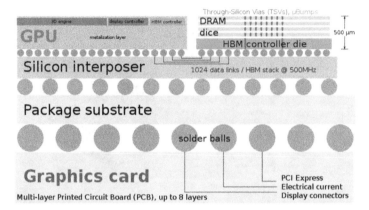

Figure 6.15. Heterogeneous integration based on TSV 3D IC technology. (This High Bandwidth Memory schematic.svg image has been obtained by the author(s) from the Wikimedia website where it was made available under a CC BY-SA 4.0 licence. It is included within this book on that basis. It is attributed to Shmuel Csaba Otto Traian.)

A successful MEMS modelling and simulation methodology should be compatible and interoperable with the IC design environments described above, in order to enable system-level design and MEMS-IC co-simulation. The feasibility of this methodology depends on the analogy between multiple energy domains. Here, three methodologies for system-level MEMS modelling and simulation are given in brief.

6.5.1 Lumped-element modelling using equivalent circuits

To implement simulation and modelling for MEMS devices that is compatible with SPICE, a simple strategy is to make similar circuits for MEMS that are purely based on lumped-component modelling [29].

6.5.2 Hierarchical abstraction of MEMS and analytical behavioural modelling

In an IC design, complex frameworks are progressively developed utilising building blocks at various levels of abstraction, starting from fundamental (or even atomic) elements, as indicated in figure 6.16. Similar reasoning has been raised and applied to MEMS design, and has proven to be feasible, though challenging, because of the uniqueness and variety of MEMS devices [29].

6.5.3 MEMS behavioural modelling based on finite and boundary element analysis

Since the geometrical structures that are uptaken by most analytical setups are discrete and constrained, MEMS developers sometimes make use of finite element analysis (FEA) and boundary element analysis (BEA) instruments. BEA/FEA instruments utilise traditional computational analysis techniques for modelling and simulation in the thermal, magnetic, electrostatic and mechanical domains. They frequently depend on automatic meshers to segment a continuous structure

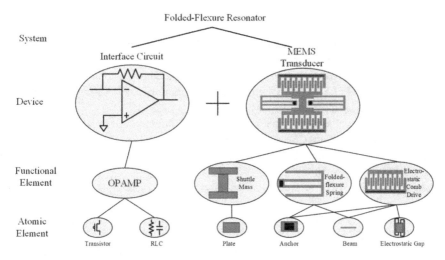

Figure 6.16. Hierarchical abstraction of a folded-flexure resonator. Adapted from [30].

into a mesh that is composed of low-order limited components. With sufficient refinement of the meshing, the designer can obtain accurate simulation results.

In summary, this chapter gave an overview of modelling and simulation tools for processes, devices, circuits and systems in brief. It also presented the difference between conventional solid-state device simulators and MEMS/NEMS device simulation tools. There are many such tools available in the market as well as in the open domain, but here we focused on the tools used to produce many publications and which are preferred by the semiconductor industry. In this chapter, we do not promote use of a particular commercial software. The sole intention of mentioning the names of these tools is to make our readers aware of some examples for further detailed exploration of the software suitable for their research/academic development work. Thus, this chapter is intended to help a newcomer embark on his/her journey in nano/microelectronics. Readers are advised to further research the availability/suitability of such modelling and simulation tools in detail with regards to their own works.

Questions

1. Mention the challenges faced in the modelling of a nanodevice.
2. What are device and circuit simulators? Name any such two simulators.
3. How to simulate a circuit with a nanodevice?
4. How and why are MEMS simulators different from conventional electronic device simulators? Name one such simulator.
5. There is no CAD environment in which to carry out the physical design of the circuit of a nanodevice. In this situation, how is it possible to implement a complete design and simulation flow for fabrication?

References

[1] Terman C J 1987 Simulation tools for VLSI *VLSI CAD Tools and Applications* ed W Fichtner and M Morf (New York: Springer) ch 3 pp 57–103

[2] Semiconductor Industry Association 2005 *International Technology Roadmap for Semiconductors (ITRS)* (https://semiconductors.org/resources/2005-international-technology-roadmap-for-semiconductors-itrs)

[3] Institute of Electrical and Electronics Engineers 2021 *International Roadmap for Devices and Systems (IRDS): Beyond CMOS* (irds.ieee.org)

[4] Tsividis Y and McAndrew C 2011 *Operation and Modeling of MOS Transistors* (Oxford: Oxford University Press)

[5] Taur Y and Ning H 2013 *Fundamentals of Modern VLSI Devices* (Cambridge: Cambridge University Press) 2nd edn

[6] Arora N D 2007 *MOSFET Modeling for VLSI Simulation: Theory and Practice* (Singapore: World Scientific)

[7] Pierret R F 1998 *Semiconductor Device Fundamentals* (New York: Pearson Education)

[8] Li S and Fu Y 2012 *3D TCAD Simulation for Semiconductor Processes, Devices and Optoelectronics* (Berlin: Springer)

[9] Synopsys, Inc. 2014 *Sentaurus TCAD v J-2014.09*

[10] SILVACO® 2016 ATLAS™ User's Manual (https://dynamic.silvaco.com/dynamicweb/jsp/downloads/DownloadManualsAction.do?req=silentmanuals&nm=atlas)

[11] Cogenda Pte Ltd 2014 *Genius, 3-D Device Simulator, Reference Manual., v1.9.3* (Singapore: Cogenda)

[12] Sanchez J E 2022 DEVSIM manual, v2.1.0, Zenodo: 10.5281/zenodo.6465427

[13] Sangiovanni-Vincentelli A L 1984 Circuit simulation *Computer Design Aids for VLSI Circuits* ed P Antognetti, D O Pederson and H Man (Dordrecht: Springer) (NATO Science Series E) 48 ch 2 pp 19–112

[14] Kundert K S and Zinke O 2004 *The Designer's Guide to Verilog-AMS* (New York: Springer) (The Designer's Guide Book Series)

[15] Cadence 2023 Hardware Description Languages: VHDL vs Verilog, and Their Functional Uses (https://resources.pcb.cadence.com/blog/2020-hardware-description-languages-vhdl-vs-verilog-and-their-functional-uses)

[16] Cannon A J 2012 Neural networks for probabilistic environmental prediction: Conditional Density Estimation Network Creation & Evaluation (CaDENCE) in R *Comput. Geosci.* **41** 126–35

[17] Parekh R 2012 Simulation and design methodology for hybrid SET-CMOS logic at room temperature operation *PhD Thesis* University of Sherbrooke

[18] Nagel L W 1996 The life of SPICE *Bipolar Circuits and Technology Meeting (Minneapolis, MN)*

[19] Tuinenga P W 1988 *SPICE: A Guide to Circuit Simulation and Analysis using PSpice* (Englewood Cliffs, NJ: Prentice-Hall)

[20] Vladimirescu A, Zhang K, Newton A R and Pederson D O 2023 *The SPICE circuit simulator, SPICE version 2G User's Guide* (https://eecg.utoronto.ca/~johns/spice/SPICE_Circuit_Simulator_Reference_Manual.html)

[21] Ma H and Liu Y 2011 Search based on the Monte-Carlo method to calculate the definite integral *3rd Pacific-Asia Conf. on Circuits, Communications and System (PACCS)*

[22] Mahalik N P 2006 *Micromanufacturing and Nanotechnology* (Berlin: Springer)

[23] PRIME Faraday Partnership 2002 An introduction to MEMS (https://lboro.ac.uk/micro-sites/mechman/research/ipm-ktn/pdf/Technology_review/an-introduction-to-mems.pdf)

[24] Comsol 2019 *Introduction to COMSOL Multiphysics: COMSOL Multiphysics® v6.0* (Stockholm: Comsol)

[25] Comsol 2023 MEMS module: analyze microelectromechanical systems with the MEMS module (https://comsol.com/mems-module)

[26] Pepin B 2011 Coventorware tutorial (https://inst.eecs.berkeley.edu/~ee245/fa12/homework/CoventorWare_Tutorial.pdf)

[27] Coventor 2023 CoventorWare (https://coventor.com/mems-solutions/products/coventor-ware/)

[28] DeSalvo G J and Swanson J A 1985 *ANSYS Engineering Analysis System User's Manual* (Houston, PA: Swanson Analysis Systems)

[29] Mukherjee T 2002 System-Level Modeling and Design of Integrated MEMS (Carnegie Mellon University)

[30] Siemens 2023 *System-level MEMS design—exploring modeling and simulation methodologies* White Paper Siemens (https://resources.sw.siemens.com/en-US/white-paper-system-level-mems-design-exploring-modeling-and-simulation-methodologies#disw-fulfillment-form)

Chapter 7

Nanofabrication

The manufacture of ultra-miniaturised systems to serve a huge variety of electronic/ optoelectronic/electromechanical applications is known as as micromanufacturing or microfabrication. Microfabrication refers to the techniques that make it possible to put a large group of transistors in a small area, which in turn makes electronic devices smaller, cheaper, faster as well as more energy efficient. By definition, microfabrication is the process by which a miniature structure of micrometre scale and smaller is fabricated. Conventionally, microfabrication processes have also been known as 'semiconductor manufacturing' or 'semiconductor device fabrication', and were used for integrated circuit (IC) fabrication. The manufacture of devices with dimensions of the order of a nanometre (10^{-9} m) is broadly termed as nanofabrication. The terms 'nanofabrication' and 'nanotechnology' are often confused with each other. However, it should be noted that nanofabrication essentially talks about fabrication techniques at the nanoscale, whereas nanotechnology is a multi-disciplinary domain that covers all the technological advancements of ultra-miniaturised systems. Thus, nanotechnology has a much greater breadth than nanofabrication. Fabricating high-density processors, high-capacity memories, and complex and sophisticated ICs is result of nanofabrication techniques. The techniques for nanoscale fabrication were initially borrowed from microfabrication. Thus, it is imperative to have a prerequisite knowledge of microfabrication. Hence, we start from the basics of microfabrication and move on to the various nanofabrication-related approaches.

7.1 Microfabrication techniques

The basic microfabrication techniques, which include doping, lithography, thin-film deposition, epitaxy and chemical vapour deposition, etching and substrate removal, and packaging, are illustrated in figure 7.1.

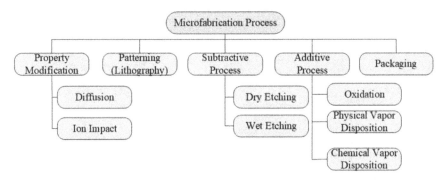

Figure 7.1. Microfabrication processes.

7.1.1 Property modification (doping)

The introduction of controlled amounts of impurities (i.e. dopants) into semiconductor materials to modify their conductivity is called doping. Impurity can be incorporated into a semiconductor by (i) ion implantation at room temperatures and high vapour pressures, or (ii) diffusion at elevated temperatures. Ion implantation is preferred for fabricating shallow junctions, whereas diffusion is usually employed for deep junctions. Until the early 1970s, doping was done mainly by diffusion at elevated temperatures [1]. In this method, the dopant atoms are placed on or near the surface of the wafer by deposition from the gas phase of the dopant or by using doped-oxide sources. The doping concentration decreases monotonically from the surface, and the profile of the dopant distribution is determined mainly by the temperature and diffusion time. Since the early 1970s, many doping operations have been performed by ion implantation. In this process the dopant ions are implanted into the semiconductor by means of an ion beam. The doping concentration has a peak distribution inside the semiconductor, and the profile of the dopant distribution is determined mainly by the ion mass and the implanted-ion energy. Both diffusion and ion implantation are used in fabricating discrete devices and ICs because these processes generally complement each other.

(a) **Doping by diffusion**

Diffusion of impurities is done by placing semiconductor wafers in a carefully controlled high-temperature quartz-tube furnace and passing through it a gas mixture containing the desired dopant. A schematic of a typical open-tube diffusion system is shown in figure 7.2. The temperature usually ranges between 800°C and 1200°C for silicon (Si) and 600°C and 1000°C for gallium arsenide (GaAs) [1]. The number of dopant atoms that diffuse into the semiconductor is related to the partial pressure of the dopant impurity in the gas mixture. For diffusion in silicon, boron is the most popular dopant for introducing a p-type impurity, whereas arsenic and phosphorus are used extensively as n-type dopants. These three elements are highly soluble in silicon, as they have solubilities above 5×10^{20} cm^{-3} in the diffusion temperature range. These dopants can be introduced in several ways, including solid sources (e.g. BN for boron, As_2O_3 for arsenic and P_2O_5 for phosphorus), liquid sources (BBr_3 $AsCl_3$ and $POCl_3$) and gaseous sources (B_2H_6, AsH_3 and PH_3).

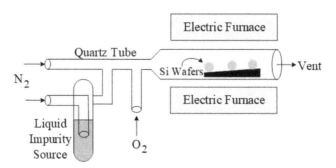

Figure 7.2. Open-tube diffusion system. [1] John Wiley & Sons. Copyright © 2007 John Wiley & Sons, Inc. All rights reserved.

However, liquid sources are most commonly used. Diffusion in a semi-conductor can be visualised as atomic movement of the diffusant (dopant atoms) in the crystal lattice by vacancies or interstitials. At elevated temper-atures, the lattice atoms vibrate around the equilibrium lattice sites. There is a finite probability that a host atom will acquire sufficient energy to leave the lattice site and become an interstitial atom, thereby creating a vacancy. When a neighbouring impurity atom migrates to the vacancy site, the mechanism is called vacancy diffusion. If an interstitial atom moves from one place to another without occupying a lattice site, the mechanism is known as interstitial diffusion. An atom smaller than the host atom often moves interstitially. Sometimes, an interstitial host atom (self-interstitially) pushes the substitutional impurity atom into an interstitial site. Subsequently, the impurity atom displaces another host atom and creates a new self-interstitial. Then the process is repeated. Interstitial diffusion is faster than substitutional diffusion. Vacancy and interstitial diffusion are considered the dominant mechanisms for diffusion of phosphorus, boron, arsenic and antimony in silicon. Phosphorus and boron diffuse via a dual (vacancy and interstitial) mechanism, with the interstitial component dominating. Arsenic and antimony diffuse predominately via the vacancy mechanism.

(b) **Doping by ion implantation**

Here, dopants are implanted in a semiconductor by means of an ion beam. The doping concentration is at its peak distribution inside the semiconductor. The dopant distribution profile mainly depends on the ion mass and the implanted-ion energy. A variety of ion implanters are available based on their current and energy rating. A schematic of a medium-current ion implanter is shown in figure 7.3 [1].

The ion source has a heated filament to break up a source gas such as BF_3 or AsH_3 into charged ions (B+ or As+). An extraction voltage, around 40 kV, causes the charged ions to move out of the ion source chamber into a mass analyser. The magnetic field of the analyser is chosen such that only ions with the desired mass-to-charge ratio can travel through it without being filtered. The selected ions then enter an acceleration tube, where they are

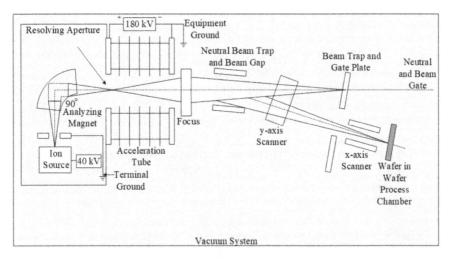

Figure 7.3. Medium-current ion implanter. [1] John Wiley & Sons. Copyright © 2007 John Wiley & Sons, Inc. All rights reserved.

accelerated to the implantation energy as they move from high voltage to ground. Apertures ensure that the ion beam is well collimated.

The pressure in the implanter is kept below 10^{-4} Pa to minimise ion scattering by gas molecules. The ion beam is then scanned over a wafer surface using electrostatic deflection plates and is implanted into the semiconductor substrate. The energetic ions lose their energies through collision with electrons and nuclei in the substrate, and finally come to rest at some depth within the lattice. The average depth can be controlled by adjusting the acceleration energy. The dopant dose can be controlled by monitoring the ion current during implantation. The principal side effect is the disruption or damage of the semiconductor lattice due to ion collisions. Therefore, a subsequent annealing treatment is needed to remove this damage. Ion implantation offers some advantages over the diffusion process, such as more precise control, lower processing temperature and reproducibility of the impurity doping. Cutting-edge CMOS processes use up to 30–40 implants.

7.1.2 Patterning

The growth of the semiconductor industry is a direct result of the capability of transferring smaller and smaller circuits onto semiconductor wafers. The two processes related to this are lithography and etching. Both require a clean room. As mentioned earlier, in chapter 2, the basic fabrication models are classified into top-down approaches, where materials are chiselled into nanoscale objects, and bottom-up approaches, where all the nanoscale objects are built from even smaller units like atoms and molecules. The ultimate goal is to calibrate for the desired functionality and properties by designing and building on a suitable scale, i.e. the nanoscale. All lithographic techniques come under the umbrella of top-down approaches.

Lithography is the process of transferring patterns of geometric shapes on a mask to a thin layer of radiation-sensitive material covering the surface of a semiconductor wafer. Patterns define various regions in an IC such as implantation regions, contact windows, bonding pads, etc. The defined resist patterns are not a permanent part of the device. The most widely used lithographic technique is optical lithography (or photolithography). This makes use of a series of process steps to transfer a layout (design) from masks to the wafer surface by means of light and photosensitive film, as shown in figure 7.4. The ability to fabricate smaller and smaller features is mainly thanks to advances in lithography. The preparation of a mask is usually the first step in the microfabrication process. This is a thin square disc of vitrified silica, which contains geometrical features of the various materials and layers of the chip to be fabricated. There are two types of photoresists: positive and negative. Positive resists consist of a photosensitive compound, a base resin and an organic solvent. Initially, the photosensitive compound is insoluble in the developer solution. After exposure to ultraviolet (UV) light, its chemical structure changes due to absorption of radiation. After development, the exposed areas become more soluble and are removed. The patterns formed on the wafer are the same as those of the masks.

In negative photoresists, polymers that are initially less insoluble in the developer solution are combined with a photosensitive compound. After exposure to UV light, their chemical structure changes due to absorption of radiation. Optical energy is converted into chemical energy, which initiates a polymer cross-linking reaction. Cross-linked polymers have more molecular weight and thus become more insoluble in the developer solution. Thus, after development, unexposed areas can removed. The patterns created on the wafer are mirror images of the mask patterns. Developer solution is absorbed here and there, and causes swelling of the resist mass. Hence, it offers a limited resolution. An illustration of pattern transfer using the same mask but opposite photoresists is shown in figure 7.5.

The major steps of the photolithography process are cleaning of the wafer, barrier layer formation, photoresist application, mask alignment, soft baking, development

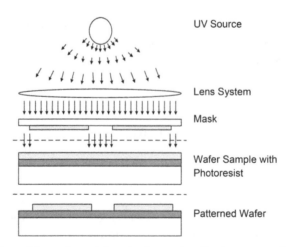

Figure **7.4.** Projection lithography. Adapted from http://optics.ansys.com/hc/article_attachments/360056057034/duv_lithography_system.jpg.

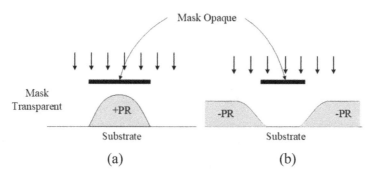

Figure 7.5. Photolithography using (a) positive, and (b) negative photoresists.

after exposure, hard baking, etching and photoresist removal [2]. A simple process of dry etching using a positive-photoresist photolithography process in semiconductor microfabrication is given diagrammatically in figure 7.6.

Preparing the wafer often involves wafer cleaning to remove native oxide and metallic and organic contaminants and surface preparation as in the application of adhesive materials. Next, the photoresist (preferably positive due to its better resolution) is applied by the process of spin-coating. The wafer is soft baked to dehydrate the photoresist at around 80°C–100°C temperature for typically 30 s. Then, the photomask is aligned with the wafer and exposed to UV light. This is followed by developing the sample and rinsing with deionised water. If required it is baked again, and then processes like etching or implantation are carried out. At last, the photoresist is stripped out using plasma or hot acids. Like any other optical projection system, optical lithography systems also suffer from the diffraction of light.

Without getting into the complicated mathematics of optical theory, for a given system the resolution limit (R) can be understood by equation (7.1). Here, the wavelength of incident light is λ, the numerical aperture of the lens used in the system is NA, and the process-dependent coefficient is k_1 (which is less than 1) [1]:

$$R = k_1 \frac{\lambda}{\text{NA}}. \tag{7.1}$$

A lens with a fixed numerical aperture can focus light with shorter wavelengths more. Similarly, a lens with a bigger numerical aperture can collect more diffracted light. The finer the detail of the feature, the further the light gets diffracted from the central optical axis. Hence, it is necessary to collect as much diffracted light as possible because it contains information of the finer details of the mask. Therefore, using light of a shorter wavelength or a lens with a large numerical aperture results in a more detailed projection on the substrate. To improve the resolution of a lens with a large numerical aperture, shorter-wavelength light and reduced k_1 factor can be used. Another very important limitation of the projection system is the depth of focus (DOF), which is shown in figure 7.7. The DOF is given by equation (7.2), with k_2 as the process-dependent coefficient (less than 1). Increasing numerical aperture improves the resolution but reduces the DOF:

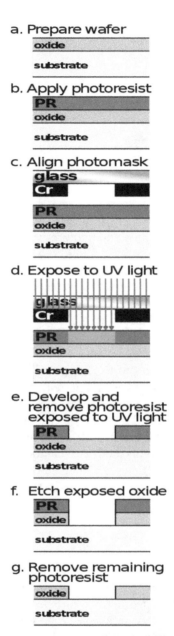

Figure 7.6. Steps involved in the positive-photoresist photolithography process. (This Photolithography etching process.svg image has been obtained by the author(s) from the Wikimedia website where it was made available under a CC BY-SA 3.0 licence. It is included within this book on that basis. It is attributed to Cmglee.

$$\mathrm{DOF} = k_2 \; \frac{\lambda}{NA^2}. \tag{7.2}$$

Usually, large-sized silicon wafers (8″ or 12″ diameter) are used to fabricate ICs, which makes it practically impossible to attain a perfectly flat wafer. There is also

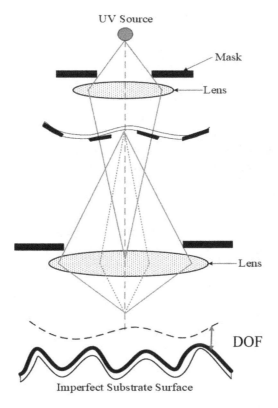

UV Source

Mask

Lens

Lens

DOF

Imperfect Substrate Surface

Figure 7.7. Defocusing due to surface imperfections in the mask and substrate. Adapted from Smart Optics Systems programme, Numerics for Control & Identification (N4CI) (http://www.dcsc.tudelft.nl/~mverhaegen/n4ci/imwacol.htm).

some topography of structure formed on the wafer in the previous steps of the process. A small DOF restricts the optical imaging to very small variations in height. Hence, many spots on the wafer will face a defocusing issue if the DOF is small, as shown in figure 7.7. Thus, increase of the numerical aperture is restricted to ensure certain DOFs at specific wavelengths. However, over the years, advancements in the achieved wafer flatness, reduced surface topography, auto-levelling systems, auto-focusing, and thinner photoresists has resulted in increased numerical aperture with lesser DOF [1, 3]. Some other limitations of photolithography include its unsuitability for curved surfaces, its expensive process, the requirement of a highly clean room and longer time required. Also, the materials and processing conditions are not suitable for biological systems.

7.1.3 Subtractive processes

Etching is an example of a subtractive process that is frequently used in the sequence of microfabrication. If the process involves cleaning of the wafer, removal of contaminants or the creation of different geometries, etching is must. There are two kinds of etching processes: wet chemical etching and dry etching. Wet chemical etching is used to remove particles, organics and metallics. It is used for blanket etches to remove native oxide, and involves treatment of a prepared surface with acid or some other chemical reagent, by which, through differential attack, a patterned structure is revealed. In wet chemical etching, then, material is dissolved when immersed in said chemical solution. It involves three steps, namely, reactant transport to the surface, selective and controlled reaction of the etchant with the film to be etched, and removal of byproducts away from the surface. Hydrofluoric, nitric and acetic acids were popular previously used etchants, but nowadays alkaline etching, i.e. the use of potassium or sodium hydroxide, is also prevalent. Once the desired shape is patterned with the photoresist, the etching process allows unprotected materials to be removed. Wet etching works very well for etching thin films on substrates, and can also be used to etch the substrate itself. The problem with substrate etching, however, is that isotropic processes can cause undercutting of the mask layer by the same distance as the etch depth. Anisotropic processes allow the etching to stop on certain crystal planes in the substrate, but still results in a loss of space, since these planes cannot be vertical to the surface when etching holes or cavities. In the manufacture of very small features on thin films (comparable to the film thickness), isotropic wet etching may cause problems since the undercutting will be at least equal to the film thickness. If these become critical limitations, one should opt for dry etching of the substrate instead. However, the cost per wafer will be 2–3 orders of magnitude higher to perform the dry etching.

With dry etching it is possible to etch almost straight down without undercutting, which provides much higher resolution. In dry etching, the material is sputtered or dissolved using reactive ions or a vapour-phase etchant. Dry-etching methods include plasma etching, reactive ion etching (RIE), sputter etching, magnetically enhanced RIE (or MERIE), reactive ion beam etching and high-density plasma (or HDP) etching. Plasma etching is the most popular etching technique. In this, it has to satisfy many stringent requirements simultaneously, including control of feature sidewall and bottom surface profiles, etch selectivity to other exposed materials, uniformity of the etch process over large substrate surfaces, and interaction with the preceding and following processing steps. Production of regular structures at nanometre dimensions of full-surface deposited homogeneous layers via etching is of the utmost interest in microelectronics, micromechanics, sensor technology, integrated optics and other fields of accurately defined nanostructures. Dry etching is the best solution for this.

A region where wet or dry etching tends to slow down drastically is called an etch stop. Such regions are important to control and stop further etching on the wafer when the desired shape and depth of the structure has been reached. Due to the nonuniformity of wafers, devices cannot be etched uniformly. Etch stops can be controlled by appropriate etchant compositions, etching agents, N_2 sparing, loading effects, etchant temperatures and diffusion effects. The depth of the etch depends on the temperature, etchant concentration and stirring control (which gives smoother surface and uniform etching). The etch rate is also controlled by the ambient light. Widely used automatic etch stops include boron etch stops, electrochemical etch stops and thin-film/silicon-on-insulator (or SOI) etch stops.

7.1.4 Oxidation

In addition to doping and deposition, oxidation is another additive process in microfabrication. In a metal-oxide-semiconductor field-effect transistor (MOSFET), the first important thin film from the oxide group is the gate-oxide layer, under which a conducting channel can be formed between the source and the drain. A related layer is the field oxide, which provides isolation from other devices. Oxides are also needed for capacitor dielectrics, local oxidation of silicon (LOCOS) pad oxides, hard masks and surface passivation. Both gate and field oxides generally are grown by a thermal oxidation process, as only thermal oxidation can provide the highest-quality oxides with the lowest interface trap densities. A very basic tube-based thermal oxidation setup, similar to the diffusion furnace shown in figure 7.2, can also be used for oxidation. The reactor consists of a resistance-heated furnace, a cylindrical fused quartz tube containing the silicon wafers held vertically in a slotted quartz boat, and a source of either pure dry oxygen or pure water vapour. The loading end of the furnace tube protrudes into a vertical flow hood where a filtered flow of air is maintained. The hood reduces dust and particulate matter in the air surrounding the wafers and minimises contamination during wafer loading. The oxidation temperature is generally in the range of $900°C-1200°C$ and the typical gas flow rate is about $1 \, l \, min^{-1}$ [1]. Oxidation systems use microprocessors to regulate the gas flow sequence, to control the automatic insertion and removal of silicon wafers, to gradually ramp the temperature up (i.e. to increase the furnace temperature linearly) from a low temperature to the desired oxidation temperature so that the wafers will not warp due to sudden temperature change, to maintain the oxidation temperature to within $\pm 1°C$, and finally to ramp the temperature down when oxidation is completed.

The following chemical reactions describe the thermal oxidation of silicon in oxygen (dry oxidation) or water vapour/steam (wet oxidation):

$Si(solid) + O_2 (gas) \rightarrow SiO_2 (solid)$————(A) Dry Oxidation
$Si(solid) + 2H_2O (gas) \rightarrow SiO_2 (solid) + 2H_2 (gas)$—— (B) Wet Oxidation

The silicon–silicon dioxide interface moves into the silicon during the oxidation process. This creates a fresh interface region with surface contamination on the original silicon ending up on the oxide surface. When silicon is thermally oxidised, the silicon dioxide structure is amorphous and quite open because only 43% of the space is occupied by silicon dioxide molecules [1].

The relatively open structure accounts for the lower density and allows a variety of impurities (such as sodium) to enter and diffuse readily through the silicon dioxide layer. Growing an oxide of thickness x consumes a layer of silicon $0.44x$ thick.

Although oxides grown in dry oxygen have the best electrical properties, considerably more time is required to grow the same oxide thickness at a given temperature in dry oxygen than in water vapour. For relatively thin oxides such as the gate oxide in a MOSFET (typically $\leqslant 20$ nm), dry oxidation is used. However, for thicker oxides such as field oxides ($\geqslant 20$ nm) in MOS ICs and for bipolar devices, oxidation in water vapour (or steam) is used to provide both adequate isolation and passivation. There are many variants of oxidation furnaces such as vertical tube furnaces, three-zone furnaces, rapid thermal oxidation furnaces, atmospheric and high-pressure oxidation furnaces, etc. Their study and comparison is left to the interest and requirements of the reader.

7.1.5 Deposition

Layering or coating thick/thin films is an important aspect of micro/nanofabrication. Most deposition processes are used to coat metals and dielectrics on a substrate. With hyper-scaling, huge demands are being made of thin-film deposition for various purposes. The techniques used for making thin films are classified into two categories: physical vapour deposition (PVD) and chemical vapour deposition (CVD). Vapour deposition is a process of coating materials that are in vapour form and condensed via methods of condensation or chemical reactivity to deposit a uniform film on the surface. Within each of these deposition methods, there are several different techniques and subcategories. These processes are usually carried out in vacuum chambers.

(a) **Physical vapour deposition**

In PVD, a solid source material is vaporised atom by atom (or molecule by molecule) and deposited on a substrate at a controlled rate, resulting in the formation of a uniform coating layer. The energy required to vaporise the source material differs for different techniques. PVD techniques are generally used to deposit metals and metal alloys (like tin, aluminium, copper, tin nitride, tantalum nitride, etc.), lines, pads, vias, contacts, and so on. PVD using a resistive heated furnace, an electron-beam furnace and thermal evaporation is shown in figures 7.8(a)–(c), respectively. In earlier generations, source material was heated above its melting point by using resistance/radio frequency (RF)/electron-beam furnaces in an evacuated chamber. Evaporated atoms would travel in a straight line at high velocities and coat a target substrate mounted in front of the sample holder. Although this was a simple and cost-effective method, it resulted in poor-quality film. In another PVD technique, called thermal evaporation, thermal energy was used to vaporise the source. The material then evaporated at a sufficiently high rate to produce deposition on adjacent surfaces. This temperature could be just a few hundred degrees in the case of volatile metals like indium or zinc, or several thousand degrees in the case of metals like molybdenum or platinum. However, there were major

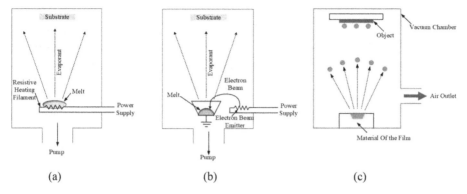

Figure 7.8. PVD using (a) resistive heated furnace, (b) electron-beam furnace, and (c) thermal evaporation. From [4]. Copyright (2016) by CRC Press. Adapted by permission of Taylor & Francis Group.

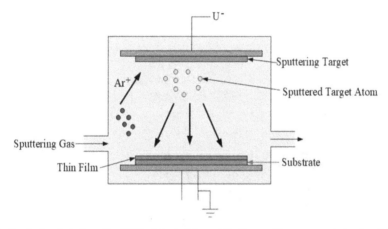

Figure 7.9. Sputtering technique for PVD. Adapted from http://www.shinmaywa.co.jp/vac/english/vacuum/vacuum_2.html.

concerns with regards to the thermal evaporation process, with its high processing temperature, inadequate step coverage, poor deposition selectivity, film quality and uniformity, and throughput.

A low-pressure, low-temperature, lower-contamination method for modern PVD is the sputtering technique. An ion source is used here. Ions are accelerated towards the target and impinge on its surface. Direct current (DC) sputtering is usually used for metal film deposition. There are two electrodes in a DC sputtering system. To implement this method, a high voltage is applied across the gas chamber and a target (substrate) is attached at one end (top) of the chamber, as shown in figure 7.9.

The source material to be coated is placed at the other end (i.e. the bottom) of the chamber. After that, inert gas (such as argon) is added inside the vacuum chamber. As a negative DC bias is applied directly on the cathode electrode of the metal target, the stray electrons accelerate and gain energy from the electric field to bombard the argon neutral atoms. If the bombarding electrons have sufficiently higher energy than the argon ionisation energy (i.e. 15.7 eV), the argon becomes ionised and plasma is created. The positive argon ions in the plasma are accelerated towards the metal target and sputter metal atoms off. The glow region of the plasma is a good conductor. At the start of the argon gas breakdown, the voltage between the two electrodes drops and hardly sustains a high field for the generation of plasma. The secondary electrons emitted from the metal target during sputtering sustain the plasma. The metal target is focused and deposited uniformly on the substrate. The sputter deposition rate directly depends on ion density. It can be improved by either providing a third electrode to provide more electrons for ionisation, by using a magnetic field and/or by using a small target for substrate separation. To minimise the effect of in-flight scattering of the sputtered target atoms, collimators are used. The ion current and energy can be independently adjusted. Sputtering is suitable for ultra-large-scale integration (or ULSI) ICs.

(b) **Chemical vapour deposition**

Chemical vapour deposition (CVD) is a method for forming a thin solid film on a substrate by the reaction of vapour-phase chemicals that contain the required constituents. In the process of CVD, the wafer being used is exposed to volatile precursors that react with the substrate surface, producing the desired deposition. The CVD process can be generalised into a sequence of steps as follows:

(1) Reactants are introduced into a reactor.
(2) Gas species are activated and dissociated by mixing, heating, plasma or other means.
(3) Reactive species are adsorbed on a substrate surface.
(4) A chemical reaction takes place with other incoming species to form a solid film.
(5) Reaction byproducts are desorbed from the substrate surface and removed from the reactor.

Although film growth is primarily accomplished at step (iv), the overall growth rate is controlled by steps (i)–(v) in series. The slowest step determines the final growth rate. As in any typical chemical kinetics, the determining factors are the concentrations of the surface species, wafer temperature, and incoming charged species and their energies. CVD is the most useful method for deposition of a wide variety of thin films in semiconductor device fabrication. CVD is used to deposit, for example, polysilicon for gate conductors, silica glass, doped silica glass such as borophosphosilicate glass (BPSG) and phosphosilicate glass (PSG), silicon nitride for dielectric films, and tungsten, tungsten silicide and titanium nitride for conducting films. Other emerging dielectrics such as high-dielectric-constant materials (e.g. hafnium silicate), low-dielectric-constant materials (e.g. carbon-doped silicate glass) and conductors (e.g. copper barrier/tantalum nitride, copper,

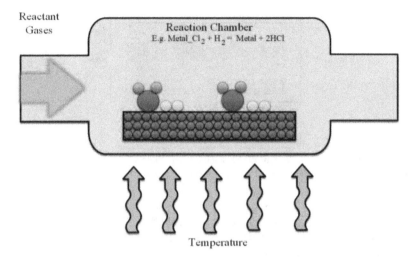

Figure 7.10. Atmospheric pressure CVD. Adapted from [7].

ruthenium) can also be deposited by CVD. There are chiefly two types of methods for CVD: atmospheric pressure methods (i.e. APCVD) and low-pressure methods (i.e. LPCVD). The low-pressure method has several variants such as thermal CVD, light-induced CVD and plasma-enhanced CVD (PECVD) [1, 5–7]. These methods differ in the intensity of the chemical reactions that take place in them.

(i) **Atmospheric pressure CVD**

In APCVD, a substrate is placed inside a reaction chamber, as illustrated in figure 7.10, and the reaction chamber is heated to some desired temperature, say 200°C–2200°C. After this, a controlled flow of mixed reactive gases is inserted in the chamber, for example in the case of a metal being deposited from metal chloride gas. In addition, hydrogen is added with the gases; after this, in controlled gas flow conditions, chlorine reacts with the hydrogen, creating hydrogen chloride (HCl) waste gas. The waste is removed from the chamber and the metal is deposited on the substrate. This coating is applied on all sides of the substrate, and its thickness can also be maintained [6, 7].

APCVD at high temperature is used mainly for the implementation of silicon films or compounded materials and for applying metals on substrates [7]. At low temperature, it is used to deposit different types of insulating films like silicon dioxide [7]. The limitations of carrying out CVD at atmospheric pressure is that it results in a nonuniform film thickness, and moreover the coating is not very robust.

(ii) **Low-pressure CVD techniques**

LPCVD is a CVD process operated at subatmospheric pressures. Reduced pressures can reduce unwanted gas-phase reactions and improve film uniformity across the wafer. However, it suffers from low deposition rates. A thermal CVD furnace is a tube furnace that is

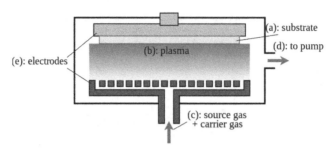

Figure 7.11. PECVD setup. (This PlasmaCVD-en.svg has been obtained by the author(s) from the Wikimedia website, where it is stated to have been released into the public domain. It is included within this book on that basis.)

operated at high temperatures, i.e. 1800°C–2000°C, for the deposition of compounds on the substrate in either hot-water or cold-water reactors [1]. A hot-water reactor is a kind of reactor in which the temperature throughout the reactor is the same (i.e. an isothermal surface). In this method, the whole surface has a uniform temperature and the substrate is applied with the material on it. The major drawback of this system is that the coating is applied to all surfaces, as every surface has the same uniform temperature. Whereas in the case of cold-water reactors, only the substrate is heated and so the coating is applied only to the substrate [1]. In light-activated (or photo) CVD, the deposition takes place at a very low temperature, controlled by optical heating so that the adsorption and subsequent decomposition takes place.

PECVD is employed to first create the gas plasma, and then the chemical reaction is carried out. Plasma is a fully or partially ionised gas composed of equal numbers of positive and negative charges and a different number of unionised molecules. A cylindrical glass or aluminium chamber, as shown in figure 7.11, contains two parallel aluminium electrodes and is sealed with aluminium endplates [1]. The upper electrode is excited with a RF signal of frequency typically equal to 13.56 MHz because of its noninterference with radio-transmitted signals. The lower electrode is grounded. The plasma is initiated by free electrons, always present in a gas, generated by cosmic rays, thermal excitation or other means. The free electrons oscillate and gain kinetic energy from the RF electric field and collide with gas molecules. The energy transferred in the collision causes the gas molecules to become ionised. The free electrons gain kinetic energy from the field again, and the process continues. When the applied voltage is larger than the breakdown potential of the gas, a sustainable plasma is generated throughout the reaction chamber. The chamber is operated at low temperatures, i.e. 100° C–400°C, by a resistance furnace. Gas flows through the plasma discharge and gets deposited. PECVD is mostly used to manufacture semiconductors, helping to deposit films on the sidewalls and on the wafers on which the metal layers are present [1, 7].

7.2 Limits of photolithography and advanced lithographic processes

A good lithography setup is characterised by high resolution, registration and throughput of its exposure tool [8]. The resolution is the minimum feature dimensions that can be transferred with high fidelity to the resist film on the semiconductor wafer. Registration refers to the measure of how accurately patterns on successive masks can be aligned with the pattern of previous masks. Throughput is the number of wafers that can be exposed per hour for a given mask level. In optical lithography, exposure techniques like contact printing, proximity printing and projection printing exist. However, owing to its better resolution, step-and-scan projection lithography is widely used. To improve resolution further, the wavelength of the exposure tool can be reduced, new resists can be developed, and resolution-enhancement techniques like phase correction masks and optical proximity corrections used. To support the aggressive scaling required in current CMOS technology, other techniques like immersion lithography have been proposed. In immersion photolithography, the air gap (with a refractive index of 1) between the lens and the wafer is filled with water (with a refractive index of 1.43), thus improving the resolution of the process. A representative setup is shown in figure 7.12. The most difficult part of this system is ensuring the presence of a completely still (i.e. completely turbulence/bubble-free) immersion puddle.

Photolithography has been successfully carried out up to feature sizes of 15 nm using a 157 nm deep-UV F_2 source, and beyond this by utilising several complex tricks. Towards the end of 2017, Global Foundries Fab8, in Malta, NY invented an extreme-UV (EUV) lithography tool that can achieve a feature size as small as 13.5 nm. However, it has its own challenges. As of 2022, ASML Holding is the only

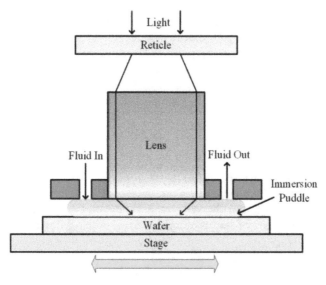

Figure 7.12. Immersion photolithography. [1] John Wiley & Sons. Copyright © 2007 John Wiley & Sons, Inc. All rights reserved.

company that produces and sells EUV systems for chip production, targeting 5 nm and 3 nm feature sizes. At the 2019 International Electron Devices Meeting (IEDM), TSMC reported the use of EUV for 5 nm in contact, via, metal line and cut layers, where the cuts can be applied to fins, gates or metal lines [9, 10] At IEDM 2020, TSMC reported their 5 nm minimum metal pitch to be reduced 30% from that of 7 nm [11], which was 40 nm in 2016 [12]. Samsung's 5 nm is lithographically the same design rule as 7 nm, with a minimum metal pitch of 36 nm [13] Thus, when transitioning from microsized features to nanosized features, there is evidently an extreme need to invent novel processes that can fabricate ultra-small features easily. Apart from optical lithography, advanced patterning processes such as electron-beam lithography, X-ray lithography, nanoimprint lithography, etc., also exist, but they are not cost effective for mass production [8].

7.2.1 Electron-beam (E-beam) lithography

E-beam lithography (EBL), inspired by scanning electron microscopes (SEMs) in the 1960s, is the most important technique in the present-day semiconductor industry as it can scale down (reduce the size) of any two-dimensional (2D) pattern to the nanometre level. Due to rapid advancements in EBL, it was able to pattern features at a scale of 60 nm as early as in the 1970s. Due to this progress, optical lithography was predicted to be doomed by the end of that decade. However, due to their high throughput and ongoing advancements of their own, optical systems continued to dominate the manufacturing process for the next 30 years. The most advanced EBL system today, with the aid of some special resist processing technology, can pattern at a scale of less than 10 nm. However, despite tremendous efforts from researchers, throughput on par with optical lithography has still not yet been achieved [3, 14]. In EBL, the wave–particle duality of electrons is exploited. Electrons in EBL work similar to the photons in optical lithography systems. Energy is transferred into the polymer-based resist material, which is also known as the electron resist. Features are developed in the exposed resist, which is then used as a mask for the transfer of features to substrate materials. A basic schematic of an EBL system and corresponding process steps are shown in figures 7.13(a) and (b), respectively. The wavelength λ of an electron with mass $m = 9.1 \times 10^{-31}$ kg and charge 1.6×10^{-19} C passing through a potential difference of V volts is given by equation (7.3):

$$\lambda = \frac{h}{\sqrt{2meV}} = \frac{1.226}{\sqrt{V}} \quad m. \tag{7.3}$$

Electrons with high energy have significantly shorter (~1000 times) wavelength than that of optical systems. Hence, one can build a very-high-resolution projection system using electrons. In fact, the wavelength is never a limiting factor in practical EBL systems. Three types of lenses can be used in e-beam devices:

    (i) Electronic gun lens: to make higher-resolution systems, field electron emission sources are used.

    (ii) Magnetic lens: to focus or deflect the moving charged particles.

    (iii) Electrostatic lens: these are less frequently used, as the lenses have more aberrations, which in turn are not used for fine focusing.

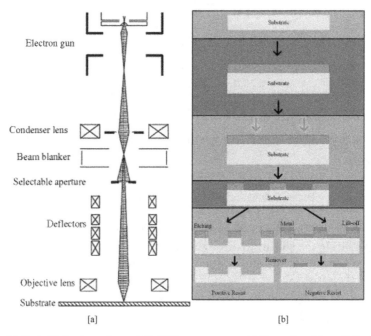

Figure 7.13. (a) Basic EBL schematic, and (b) EBL process steps. (a) Reproduced from [15]. Copyright (c) 2004 The Japan Society of Applied Physics. All rights reserved. (b) Adapted from http://d1otjdv2bf0507. cloudfront.net/images/Article_Images/ImageForArticle_3219(1).jpg.

The e-beam process proceeds as follows:
- (i) First, the sample is cleaned to remove any debris like oils, organics, etc.
- (ii) Spin coating of the electron resist is applied on the surface of the film (positive or negative resist).
- (iii) The e-beam is applied to remove the uncovered deposited film.
- (iv) Steps (i)–(iii) are repeated again until the desired pattern is obtained.
- (v) Finally, the electron resist can be removed.

EBL is preferred over photolithography for the following reasons:
- (i) In an e-beam, the speed is high enough for complex patterns, whereas in photolithography the speed is better suited for larger shapes.
- (ii) EBL offers point by point exposure, which limits the speed compared to photolithography, which has high speed but gives parallel exposure.
- (iii) EBL is not diffraction-limited, with exposure up to 10 nm, as compared to photolithography, which has light diffraction limits with exposure only up to 50 nm [3].

However, again, the proximity effect due to electron scattering and electron aberrations are the limiting factors to the resolution. This results in underexposure of isolated features and overexposure of densely distributed features. Electron scattering results due to both forward-scattering and back-scattering. In forward-scattering, electrons are deflected due to electron–electron interaction, similar to the

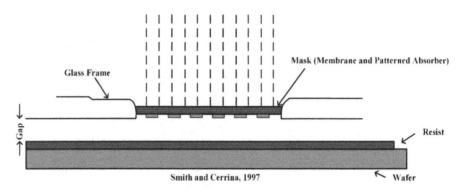

diffraction of light; this, in turn, broadens the incident beam. In back-scattering, electrons penetrate the substrate through the resist. These electrons then collide with the heavy substrate nuclei and scatter back into the resist, causing unintended exposure of the resist [3].

7.2.2 X-ray lithography

X-rays have a very small wavelength (less than 1 nm) and hence can overcome the diffraction-related issues posed by photolithography with an added advantage of a faster process. Deep X-ray lithography (or DXRL) procedures use wavelengths of the order of 0.1 nm to fabricate uniform three-dimensional (3D) structures with high aspect ratios.

As shown in figure 7.14, an X-ray lithography (XRL) mask consists of an X-ray absorber made up of gold or compounds of tantalum or tungsten, on an X-ray-transparent high-cost membrane like silicon carbide or diamond. This mask is patterned by direct EBL writing. The membrane can be stretched for overlay accuracy. However, due to repeated exposure, the membrane mask may degrade. X-rays generate secondary electrons just as in the cases of EUV lithography and EBL. The sidewall roughness and slopes are influenced by these secondary electrons as they can travel a few micrometres in the area under the absorber, depending on the exposure X-ray energy [5]. XRL finds its dominant use in the basic steps of the LIGA process (a German acronym for *Lithographie Galvanoformung Abformung*, meaning lithography, electrodeposition and moulding), which includes the electrodeposition of metal in a developed resist structure in order to obtain a mould or an electrode for a subsequent replication process like moulding or electrical discharge machining. Expensive membranes and X-ray sources are two limitations for widespread use of XRL in the microfabrication industry. However, XRL is a popular technique in the manufacture of high-aspect-ratio (HAR) micro-electromechanical (MEMS) ICs [8, 14].

7.2.3 Nanoimprint lithography

Nanoimprint lithography is another prominent top-down technique for nanostructuring in large areas with high throughput. It is also regularly known as stamping. One of its major advantages is that the fabricated nanostructures are a lot smaller than the

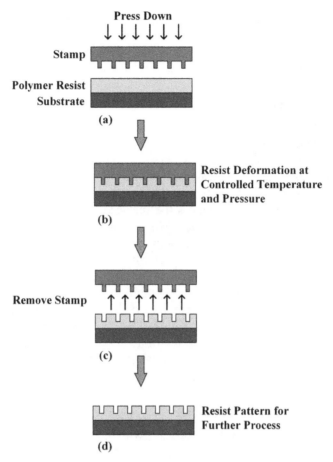

Figure 7.15. Nanoimprint lithography. Adapted from [3]. Copyright (2017), with permission from Springer Nature.

wavelength of light. The process of nanoimprint lithography starts by making a polymeric material, which is resistant and substrate coated, from pressing a reusable stamp, as shown in figures 7.15(a) and (b). The stamp is removed after deformation of the resist to the shape of the stamp, which is complementary, as shown in figure 7.15(c). There are many different types of resists, in general. This also includes resists that can be solidified after stamping in the exposure of UV light. Figure 7.15(d) shows an example of a patterned resist to be utilised in further processing [2].

7.3 Self-assembly processes

Conventional lithographic techniques essentially fall under top-down fabrication approaches. Even after many years of advancement, every microfabrication technology intrinsically uses some variation of lithographic patterning processes. This therefore limits the smallest feature size. In contrast, if one observes natural processes, it is possible to see that many complex organisms are produced by self-organisation.

Self-assembly occurs when nanoscale objects interact with each other by a balance of attractive and repulsive forces. There are two particular self-assembly techniques that are interesting from the perspective of development in nanofabrication, molecular self-assembly and particle self-assembly, which we detail as follows.

7.3.1 Molecular self-assembly

The best known method from the category of bottom-up approaches is molecular self-assembly. This consists of the spontaneous organisation of molecules into structurally precise and unambiguous stable aggregates at the nanometre scale. The particles or molecules are transported to a surface area and form self-assembled monolayers (SAMs) by liquids. A typical SAM contains a tail group that adheres to the substrate, a head group that defines the surface characteristics and a backbone that gives the required thickness to the SAM. A graphic depicting the self-assembly of a monolayer of thiolated carbohydrates on gold is shown in figure 7.16. Details of SAM formation are also presented in chapter 4.

Molecular self-assembly processes, which are relevant to nanotechnology, involve surface-active reactions wherein molecules combine in order to approximate lattices or chains by noncovalent bonds. Compared to the structures formed by covalent bonds, these structures are easier to create and control. Moreover, the molecules of these structures are well ordered and robust under varied conditions. These molecules form spontaneously, possess intrinsic thermodynamic stability, reject defects and permit control of film thickness up to approximately 1 nm.

7.3.2 Self-assembly for colloidal particles

Self-assembly of particles (likely spherical) with diameters ranging from a few nanometres to a few micrometres in a liquid suspension is known as colloidal self-assembly. Particles of these dimensions have completely different properties compared to their intrinsic atoms as well as their bulk form in a solid state. The self-

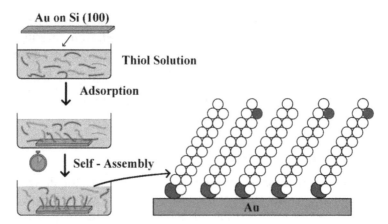

Figure 7.16. Self-assembly of a monolayer of thiolated carbohydrates on gold. Adapted from [14]. Copyright (2006), with permission from Springer Nature.

assembly process of colloidal particles, being physical in nature, can be promoted by various forces such as tension force or capillary force to induce the molecules to form colloidal crystals [14]. Moreover, electrostatic force and van der Waals forces can also be used to generate a force of attraction between such particles.

Self-assembly has shown potential of being the future of fabrication because of two reasons: first, is in the design of true nanostructures, at which current technology is failing; second, is in the cost-effectiveness of this solution [14]. In order to be useful across a broad spectrum of applications, self-assembly processes need to cross over with conventional nanofabrication technologies to form planar or solid structures. More precisely, the self-assembly processes need to be guided and, hence, cannot work single-handedly. As such, almost all self-assembly processes are assisted by conventional nanofabrication processes. As such, an integration of top-down and bottom-up approaches will always be a primary aspect in the development of self-assembly-based nanofabrication. The lattice-like structures that result from self-assembly are heavily affected by external energy and environmental conditions of the experiment. Therefore, these factors can be exploited in such a way that they guide self-assembly processes towards a predefined end. Fabricated nanopatterns are able to form physical boundaries in order to inhibit self-assembly from spreading further to regions with a particular surface energy. These regions either promote or prohibit the further self-assembly process [14]. It is possible to control these conditions using five forces found in nature: surface topography, surface energy, electrostatic force, magnetic force and thermal energy [14].

7.4 Nano measurement and characterisation tools

For industrial and academic studies at the nanoscale, it is necessary to adopt advanced instrumentation platforms. To achieve fruitful research and further development in micro/ nanostructures, material science, sensors, devices, integration, modelling and simulations at the nanoscale, the instruments used must offer high levels of accuracy, sensitivity, resolution, fidelity and repeatability. Nanotechnology, being a multidisciplinary field, demands the measurement of physical, chemical, electrical, thermal, optical as well as biological parameters. Optical measurement units, electron microscopes, spectrometers, and so on, play a very significant role at this scale: they not only characterise materials but also help in synthesising novel materials and nanostructures. These instruments offer much higher resolution as they use ultra-low-wavelength electrons to form images instead of light (1000 times more resolvable). An understanding of this equipment is therefore necessary not only to study existing nanosystems but also to develop new ones.

7.4.1 Energy-dispersive X-ray spectroscopy

Kai M Siegbahn won the Nobel Prize for developing high-resolution electron spectroscopy in 1981. Different kinds of emission, including X-rays, are produced when an electron beam interacts with a target. Subsequently, the characteristic X-rays of different elements can be segregated into an energy spectrum using an energy-dispersive (EDS) detector, and the spectrum analysed using EDS system software to determine the abundance of certain elements. The chemical composition of substances

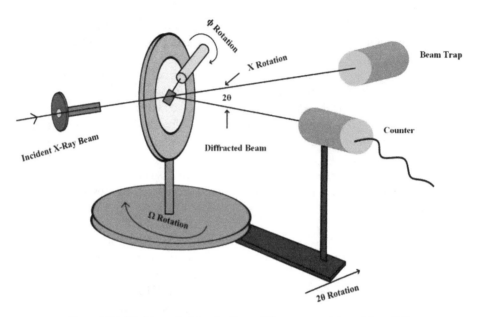

Figure 7.17. Working principle of a X-ray diffractometer. Adapted from [17].

can be ascertained to a precision of a few micrometres using EDS. Usually, a SEM instrument is integrated with EDS systems. The constituent parts of a typical EDS system are liquid nitrogen for cooling, software for assimilating and analysing energy spectra, and a sensitive X-ray detector. The detector is set up at the extremity of a long arm in a chamber, which is cooled using the liquid nitrogen and contains the sample.

Sir William Henry Bragg and William Lawrence Bragg jointly contributed to analysing crystalline structures with the help of X-rays. They were awarded the Nobel Prize for the same in 1915. X-ray diffraction (XRD) (figure 7.17) is a powerful method to study nanomaterials with at least one dimension in the range of 1–100 nm [17]. Another primary reason for the use of XRD is because the wavelength of X-rays is in the range of the atomic scale. This technique is the default choice for determining strain states in thin films because of its unparalleled accuracy [17]. Also, it is very economical compared to alternative electron bombardment methods. To study characteristic properties like bond angles, chemical composition, and so on, of crystalline substances, both natural and man-made, X-ray spectroscopy is used as a nondestructive technique.

Each crystalline substance has its own characteristic X-ray powder pattern that can be used for identification. Waves with wavelengths of a crystal lattice spacing are highly scattered. The sample should be a crystalline material. X-rays from a cathode ray tube are filtered and directed towards the sample, then constructive interference is formed because of interaction. Here, Bragg's law (figure 7.18) gives the relationship between the wavelength of the incident radiation to the diffraction angle and lattice spacing as $n\lambda = 2d\sin\theta$.

XRD is used to achieve a magnification of 0.1–10 nm. It is used in the identity phase, atomic spacing and cell dimension of crystalline materials [18]. Its application

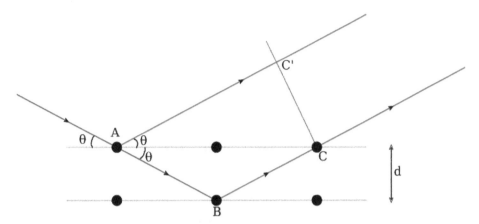

Figure 7.18. Conceptualisation of Bragg's law. (This Bragg's law.svg image has been obtained by the author(s) from the Wikimedia website where it was made available under a CC BY-SA 3.0 licence. It is included within this book on that basis. It is attributed to M. Hadjiantonis.)

includes investigation of high- and low-temperature phases, identification of solid solutions and determination of unit parameters. Another application includes calculating the ratio of diffraction peaks in amorphous and crystalline materials as crystalline materials have sharp and narrow diffraction peaks while amorphous materials have broad and halo-like peaks.

7.4.2 Electron microscopy

Electron microscopy is necessary for the development of nanotechnology and nanodevices, in transforming a material's atoms from ideal representations to real-life objects and in enabling the engineering of materials atom by atom [19]. Electron microscopes emit electrons and create images that are magnified by a million times, so that structures as small as single atoms can be visualised [19]. By imaging nanoscale structures, electron microscopy provides a way to understand material composition, structure and performance. This has led to many technological advances.

(a) **Transmission electron microscope**

In this method, an image is formed by transmitting an electron beam through a sample on a fluorescent screen or photographic film layer via magnifying and focussing electrodes. Even a single column of atoms can be captured using a transmission electron microscope (TEM) [8, 19, 20]. Figure 7.19 shows the main components of a typical TEM.

TEM microscopes project a stream of electrons through an object to acquire an image of the object. A heated filament that makes up the cathode is used to generate a beam of electrons, which works like a ray of light. There are two electromagnetic coils, one which concentrates the beam and the other which is used to focus the beam on a certain section of the object. There is also a third lens that magnifies the image, which is known as the projector image. The image is made visible on the fluorescent screen at the bottom of

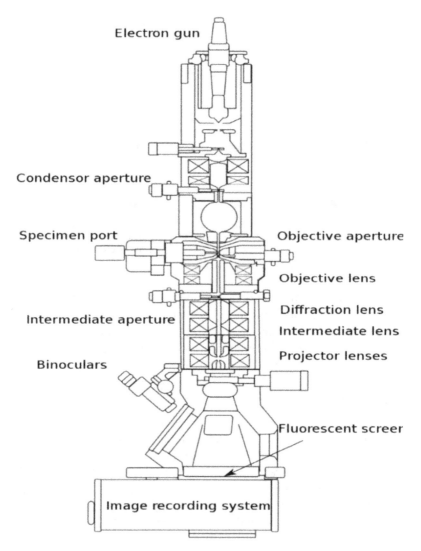

Figure 7.19. Main TEM components. (This Scheme TEM en.svg image has been obtained by the author(s) from the Wikimedia website where it was made available under a CC BY-SA 3.0 licence. It is included within this book on that basis. It is attributed to Gringer.)

the machine. The electrons that are bombarded on the surface of the specimen are either absorbed, scattered elastically or scattered inelastically. A contrast will be created in the image due to the absorption of electrons in the object. Regions that have higher thickness will appear dark, and where the object is absent the image will look bright. The formation of contrast takes place due to the diffraction of electrons from their original path without any loss of energy. In the case of Bragg scattering, the electrons will be dispersed into the back focal plane. The inelastic scattering electrons give information about the element they collided with via the loss of energy observed in the collision. It is possible to gain details like their composition and bond structure from this mechanism.

A TEM can be used to capture high-magnification images of the internal structure of a sample, to observe the crystalline structure of an object and evaluate the stress or internal fractures of a sample, or to view contamination within a sample through the use of diffraction patterns.

(b) **Scanning electron microscope**

In 1986, the Nobel Prize in physics was shared between Ernst Ruska, for the design of the first electron microscope and his fundamental work in electron optics, and Gerd Binning and Heinrich Rohrer, for their design of a scanning tunnelling microscope (STM). An electron microscope utilises a light emission electron as an illumination source. A SEM is a kind of electron microscope that produces an image by checking with the assistance of a beam of accelerated electrons [19, 20]. The electron in focus would either be assimilated or reflected. Another possibility is the discharge of auxiliary electrons, energised because of the bombardment of electrons. This would bring about the generation of a mixture of signals at the surface of solid specimens. These electron–sample interactions can uncover information such as the surface/texture and chemical composition of a material alongside its crystalline structure. A typical structural schematic of a SEM is illustrated in figure 7.20.

In a SEM, the bombarding electrons carry a significant amount of kinetic energy, which results in the production of a variety of signals due to the

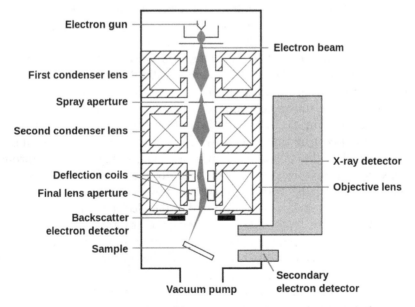

Figure 7.20. Schematic of a typical SEM. (This Schema MEB (en).svg image has been obtained by the author (s) from the Wikimedia website where it was made available under a CC BY-SA 3.0 licence. It is included within this book on that basis. It is attributed to Steff.)

interaction of electrons and atoms at various depths of the sample [19, 20]. The various signals produced are as follows:
 (i) Secondary electrons.
 (ii) Reflected or back-scattered electrons (BSEs).
(iii) Diffracted BSEs.
(iv) Characteristic X-rays and light.
   Electrons that are emitted from very close to the surface are known as secondary electrons. Therefore, secondary electron imaging (or SEI) enables the production of images of the surface of a sample to a high resolution. Back-scattered electrons (BSEs) are emitted from a deeper location within a sample, which results in a lower resolution of the images as compared to those produced by secondary electrons. The intensity of signals produced by BSEs depends on the atomic number of the sample. The excitation of inner-shell electrons of the sample leads to the emission of X-rays. This results in a release of energy as the higher-shell electrons move to a lower shell. This allows calculation of the wavelength, and the identification and measurement of abundance of that material in the substance.
   In this kind of electron microscope, the sample is scanned in a raster pattern by the emitted electron beam. Firstly, the electron source generates electrons at the top of the column. When the thermal energy of the electrons surpasses the work function of the sample material, they are emitted, resulting in acceleration and attraction by the positive charge on the anode. Vacuum-like conditions are required for the entirety of the electron column. In order to protect against vibration, noise and contamination, the electron source and other components of the electron microscope are placed in a special chamber to maintain a vacuum. The vacuum enables the capturing of high-resolution images while safeguarding the electron source from contamination. Atoms and other molecules will prevail in the electron column without a vacuum. These will interact with the electrons, leading to the deflection of the electron beam and the reduction of image quality. Further, the detectors' electron collection efficiency in the column is increased in a high vacuum. The electrons' path is controlled using lenses. Electromagnetic lenses are used here since electrons cannot penetrate glass. They are formed by wire coils within pieces of a metal pole. The passing of current through the coils generates a magnetic field. Due to the sensitivity of electrons to magnetic fields, the current applied to the electromagnetic lenses can be changed to manipulate their path within the electron column.
   Typically, two types of electromagnetic lenses are used. As the electrons travel towards the specimen, the first lens that they meet is the condenser lens. The beam is converged by this lens before the opening of the electron beam cone and further convergence again by the objective lens, before finally reaching the specimen. The size of the electron beam, and thus the resolution, is defined by the condenser lens, while the objective lens focuses the electron beam onto the specimen. The beam is rastered onto the material using scanning coils, which are also a part of the lens system of the SEM.

The lenses are used, along with apertures in many cases, to control the beam size. SEMs are typically used for imaging objects' shapes in high resolution, detecting spatial variation in chemical configurations, analysing crystalline structure and/or qualitative chemical composition for phases and discriminating phases based on the mean atomic number using BSE.

(c) **Scanning tunnelling microscope**

As mentioned, Gerd Binning and Heinrich Rohrer invented the first STM in 1982 while working at an **IBM** research laboratory. A structural schematic of an STM is shown in figure 7.21. It contains a sharp probing tip, a piezoelectric scanning unit, a coarse positioning unit, vibration-isolation stage data processing and a display unit. It operates on the basis of the tunnelling effect of electrons. A sharp probing tip moves at a certain distance over a specimen. Due to the spherical shape of electrons, the distance between the tip and the probe varies as the tip moves, and therefore the tunnelling current varies continuously. To maintain a current constant, the tip moves vertically up and down, and the vertical displacement of the tip gives an image of the scanned sample.

Alternatively, the tip can be kept fixed vertically and the current change can be observed to scan the surface of the specimen. STMs find typical use in

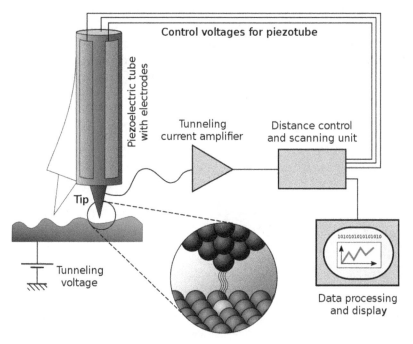

Figure 7.21. Construction of a STM. (This Scanning Tunnelling Microscope schematic.svg image has been obtained by the author(s) from the Wikimedia website where it was made available under a CC BY-SA 2.0 licence. It is included within this book on that basis. It is attributed to Michael Schmid and Grzegorz Pietrzak.)

mapping structures in the human body, investigating chemical reactions, improving properties of materials and device fabrication.

(d) **Atomic force microscope**

The first experimental implementation of an atomic force microscope (AFM) was demonstrated by Binnig, Quate and Gerber in 1986, and the first commercial AFM was made available in 1989. As in any technique of scanning probe microscopy, in AFMs too a sharp probe is scanned across a surface and certain probe–sample interaction or interactions are monitored [1, 19]. An AFM senses ithe nteratomic forces that occur between a probe tip and a substrate. A structural schematic is illustrated in figure 7.22.

The major components of an AFM include a sensing surface, a sensing tip, a laser source and a position detector. It has three prime mechanisms: surface sensing, detection and imaging. During surface sensing, a nanoscale needle makes contact with a specimen to be scanned. As the needle moves, the needle is deflected or attracted by the specimen due to atomic forces. This deflection of the needle is captured using a laser light and a photodetector. Deflection of the laser light gives the z-depth of the specimen, and it is converted into an image. AFMs find applications in studies dealing with the unfolding of proteins, the imaging of biomolecules, antibody antigen binding studies, in examining the binding forces of complementary DNA strands, studies of surface frictional forces, ion channel localisation, etc.

Table 7.1 presents a comparison of these electron microscopes with respect to their performance metrics. A quick glance at the table will make the reader aware of the advantages and disadvantages related to each device. It is hoped that this analysis will help the reader to make an informed choice about the tools they wish to use.

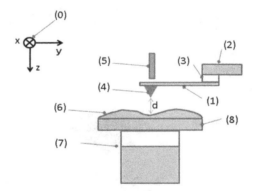

Figure 7.22. Typical AFM system: (1) cantilever, (2) support for cantilever, (3) piezoelectric element (to oscillate cantilever at its eigenfrequency), (4) tip (fixed to open end of a cantilever, acts as the probe), (5) detector of deflection and motion of the cantilever, (6) sample to be measured by the AFM, (7) xyz drive (moves sample (6) and stage (8) in the x-, y-, and z-directions with respect to a tip apex (4)), and (8) stage. (This AFM conf.jpg image has been obtained by the author(s) from the Wikimedia website where it was made available under a CC BY-SA 4.0 licence. It is included within this book on that basis. It is attributed to Tom Toyosaki.)

Table 7.1. Comparison of different electron microscopes.

S. No.	Parameters	TEM	SEM	STM	AFM
1.	Resolution	1 Å	0.5 nm	x–y: 0.1 nm; z: 0.01 nm	x–y: 1 nm; z: 0.1 nm
2.	Maximum field of view	100 nm	1 mm	10–100 nm	100 μm
3.	Measurements	2D image	2D image and material properties	3D image	3D image
4.	Sample surface	Conductive coating is required	Conductive coating is required	Conductive coating is required	Irrelevant
5.	Relative cost	High	Medium	Low	Low
6.	Cons	• Requires extensive sample preparation • Small field of view • Resolution limitation • High capital and running cost • E-beam can damage the sample	• Requires sample preparation • Limited to solid, inorganic materials • Isolated vacuum environment required • Risk of radiation exposure • Expensive	• Requires conductive coating over specimen surface • High tunnelling current can alter the surface • Slow operation of feedback can result into the tip scraping on the surface	• Large size of acquired data • Maximum lateral movement is limited to 10–20 μm • Lower operation speed compared to the others • Physical contact with specimen can alter surface properties

7.	Pros			
	• Real (image) and reciprocal space (diffraction pattern) • Chemical information can be obtained • Energy filtered images possible via EELS filter • High resolution imaging possible • Possible to obtain amplitude and phase contrast images	• Continuously variable magnification • High resolution • Good depth of focus • Capable of scanning at real time	• Operates on a variety of surfaces and temperatures • Gives a 3D profile of a surface of the specimen • High resolution • Specimen is not damaged during electron tunnelling	• Uses samples with just minor preparation, over a large range of temperatures and repetition • Can image a 3D surface profile • Special treatment such as metal/carbon coating are not required • It provides higher resolution than SEM • Allows topographical imaging of samples like DNA molecules, protein absorption, crystal growth, and living cells

7.5 Thin-film technology and synthesis

A thin film is a thin layer of a material on a substrate. The thickness of a thin film can range from a few nanometres to a few micrometres, which means it has a high surface-area-to-volume ratio. Owing to this, many material properties like hardness (e.g. carbon nanotubes are harder than diamond), corrosion resistance (e.g. a thin-film coating on iron can make it noncorrosive), optical absorption, etc., are found to be improved manifold. The term 'thin film' is, however, relative to the sense of its application. The same substance with the same chemical composition and same thickness may be categorised as a thin/thick film depending on the application where it is being used. Thin-film technology has proven helpful in many areas of nano-science and nanotechnology. In optical coatings, antireflection coating and high-reflection coatings are the simplest example of thin-film technology [21]. Antireflection coating is useful in cases like antiglare features for spectacle lenses and camera lenses. Thin-film solar cells are used in calculators and watches. They are also used in building-integrated photovoltaics, i.e. to replace conventional building materials like roofs, facades and skylights, because of their flexibility and light weight. They are also used in photolithography to reduce the image distortions that occur while capturing images. High-reflective coatings are used in dielectric mirrors, which are very useful in the production of inexpensive and high-quality mirrors for optics experiments in laboratories. Optical coatings are usually made of alternating thin layers of aluminium oxide (Al_2O_3) and tantalum pentoxide (Ta_2O_5). Protective thin films are used to shield an object from any reaction from outside. The thin-film coating on the object reduces the amount of surface open to the external atmosphere and hence reduces the chances of any unwanted chemical reactions [21]. For example, bottles of carbonated drinks are often given a coating to avoid any reactions with atmospheric CO_2. Metals like copper, aluminium, gold or silver are good conductors of electricity, hence they are used in the form of thin layers in many electrical devices. For example, copper layers are used in printed circuit boards and as the outer ground layer among all the layers in coaxial cables. They are also used in ICs to build a network between transistors and capacitors in the form of thin aluminium or copper layers. Thin-film technology, when combined with the advantages of solid-state batteries, can also be used to build thin-film lithium-ion batteries, making the battery usable in devices that are very small in size. Some of their main uses are in wireless sensors, implantable medical devices, RFID tags, smart cards and renewable-energy storage devices.

Frequently, thin films are synthesised by CVD/PVD techniques [21]. Additionally, atomic layer deposition (or ALD) and molecular layer deposition (or MLD) processes, which use two or more gas-phase precursors, can be used. One by one, these precursors react with the substrate. The process for depositing each precursor is divided into two half-reactions: the first part includes the deposition of the precursor, and the second deals with the evacuation of the reaction chamber to prepare for the next precursor. The process is repeated multiple times until a film of desired thickness is formed. This method takes more time than CVD but can be carried out at lower temperatures. Other popular techniques for thin-film deposition are the spin-coating and dip-coating methods, which use a liquid or sol–gel precursor [21]. In the spin-coating method, the sol–gel or liquid precursor is put on the smooth, flat surface of the substrate, and then the surface is rotated at a high velocity so that the precursor is spread evenly on the surface of the substrate. Thin films of required thickness can be formed by repeated depositions. The amorphous thin-film layer formed is crystallised using thermal treatment. The dip-coating method is similar to spin coating, except here the substrate is first dipped in the precursor and then rotated in monitored conditions. By supervising the speed of rotation, the evaporation conditions (like the moisture and temperature) and the volatility of the solvent, the thickness of the film, homogeneity and microscopic morphology can be controlled. Deposited thin films may have intrinsic compressive or tensile stresses, which can lead to deviations in electrical properties, cracking, corrosion and adhesion problems. Hence, a good deposition method is very important [21].

During the deposition process, the nucleation and coalescing result in the formation of thin films. While keeping the incident energy low, the material or component to be deposited loses its velocity component perpendicular to the surface of the substrate and is adsorbed physically on the surface.

As there is a temperature difference between the species and substrate, the species move over the surface of the substrate. In the meantime, they interact with each other, effectively forming clusters of increasing size. These primary clusters are called nuclei. During the initial nucleation, the clusters are thermodynamically unstable until they reach a particular critical size. After that, a cluster grows when it collides with other adsorbed species before getting desorbed. Now the clusters become thermodynamically stable. At this point, the nucleation barrier has been broken. The critical nuclei grow and form an island until a specific saturation nucleation density is achieved. In the coalescence stage, the small islands start coming together to form a whole mass. Large islands start to develop, leaving small parts of the substrate uncovered. Here, disconnected island-type structures change to a percolative type of network. This process continues until the result is the formation of a consistent film covering the whole surface [21].

As shown in figure 7.23, three models have been proposed for the exact growth process of thin films, namely, the Volmer–Weber [22], Frank–van der Merwe [23] and Stranski–Krastanov [24] models. They are also referred to as modes of thin-film growth. In Volmer–Weber or island-type development, the attraction between atoms of film material is higher than their bonding force to the substrate, prompting the arrangement of 3D islands or atom groups. This results in the formation of rough surfaces. In Frank–van der Merwe or layer-type development, the attraction between atoms of film material is lesser than their bonding force to the substrate. Hence, initially a single layer forms and then another layer is deposited. This 2D development happens layer by layer, demonstrating that an initial total film structure precedes the development of resulting layers. Stranski–Krastanov or mixed-type development is also referred to as 'layer in addition to island development'. As this would suggest, it is a combination of the above two growth modes. First, layer-by-layer film development occurs and then the formation transitions to

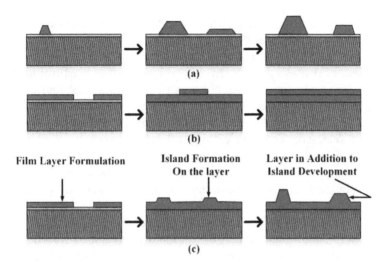

Figure 7.23. Growth modes. (a) Volmer–Weber, (b) Frank–van der Merwe, (c) Stranski–Krastanov.

island type. This transition from layer-by-layer to island-type formation depends on the physical and chemical properties of both the substrate and film material.

Surface mobility, nucleation barrier and film density affect the growth rate of the film. A high surface mobility tends to increase the surface smoothness. A low nucleation barrier results in smooth, finely grained deposits, which then start to become continuous as the thickness reduces. A large nucleation barrier results in rough, coarse-grained films, which then become continuous as the thickness increases. The bonding between substrate and film is known as adhesion, which is subject to the nanoscopic topography of the surface of the substrate, its cleanliness and chemical nature. This adhesion property can be advantageous for increasing the values of initial nucleation density, the adsorption energy of the deposit and the kinetic energy of incident species. On the surface of the substrate, more nucleation centres can be provided to improve the adhesiveness of the film to the substrate. Similarly, contaminants can decrease the film's adhesion with the substrate [21].

7.6 Microelectromechanical, microoptoelectromechanical systems and nanoelectromechanical technologies

MEMS, MOEMS and NEMS technologies represent an amalgamation of optics, electronics and mechanics and are subsets of microsystems technology. They all use almost similar batch-processing techniques for their fabrication and design but differ only in their applied engineering fields and physical dimensions [8, 14, 20]. When the technology is scaled down so that the components are manufactured at the nanoscale, electromechanical systems are referred to as NEMS. These are currently still mostly in the research phase and not many NEMS devices are available on the market. They not only incorporate all the advantages of MEMS devices, but also, due to their reduced size and greater efficiency, offer lower power consumption and lower production costs. MEMS surfaces become nonplanar after very few iterations of film deposition, patterning and etching due to the large thicknesses involved. In MEMS, movable or suspended items (i.e. released structures) can be manufactured. Realising these released structures, surrounded by voids, cavities and gaps, demands a differentiated fabrication process. In such a process, the structural material is the material that makes up the operational structure; the material that produces gaps and spaces in the mechanical structure is called the sacrificial material. The sacrificial material is removed at the end by a chemical process when the mechanical structures are released. The fabrication technology consists of microelectronic/ optical structures and responders, as well as mechanical designs built onto non-conducting materials. The techniques used for fabrication are divided into bulk micromachining, surface micromachining and HAR processes [8]. They primarily involve deposition, patterning and etching.

Micromachining is a broad term, describing all precision techniques used to build small structures. These techniques were originally developed for the IC industry (circa 1960). There two main types of micromachining: bulk micromachining (a subtractive process) and surface micromachining (an additive process). Bulk micromachining refers to the formation of microstructures via the removal of materials from bulk substrates [8]. Surface micromachining allows designers to fabricate free-form complex and integrated electromechanical structures [8]. The wafer itself is the substrate on which multiple alternate layers of structural and sacrificial material are deposited and etched. Layering these two materials properly produces movable, suspended or interlocked structures. Whereas bulk micromachining creates devices by etching into a wafer, surface micromachining builds devices up from the wafer layer by layer. It involves a repeated sequence of thin-film deposition on a wafer, photo-patterning of the films, and then etching of the patterns into the films. While micromachining techniques can be utilised to fabricate normal 3D structures, special techniques such as LIGA are required to realise features with high aspect ratios (height of the feature/width of the feature) [8].

7.6.1 Bulk micromachining

Bulk micromachining is currently by far the most commercially successful technique, helping to manufacture devices such as pressure sensors and inkjet print heads. This allows the creation of various micromechanical components such as beams, plates and membranes that can be used to fabricate a variety of sensors and actuators. The methods commonly used to remove excess material are wet and dry etching, allowing varying degrees of control on the profile of the final structure. Figure 7.24 shows various structures fabricated by bulk micromachining including a simple cantilever and complex pillar. For simpler microstructure fabrication tasks, etch masks and stops carry out selective etching with anisotropic silicon etchants, which etch single-crystal silicon along a single plane. Generally, good etch masks are provided by SiO_2, Si_3N_4 and some metal films like chromium and gold. Etch masks define the initial geometry of the region to be etched, and etch stops are used to define microstructure thickness. The final structure has one side exposed to the environment and the other side enclosed in a package.

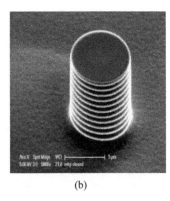

(a) (b)

Figure 7.24. Various structures fabricated by bulk micromachining. (a) Bulk micromachined cantilever fabricated by p+ etch stop and anisotropic etching, (b) complex shapes patterned using deep reactive ion etching. (a) Reproduced from [25]. © IOP Publishing Ltd. All rights reserved. (b) This Bosch process PILLAR. jpg has been obtained by the author(s) from the Wikimedia website, where it is stated to have been released into the public domain. It is included within this book on that basis.

Bulk micromachining is a relatively simple way of fabricating robust yet highly sensitive MEMS. It is well suited for applications that do not require a high level of complexity. However, it is an expensive method since a lot of material is wasted. Bulk micromachined pressure sensors offer several advantages over traditional pressure sensors. They cost less, are highly reliable, manufacturable, and there is a very good repeatability between devices [8]. All new cars on the market today have several micromachined pressure sensors, typically used to measure manifold pressure in the engine. The small size and high reliability of micromachined pressure sensors make them ideal for a variety of medical applications, as well.

7.6.2 Surface micromachining

While bulk micromachining creates devices by etching into a wafer, surface micro-machining builds devices up from the wafer layer by layer. It involves a repeated sequence of thin-film deposition on a wafer, photo-patterning of the films, and then etching the patterns into the films [14]. In order to create moving, functioning machines, these layers alternate thin films of a structural material (typically silicon) and a sacrificial material (typically silicon dioxide). Figure 7.25 demonstrates the basic steps of surface micromachining for fabrication of a cantilever. The structural material will form the mechanical elements, and the sacrificial material creates the gaps and spaces between the mechanical elements. In the case of the structural level being made from silicon and the sacrificial material being silicon dioxide, the final 'release' process is performed by placing the wafer in hydrofluoric acid. The hydrofluoric acid quickly etches away the silicon dioxide, leaving the silicon undisturbed. Surface micromachining requires more fabrication steps than bulk micromachining, and hence is more expensive. It is able to create much more complicated devices, capable of sophisticated functionality. Thus, surface micro-machining is suitable for applications requiring more complex mechanical elements.

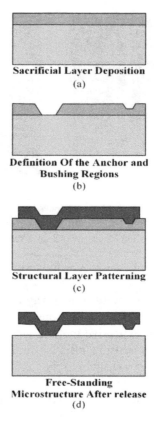

Figure 7.25. Demonstration of surface micromachining. (a) Sacrificial layer deposition, (b) definition of the anchor and bushing regions, (c) structural layer patterning, and (d) the free-standing microstructure after release. Adapted from [14]. Copyright (2006), with permission from Springer Nature.

A summary comparison between bulk and surface micromachining shows that bulk micromachining etches the silicon substrate, whereas surface micromachining etches away layers deposited on top of the silicon substrate. Bulk micromachining typically uses wet-etching techniques, while surface micromachining primarily uses dry-etching techniques. While both techniques are similar from a processing complexity perspective, there is a tremendous difference in their outcomes. Applications of bulk micromachining usually include the fabrication of microstructures such as trenches and holes of lateral dimensions of the order of 1–5 mm and vertical dimensions of the order of 100–500 μm. In contrast, surface micromachining is useful for forming free-moving microstructures, including basic rotating structures that cannot be realised using bulk micromachining techniques, of typical lateral dimensions of the order of 100–500 μm and vertical dimensions of the order of 0.5–2 μm.

7.6.3 Material aspects of micromachining

The most commonly used structural material in surface micromachining is polysilicon. It offers process compatibility with conventional IC fabrication technology, and offers very good mechanical properties. In addition, it can easily be patterned with wet/dry etching. For wet etching, standard silicon etchants are used; for dry etching, RIE is preferred. Some other materials that can be used in surface micromachining are metals, silicon nitride, polymers and single-crystal silicon [14, 20].

Polysilicon is usually deposited via LPCVD. This involves breaking up silane (silicon hydride) at temperatures above 600°C to deposit polysilicon on oxide substrates. The electrical, structural and thermal properties of polysilicon depend on the deposition conditions such as substrate temperature, pressure, flow rate, etc. Below 580°C, it is amorphous. At 600°C, its grains are very fine; at 625°C, they are little larger; and at 645°C, they are columnar. For MEMS applications, deposition is done at 620°C, where the grain size is small. This ensures isotropic properties since the number of grain boundaries are the same in all directions. The typical thickness of a structural layer ranges between 2–4 μm. Deposited polysilicon layers have compressive stress. In MEMS applications, where the structural layer has to be free standing, residual stress can be very dangerous. Hence, they are annealed at high temperatures (e.g. 900°C–1500°C) [14, 20].

For the sacrificial layer, the most commonly used materials are silicon dioxide, phosphosilicate glass, oxidised porous silicon and photoresists. The sacrificial oxide is 1.5–2 μm thick and easy to etch. Since the sacrificial oxide etchant has to go through the patterned structural layer for the release, this etching step is usually long. In many instances, a wet chemical etch is preferred. Usually, a low-temperature oxide that etches easily in buffered hydrofluoric acid is preferred. Etching times can be as long as 100 min [14, 20].

7.6.4 Wafer bonding

Substrate (wafer) bonding (silicon–silicon, silicon–glass and glass–glass) is among the most important fabrication techniques in microsystem technology [14]. As the name suggests, it is a technique by which two wafers are joined together. It is frequently used to fabricate complex 3D structures both as a functional unit and as part of a final microsystem package and encapsulation. The addition of material can be done by wafer bonding under a high-pressure, high-temperature and high-voltage environments. Two or more bulk micromachined wafers, e.g. Si to Si, Si to quartz, can be bonded together. Wafer bonding is used heavily in the production of pressure sensors, valves, regulators, inkjet nozzles, pumps, chemical sensors and miniature analytical instruments. The two most important bonding techniques are silicon–glass electrostatic (or anodic) bonding and silicon–silicon fusion (or silicon direct bonding). In wafer bonding, a silicon wafer and silicon/glass substrate are brought together and heated to a high temperature. Then, an electric field is applied across the joint, which develops a strong bond between the two materials. The basic steps of wafer bonding include surface preparation, fusion, pressurisation and annealing. The quality of the bond depends on the surface conditions. A surface contaminant of the order of 1 μm can damage an area as large as 1 cm in diameter. Two major characterisation techniques for substrate bonding evaluation are visual inspection

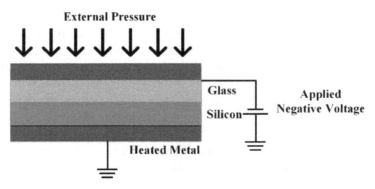

Figure 7.26. Anodic bonding. Adapted from [14]. Copyright (2006), with permission from Springer Nature.

and blade tests. These essentially evaluate cracks/gaps in the bond to estimate the resulting bonding strength. Silicon–glass anodic bond interfaces do not require any particular imaging instrument and can be inspected with an unaided eye or under a light microscope. However, infrared imaging, X-ray photography and ultrasonic methods are necessary for opaque substrates (mainly silicon–silicon).

(a) **Anodic bonding**

Anodic bonding is applied to silicon wafers and glass substrates with a high content of alkali metals (around 3.5% sodium). It is also known as electrostatic bonding. As shown in figure 7.26, a high negative voltage is applied between two mechanical supports of the glass and silicon assembly and the composition is heated upto 500°C. The voltage creates a strong electric field, pulling the two surfaces together. Positive sodium ions from the glass become attached to the negative electrode and are neutralised, creating a strong bond between them.

During bonding, oxygen from the glass is transported to the glass–silicon interface. SiO_2 is formed in this process and permanent bonds are created. Other techniques include adhesive layers like wax, epoxy, etc. Furthermore, this wafer bonding technique can be combined with basic micromachined structures to design complex microdevices such as valves and pumps.

(b) **Fusion bonding**

Fusion bonding is also known as silicon direct bonding. Silicon wafers are bonded through formation of oxides on their surfaces. As the thermal mismatch between the bond materials is low, it is used in a wide range of applications in microelectronics and microsystems technology to bond two silicon wafers together. A fusion bonding setup is depicted in figure 7.27.

In fusion/direct bonding, two wafers are cleaned and made hydrophilic in an acid mixture. They are then rinsed and spin dried. The wafers are then wetted with silicate solutions. The wafers are again rinsed and dried. Following this, they are joined by applying an external pressure, initially at the centre and then uniformly throughout the surface. Bonding is controlled by the viscosity and pressure of

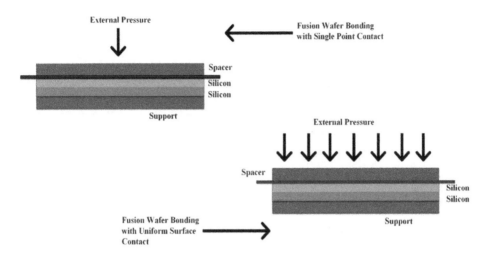

Figure 7.27. Direct wafer bonding. Adapted from [14]. Copyright (2006), with permission from Springer Nature.

ambient gas and surface energy. To dehydrate the surface from moisture, the wafers are heated. Spacers are provided to facilitate mechanical/electrical contact. At high temperatures, the bond strength is almost equal to silicon itself. This method is economical, offers ease of bonding and a high bond strength due to the homogeneity. Bonding can be carried out at high or low temperatures depending on the materials used. Maximum bond strengths are achieved at temperatures between 700°C–1100°C. (Thermally sensitive devices can be bonded at 200°C–400°C by using separate surface activation methods.)

In addition to these techniques, several other alternative methods that utilise an intermediate layer (eutectic, adhesive and glass frit) have also been proposed; these are left to the interest of the reader.

7.6.5 High-aspect-ratio micromachining

Bulk and surface micromachining technologies fulfil the requirements of a broad range of applications. Certain applications, however, require the fabrication of HAR structures, which is not possible with the aforementioned technologies. To achieve this, three main technologies exist: LIGA, HEXSIL (short for HEXagonal honeycomb poly SILicon) and HARPSS (which denotes high-aspect-ratio micromachining combined with poly- and single-crystal silicon). These are capable of producing structures with vertical dimensions much larger than the lateral dimensions by means of X-ray lithography (LIGA) and deep reactive ion etching (HEXSIL and HARPSS). They allow fabrication of tall, precision MEMS structures with vertical sides of the order of centimeters against horizontal dimensions of the order of a few millimetres. The most popular technique is LIGA process. As

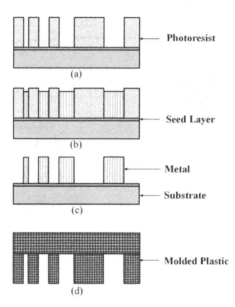

Figure 7.28. Outline of X-ray LIGA process. (a) Photoresist patterning, (b) electroplating of metal, (c) resist removal, and (d) moulded plastic components. Adapted from [14]. Copyright (2006), with permission from Springer Nature.

aforementioned, LIGA is a German acronym for *Lithographie, Galvanoformung, Abformung*, which utilises a three-step process involving lithography, electroplating and moulding. It usually uses UV or X-ray photolithographic processes with thick photoresist layers of the order of 100 μm. Figure 7.28 illustrates the X-ray LIGA process [14].

It is very popular as it can yield high aspect ratios on the order of 100:1. In this process, an X-ray-sensitive polymer photoresist, typically polymethyl methacrylate (or PMMA), bonded to an electrically conductive substrate, is exposed to parallel beams of high-energy X-rays from a synchrotron radiation source through a mask partly covered with a strong X-ray-absorbing material. Chemical removal of exposed (or unexposed) photoresist results in a 3D structure, which can be filled by the electrodeposition of metal. The resist is chemically stripped away to produce a metallic mould insert.

To sum up, it can be inferred that in bulk micromachining dry etching should be employed to produce 2D free-form geometries with vertical sidewalls in substrates. Anisotropic wet-etching systems with protection for wafer frontsides should be preferred during etching, bonding and aligning of systems to join wafers and perform photolithography on the stacked substrates. In surface micromachining, release and drying systems should be employed to realise free-standing micro-structures. In the LIGA process, batch-plating systems are used to create metal moulds, and plastic injection moulding systems are used to create components from metal moulds.

7.7 Process integration

The industrial processes involved in the fabrication of chips are quite sensitive and require specialised production facilities. In general, chip production takes a time of between 6–8 weeks; but this time increases as the size of the chips is reduced [1, 6]. ICs on a 10 μm technology node were initially fabricated in 1971. This node has been gradually downscaled to 7 nm. Furthermore, a 5 nm node is currently in research and will hit the market soon. When the technology node is reduced, the chip becomes more integrated, and hence requires better cleaner rooms to avoid chip failures and improve yield. In addition, special care has to be taken to tackle the hazardous materials and complex sophisticated processes. Finally, amalgamation of all industrial processes results in an IC as a product. Process integration can be divided into three basic parts: device fabrication, isolation and interconnection. The device fabrication step determines the technology with which the IC will be realised. In this, individual devices need to be isolated from each other on a single substrate. After isolation, devices will be connected through specific electrical paths to implement the desired functionality.

As discussed earlier, device fabrication consists of etching, oxidation, ion implantation, diffusion, lithography, metallisation, etc. At the nanoscale, manufacturing is undertaken by top-down and bottom-up approaches. In top-down approaches, work commences by forming patterns on a large scale, and then gradually reducing to the nanoscale, just as a sculpture is made from a piece of rock/stone. These methods are expensive and use a larger amount of materials, as well as generating waste in the form of the discarded excess material. Thus, they are not suitable for large-scale production. In bottom-up approaches, elementary forms such as atoms or molecules are used to build up nanostructures. They are less expensive, but time consuming.

To achieve the necessary functionality, many devices may need to be fabricated on single substrate. If these are not isolated, they will just create a short circuit throughout the chip area. Thus, lateral isolation is needed. Use of field oxides, LOCOS, shallow trench isolation (STI), etc. are popular device isolation techniques [1, 6, 7].

The last step of process integration is interconnection. In order that two devices can communicate with each other, it is necessary to connect them with an interconnect. Earlier interconnects were made from aluminium and a layer of dielectric was deposited above them. With increasing levels of device integration, and in order to achieve the desired small areas, multilevel metallisation schemes were deployed. However, now, due to increased scaling, circuit density and interconnect delay (i.e. parasitic time constant) are gaining importance. In order to reduce this delay, today's interconnects are made of copper [1, 6, 7]. Interconnect design is one of the hottest research areas in today's context. Many emerging nanostructures are being looked upon as possible replacements for copper interconnects. In table 7.2, potential future interconnects are listed along with their advantages and concerns [14].

Table 7.2. Potential futuristic interconnects [26].

Application	Option	Potential advantage	Primary concerns
Cu replacements	Other metals (Ag, silicides, stacks)	Potential lower resistance in fine geometries	Grain boundary scattering, integration issue, reliability
	Nanowires	Ballistic conduction in narrow lines	Quantum contact resistance, controlled placement, low density, substrate interactions
	Carbon nanotubes	Ballistic conduction in narrow lines, electro migration resistance	Quantum contact resistance, controlled placement, low density, chirality control, substrate interactions, parametric spread
	Graphene nanoribbons	Ballistic conduction in narrow films, planar growth, electro migration resistance	Quantum contact resistance, control of edges, deposition, etch stopping and stacking, substrate interactions
	Optical (interchip)	High bandwidth, low power and latency, noise immunity	Connection and alignment between die and package, optical (electrical) conversion efficiencies
	Optical (intrachip)	Latency and power reduction for long lines, high bandwidth with wavelength-division multiplexing (WDM)	Benefits only for long lines, need compact components, integration issues, need WDM, energy cost
	Wireless	Available with current technology, parallel transport medium, high fan out capability	Very limited bandwidth, intra-die communication difficult, large area and power overhead
	Superconductors	Zero resistance interconnection, high Q passives	Cryogenic cooling, frequency dependent resistance, defects, low critical current density, inductive noise and crosstalk
Native device interconnects	Nanowires	No contact resistance to device, ballistic transport over microns	Quantum contact resistance to Cu, substrate interconnection, fan out/branching and placement control
	Carbon nanotubes	No contact resistance to device, ballistic transport over microns	Quantum contact resistance to Cu, fan out/branching and placement control
	Graphene nanoribbons	No contact resistance to device, ballistic transport over microns, support for multi fanouts	Quantum contact resistance to Cu, deposition and patterning processes.
	Spin conductors (Si (Mn), Ga(Mn)As)	Long diffusion length for spin excitons	Low temperature requirements, low speed. surface magnetic interactions

This chapter summarises the prevailing and future processes for cutting-edge semiconductor manufacturing. Solid-state as well as MEMS-based IC fabrication is presented in a brief but comprehensive manner. The significance of top-down and bottom-up approaches is explained in detail. It is hoped that the contents of this chapter will equip the reader with an overview of micro- and nanofabrication methodologies.

Questions

(1) What is the meaning of nanofabrication?
(2) List the challenges in fabricating a device at the nanoscale.
(3) What is top-down nanofabrication and bottom-up nanofabrication? Which approach, in your opinion, is more suitable?
(4) Which type of nanofabrication is the fastest and therefore the cheapest (though also the least reliable)?
(5) List the processes in the order of their implementation in order to develop a nanotransistor of your choice.
(6) What is the future of nanofabrication?
(7) What is the fundamental physics behind the working of a STM? Does it aid in the movement of the atoms?
(8) During lithography, at the nanoscale structures do not appear properly due to quantum limits and require correction. How can this be achieved?

References

[1] Sze S M 2002 *Semiconductor Devices, Physics and Technology* 2nd edn (New York: Wiley)
[2] Phelps G J 2004 Dopant ion implantation simulations in 4H-silicon carbide *Simul. Mater. Sci. Eng.* **12** 1139
[3] Zheng C 2017 *Nanofabrication: Principles, Capabilities and Limits* (Berlin: Springer Nature) pp 365–88
[4] Sarangan A 2016 *Nanofabrication: Principles to Laboratory Practice* (Boca Raton, FL: CRC Press) 1st edn
[5] Campbell S A 2001 *The Science and Engineering of Microelectronic Fabrication* 2nd edn (Oxford: Oxford University Press)
[6] Sze S M (ed) 2017 *VLSI Technology* (New York: McGraw-Hill) 2nd edn
[7] James Plummer M, Deal and Griffin P 2003 *Silicon VLSI Technology* (Englewood Cliffs, NJ: Prentice-Hall Electronics)
[8] Rogers P and Adams 2008 *Nanotechnology: Understanding Small Systems* (Boca Raton, FL: CRC Press)
[9] Yeap G *et al* 2019 5nm CMOS production technology platform featuring full-fledged EUV, and high mobility channel FinFETs with densest 0.021 μm 2 SRAM cells for mobile SoC and high performance computing applications *EEE International Electron Devices Meeting (IEDM)* (Piscataway, NJ: IEEE) pp 10.1109/IEDM19573.2019.8993577.
[10] Adan O and Houchens K 2019 On device EPE: minimizing overlay, pattern placement, and pitch-walk, in presence of EUV stochastics and etch variations *Proc. SPIE* **10959** 1095904
[11] Liu J C *et al* 2020 A Reliability Enhanced 5nm CMOS Technology Featuring 5th Generation FinFET with Fully-Developed EUV and High Mobility Channel for Mobile

SoC and High Performance Computing Application *2020 IEEE International Electron Devices Meeting (IEDM)* (Piscataway, NJ: IEEE)

[12] Wu S-Y *et al* 2016 A 7nm CMOS platform technology featuring 4 th generation FinFET transistors with a 0.027 um 2 high density 6-T SRAM cell for mobile SoC applications *2016 IEEE International Electron Devices Meeting (IEDM)* (PIscataway, NJ: IEEE) 2–6

[13] Samsung 5 nm update https://en.wikichip.org/wiki/5_nm_lithography_process

[14] Mahalik N P 2006 *Micromanufacturing and Nanotechnology* (Berlin: Springer)

[15] Yamazaki K and Namatsu H 2004 5-nm-order electron-beam lithography for nanodevice fabrication *Jpn. J. Appl. Phys.* **43** 3767

[16] Cerrina F 2000 X-ray imaging: applications to patterning and lithography *J. Phys. D: Appl. Phys.* **33** R103

[17] XRD-6100 X-ray diffractometer from Shimadzu https://serc.carleton.edu/research_education/geochemsheets/techniques/SXD.html

[18] Henry D, Eby N, Goodge J and Mogk D 2023 X-ray reflection in accordance with Bragg's Law (https://serc.carleton.edu/research_education/geochemsheets/BraggsLaw.html)

[19] Exploring uncharted realms with electron microscopy https://assets.thermofisher.com/TFS-Assets/MSD/Handbooks/Exploring-uncharted-realms-with-electron-microscopy.pdf

[20] Kohler and Fritzsche 2004 *Nanotechnology: An Introduction to Nanostructuring Techniques* (New York: Wiley-VCH) 1st edn

[21] Venables J A 2000 *Introduction to Surface and Thin Film Processes* 1st edn (Cambridge: Cambridge University Press)

[22] Volmer M and Weber A 1926 Keimbildung in übersättigten Gebilden *Z. Phys. Chem.* **119U** 277–301 (in German)

[23] Frank F, van der Charles and Merwe J H 1949 One-dimensional dislocations. I. Static theory *Proc. Royal Soc. London: Series A, Math. Phys. Sci.* **198** 205–16

[24] Stranski I N and Krastanov L 1938 Zur Theorie der orientierten Ausscheidung von Ionenkristallen aufeinander *Monatsh. Chem. Verw. Teile Anderer Wiss.* **146** 351–64

[25] Pal P and Chandra S 2004 Bulk-micromachined structures inside anisotropically etched cavities *Smart Mater. Struct.* **13** 1424

[26] SIA 2013 *International Technology Roadmap for Semiconductors (ITRS)* Semiconductor Industry Association https://www.semiconductors.org/resources/2013-international-technology-roadmap-for-semiconductors-itrs/

IOP Publishing

Nanoelectronics
Physics, technology and applications
Rutu Parekh and Rasika Dhavse

Chapter 8

Emerging nanoelectronic architectures

Many efforts have been made at designing, developing, simulating, modelling, fabricating and characterising novel nanoelectronic devices. However, the development of system architectures employing these devices is still in its infancy. A large amount of resources need to be deployed to this end because any given novel nanoelectronic device will not be useful in isolation. Rather, they will be put to good use when they form a system-level architecture with very high density (around 10^{15} devices per cm^2). Architecture-level considerations such as circuit/system-level performance, reliability, interconnection of smaller devices to larger ones, increasing power dissipation with increasing packaging density and, most importantly, fault tolerance (i.e. system failure) play a vital role. Presently, different nanoelectronic devices are at different levels of maturity. That makes the continuing emergence of nanoelectronic architectures interesting and equally challenging. This chapter is intended to stimulate a thought process in the reader regarding the performance of future nanoelectronic architectures. The architectures illustrated here are just a glimpse of what is possible.

As shown in figure 8.1, computing architectures are classified into program-centric computing and data-centric computing. A device is said to have a program-centric architecture when computation is performed in the form of set of stored instructions. The device is created in such a way that it performs a set of predetermined functions based on instructions stored into the device. Stored-program digital computers—also known as (classical) von Neumann architecture —is an example of program-centric architecture. The first analogue 'computers' were invented in 100 BC and were superseded later by the emergence of digital computers. With the development of very-large-scale integration (VLSI) technology, more and more research commenced on analogue computers. Several complementary metal-oxide-semiconductor field-effect transistors (MOSFETs) can be used to integrate analogue as well as digital computers to give rise to hybrid computers. In previous generations, there has been comparatively less data, and it has been

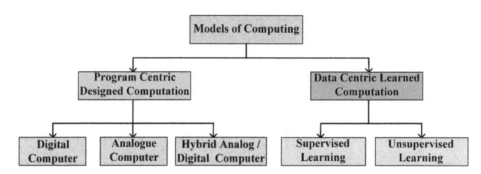

Figure 8.1. Various computing architectures. © 2010 IEEE. Adapted from [1].

structured in one way or another. Over the past two decades, however, there has been a phenomenal increase in the quantities of data, most of which is unstructured. This has given rise to data-centric architecture in devices. In data-centric computing, computation is mainly based on data rather than a particular 'program' with predefined instructions. Neural networks and machine learning is an example of data-centric computation. Machine learning itself is broadly divided into two fields: supervised learning and unsupervised learning. Supervised learning is when the result dataset is already known and computation is carried out to come up with a function that maps input data to the given output data. On the other hand, unsupervised learning is when the computer learns from the input data itself. The taxonomy of architectures based on their 'computational efficiency' is as follows [1–3]:

- **von Neumann architecture** uses memory storage separated from different resources to store programs as well as data.
- **More-Neumann architecture** uses the stored memory concept but some multi-core systems are also present.
- In **More-than-Neumann architecture**, the resources are integrated to a higher degree such that their combination elevates the system performance to a higher level. Thus, in such an architecture, precise configurations of individual elements are combined to perform certain functions.
- **Beyond-Neumann architecture** can solve complex computational problems much faster than the ones discussed above.

Figure 8.2 illustrates a hierarchy of recent extended memory architectures from a computational perspective.

It is very difficult to identify applications of emerging memory devices due to the lack of architectural models or appropriate interconnect systems in order to incorporate novel devices in existing architectures. Thus, emerging memory devices can be used as a replacement for a particular part of a conventional circuit; or as an augmentation to existing conventional circuit, increasing performance; or as a device that adds additional functionalities to information processing by its unique properties. Memory systems, which may consume almost 30% of a system's power, vary in size from gigabytes to as high as exabytes. In every case, data read/write

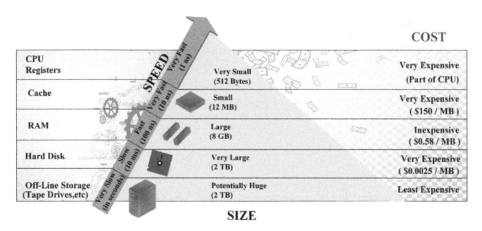

CPU Registers	SPEED	Very Small (512 Bytes)	COST Very Expensive (Part of CPU)
Cache		Small (12 MB)	Very Expensive ($150 / MB)
RAM		Large (8 GB)	Inexpensive ($0.58 / MB)
Hard Disk		Very Large (2 TB)	Very Expensive ($0.0025 / MB)
Off-Line Storage (Tape Drives,etc)		Potentially Huge (2 TB)	Least Expensive
		SIZE	

Figure 8.2. Extended memory hierarchy. Adapted from [4] with permission.

speeds, cost and power consumption are critical considerations. Volatile dynamic RAM (DRAM) consumes a major percentage of this power for refresh purposes, leading to high power consumption in today's servers [2]. Thus, persistent memory that avoids continuous refreshing is more valuable. The following sections discuss some typical emerging memory architectures for computing systems like storage class memory (SCM), neuromorphic memory, quantum cellular automata (QCAs) and corticular memory.

8.1 Storage class memory

SCM is a technology that enables new levels in the memory hierarchy containing the different capabilities of solid-state drives (SSDs), hard-disk drives (HDDs) and magnetic drives. Compared to other storage technologies, it can provide robustness and offers a low cost per bit. There are two classes of SCM [3]: M-class SCM (M-SCM), which has very low energy consumption with speeds equivalent to DRAM, and S-class SCM (S-SCM), which can achieve high-capacity and durable data storage similar to SSDs. The desirable properties of M-SCM and S-SCM are listed in table 8.1. To implement SCM, emerging memory devices and correct interfaces with new architectures are strongly needed. As shown in figure 8.3, SCM architecture is normally created out of multilevel NAND-based flash cells with a three-dimensional (3D) stacking methodology. In the pursuit of large memory capacity, it offers dense packing and individual bit access by employing a crosspoint structure with selector lines. 3D stacking allows upscalability. Novel memory cell materials can be explored to offer better switching at the cell level and high-performance array architecture (approximately 1000 times faster than conventional NAND arrangements).

The important desired properties of memory devices for SCM applications are reliability, fast access, significant endurance, low cost per bit and huge storage capacity. The most important advantage of emerging SCM technology is that it can help in increasing overall system performance by improving the availability/delivery

Table 8.1. Desirable properties of M-SCM and S-SCM [4].

S. No`.	Properties	M-SCM	S-SCM
1.	Cost per bit	Comparable to DRAM or better	Approaching HDD
2.	Read/write latency	<100 ns	~1 ms or longer
3.	Persistence	Sufficient to survive power failure and eliminate refresh power	Long term
4.	Interface	TBD, likely an extension of main memory address space, word addressable	TBD, either memory extension, page based or disk like
5.	Fault management	Hardware-based failure remapping and error correction code (ECC)	Hardware-based failure remapping and ECC
6.	Architecture	Merged with DRAM L4 cache	Serves as cache for disks

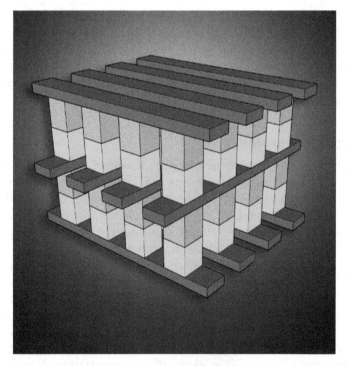

Figure 8.3. 3D structure of SCM. (This 3D XPoint.png image has been obtained by the author(s) from the Wikimedia website where it was made available under a CC BY-SA 4.0 licence. It is included within this book on that basis. It is attributed to Trolomite.)

of the most needed data. Both M-SCM and S-SCM have the capability to increase overall system performance as they avoid memory corruption caused by errors or leaks. SCM decreases processing and boosts throughput by reducing the amount of data that must be sent back and forth between memory and storage. SCM presents a novel and high-performance storage type that is thus suitable for various perform-ance requirements, and it can be used to replace conventional technologies like HDDs and SSDs. If different types of memory like M-SCM and S-SCM are combined in the same package, then modern mobile devices, for instance, can become more compact.

The purchasing and maintenance costs of a technology are accounted for in calculating the total cost. The average price of SCM drives is a drawback to broader adoption, may be 10 times more expensive than a similar non volatile memory configuration. The purchasing costs of HDDs are lower than flash memory, which are in turn lower than DRAM. However, purchasing costs should not only be considered, as SCM has lower power consumption than HDDs and further consumes less space. The data centers can benefit from SCM as it offers persistant and high performance storage workloads. With SCM, data and program code are retained during sysem crash or power failure in presence of onboard power source. In this, it is helpful for insurance companies and financial companies to analyse data more easily and to detect whether there are any suspicious transactions. It is also useful for companies to detect and defend against any cybercrime activities. SCM can be used by trading companies for better execution of financial transactions. It promotes data transfer at much higher rates and also supports next-generation complex computing [5]. HDDs and DRAM can sustain a large number of read/write operations successfully. However, in emerging SCM, wear levelling (uniform spread of the write operations) is a major issue. In certain scenarios data and metadata are written with greater frequency, and so one must see that the storage locations do not get worn easily

The architecture, read/write speed, endurance and interface of M-SCM should have a resemblance with that of DRAM. In M-SCM, use of write buffers ensures that all writes are completed by the device even in the case of any power loss. Here, the computations are done near to the data storage and hence there is no need to load data to the processor. Caches are used so that data with high requirement probability is easier to access. However, the endurance of emerging devices is inferior as compared to that of DRAM, and so some innovation is needed to overcome this. This should try to optimise the performance as well as power use by careful choice of the right memory technologies, and should also maximise the lifespan. Further, there is a need for an interface that can support a number of generations of M-SCM to come, because otherwise the cost to adapt new generations would be considerable. For a number of systems, M-SCM must be able to use existing power supplies and controllers to the fullest possible extent. M-SCM must be power efficient and also give an indication whenever read/writes are completed without failure. Also, it must support multiple data rates.

S-SCM is expected to replace the HDD, and to integrate it as the primary storage device. This technology has a number of advantages such as decent speed and less of penalty for seeking time. Its purchase cost and maintenance cost must be similar to HDDs.

From a research and development point of view, computational efficiency is a prime concern. Hence, with respect to benchmark memory technologies, certain

Table 8.2. Parameters of benchmark vs target memories [4].

S. No.	Parameter	Benchmark			Target	
		HDD	NAND Flash	DRAM	M-SCM	S-SCM
1.	Read/write latency	3.5 ms	~100 μs (block erase ~1 ms)	<100 ns	<100 ns	1–10 μs
2.	Endurance (cycles)	Unlimited	10^4–10^5	Unlimited	$>10^9$	$>10^6$
3.	Retention	>10 years	>10 years	64 ms	>5 days	>10 years
4.	ON power (W/GB)	~0.04	~0.01–0.04	0.4	<0.4	~0.04
5.	Standby power	~20% of ON power	<10% of ON power	~25% of ON power	<1% of ON power	<1% of ON power
6.	Density	~10^{11} bit cm^{-2}	~10^{10} bit cm^{-2}	~10^9 bit cm^{-2}	$>10^{10}$ bit cm^{-2}	$>10^{10}$ bit cm^{-2}
7.	Cost ($/Gb)	0.1	2	10	<10	<3–4

targets for SCM can be set, as indicated in table 8.2. A hybrid version of these architectures is envisaged to be very useful in the global market. Similar to that of M-SCM, S-SCM faces challenges such as its interface, endurance, cost efficiency and architecture. Efforts to design SCM using 3D-Xpoint technology, resistive RAM (or ReRAM), magnetoresistive RAM (or MRAM), spin-transfer torque RAM (or STT-RAM), nanotube-based RAM, etc. are ongoing. Some may lead to revolutionary new architectures.

8.2 Morphic computing: the architectures that can learn

IBM Watson is a famously successful presentation of a data-centric computation system. It was developed for, and went on to win against the human champion of, the well-known TV show 'Jeopardy'. Presently, the development of IBM Watson is being tailored towards medical applications. Watson is passed through an intensive training process to gain knowledge, making it capable of answering a collection of difficult and wide-ranged questions. However, throughout its entire training process, Watson is built without any morphic features (meaning features that mimic the human brain's learning capabilities) [4]. Another kind of data-centric architectures, then, are morphic architectures. 'Morphic' is a word used in electronics to refer to circuits or devices that can adapt to a given problem and handle it efficiently. These architectures take inspiration from biological structures as well as scientific computational models. It was first introduced in the 'Emerging Research Architecture' section of ITRS 2007 [6], referring to architectures, inspired by biology, that embody a new computation paradigm in which adaptation serves a key function in tackling the particulars of problems. This perspective is clearly indicated in figure 8.4, which shows the roadmap beyond Moore's law.

Morphic architectures include a broad range of mixed-signal structures that are centred on a specific application and which draw inspiration for their structure from the application. In some cases, processing is carried out in the analogue domain,

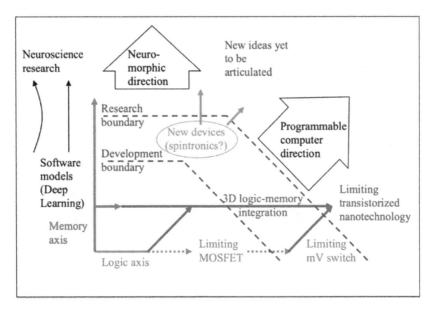

Figure 8.4. Morphic architectures as indicated in the electronics roadmap beyond Moore's law. Adapted from [7]. CC BY 4.0.

providing orders-of-magnitude enhancement in overall performance and strength dissipation, albeit with reduced accuracy. As an example, biologically inspired inference networks for cognition may also yield to a partially analogue implementation and report good performance metrics relative to their digital counterparts.

In recent years, many studies have been undertaken to create architectures that can learn with or without training from existing data. Systems exist, deployed on classical von Neumann architecture, that can learn against biological counterparts. However, they are inefficient with regards to noisy data and encounter problems involving pattern recognition, thus requiring self-configurable devices. A dense network resembling the biological synapses that help in communication can be built using $TiN/HfO_2/Ti/TiN$ memristors [8]. There are three distinct emerging architectures under this category, i.e. neuromorphic, cortical and cellular automata. They are expected to provide considerable opportunities for development of new high-performance data-centric computers. We will here look at the visualisation and implementation of these architectures through different circuits. We also detail the current advancements in such technologies and, lastly, try to indicate where these architectures will be heading in the future.

8.2.1 Neuromorphic architecture

Neuromorphic architecture was first introduced by Carver Mead in 1989 [9]. It is also known as neuromorphic processing or neuromorphic engineering. These hardware systems tend to use a basic neural computation function as their primary

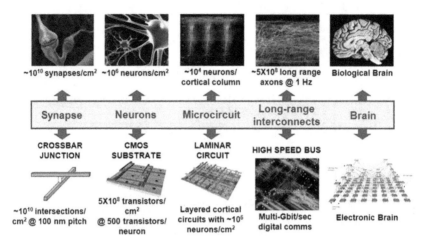

Figure 8.5. Correspondence between biological structures and hardware-based neuromorphic architecture. Reproduced from http://thenewstack.io/ai-hardware-software-dilemma/.

operation. Their working principle utilises concepts such as weighted connections, long-term potentiation, nonlinear activations and inhibitions from computational neuroscience to solve real-world problems. As compared to existing solutions such as classical von Neumann architecture, neuromorphic architecture exhibits low-power operation and fault-tolerant hardware solutions, which are extremely useful in the distributed and computation-intensive tasks commonly seen in today's embedded hardware. It is essentially a computer architecture designed to mimic biological neurons. The relevant functional units in biology consist of neurons, axons, synapses and dendrites, wherein neurons constitute the basic building blocks, connections between two neurons are called synapses, and axons and dendrites connect lots of synapses in long-range structures that resemble computer buses. Typical analogies between biological and neuromorphic structures are shown in figure 8.5.

There are two phases of program execution, that is, learning and operation. During the learning phase, the synapse weight is defined and determined. In the network, when a threshold voltage is obtained, the neuron sends a spike. This is used as a base for neural networks. The voltage is added to or deducted from neurons according to what voltage is passed through the synapses. Dendrites are used to send signals to neurons and axons are used to send them out. In the operation phase, the computation is completed. Present data-centric architectures face problems of scaling and power, which can be solved by neuromorphic computing. The design of neuromorphic architecture involves a combination of various disciplines such as physics, biology, mathematics and computer science to build the artificial neural systems, as used in object recognition systems, autonomous rovers and audio processors, which all can work on the principles of biological neural systems. The characteristic features of neuromorphic computing include the following [9]:

(a) Exhibit human-like intelligence, which can be specifically found in the neurons of humans.
(b) Demonstrate the ability to handle anomalies, faults and detect noise in order to increase performance at certain tasks.
(c) Operate at very low input power.

The human brain is divided into areas (or cortexes) that each have specific functions like visual processing, memory, and so. Similarly, neuromorphic machines are also application specific. The performance of existing von Neumann devices can be increased using these neuromorphic machines as supplemental CMOS, as can be seen in figure 8.6. As a result of their variety of applications, they are referred to as 'More [than] Moore' candidates [10]. In real scenarios, one of the main issues is the efficient implement of the artificial neurons. As the actual neuronal system in humans is too complex, the level of abstraction of a neuron is an important consideration in this regard. This implementation may be as simple as the addition and subtraction of signals entering a neuron. Existing technologies can be used to represent a neuron, such as single-electron transistors (SETs), memristors, resistance temperature detectors (RTDs), etc. Depending on this technology, the potential and level of abstraction can vary. In this way, it may be possible to create new opportunities in emerging nanoelectronics devices.

Another problem is in how signals are handled along the neurons and synapses. In a biological neuron, the computation occurs in a digital–analogue–digital

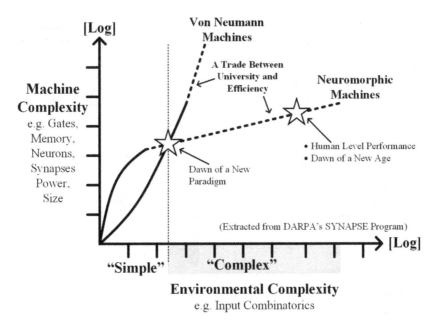

Figure 8.6. Comparison between the design complexity of von Neumann and neuromorphic architectures with increasing number of inputs. Adapted from http://www.nextbigfuture.com/2011/07/darpa-synapse-phase-2-targets.html.

fashion. Researchers have tried various techniques to develop analogue synapses using flash memory, but all suffer a limitation of designing a controller for the electrons and of limited rewriting. In contrast, memristors (resistive RAMs and atomic switches) offer a better solution for implementing such synapses. CMOS neurons are often used with memristive nanojunctions and additional controllers to constitute hybrid CMOL architectures. CMOL (CrossNets) was introduced in ITRS 2007 to refer to structures wherein nanogrids of a single molecule are fabricated on a traditional CMOS. These architectures pave the way for the development of large-scale neural networks that can have many layers [10].

A final significant issue is to incorporate the noise-tolerant property of artificial neural networks into electronics. The most common technique to deal with noise in analogue and digital systems is suppression. Neural networks, on the other hand, use this noise to improve their performance. Some noise-sensitive devices may use this concept to design computation models. Figure 8.7 illustrates the difference between von Neumann and neuromorphic architecture [10] based on classic logic circuit examples.

It is further interesting to look at some examples of neuromorphic architecture. TrueNorth is a neuromorphic CMOS-incorporated circuit created by IBM in 2014 [11]. It is a manycore processor prepared on a chip structure, with 4096 cores, each having 256 programmable simulated neurons for a total of over one million neurons. Each neurons has 256 programmable 'synapses' that pass the signal between them. Hence, the total number of programmable synapses is just over 268 million (2^{28}). Its basic transistor count is 5.4 billion. TrueNorth avoids the von Neumann-architecture bottleneck and is extremely energy-efficient, with IBM claiming a power

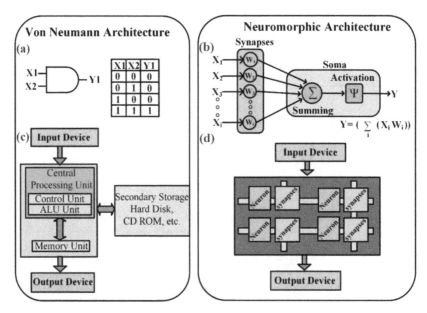

Figure 8.7. Contrast between von Neumann and neuromorphic architectures. (a) Traditional AND gate with its logic table, (b) basic computation model of an artificial neuron, (c) schematic of von Neumann architecture, and (d) schematic of neuromorphic architecture. Adapted from [10].

consumption of 70 milliwatts and a power density that is 1/10,000th of conventional microprocessors. Memory, computation, and communication are handled in each of the 4096 neurosynaptic cores. Because it only uses the power required for computation, the SyNAPSE chip runs at lower temperatures and with less electricity. Its utility in a board is shown in figure 8.8.

Figure 8.9 shows the fine-grained mesh architecture of Loihi, which is a neuro-morphic chip planned by Intel Labs that uses an associated unconventional spiking neural system (SNN) to realise versatile self-changing occasion-driven fine-grained

Figure 8.8. DARPA SyNAPSE board with 16 TrueNorth chips. (This DARPA SyNAPSE 16 Chip Board.jpg has been obtained by the author(s) from the Wikimedia website, where it is stated to have been released into the public domain. It is included within this book on that basis.)

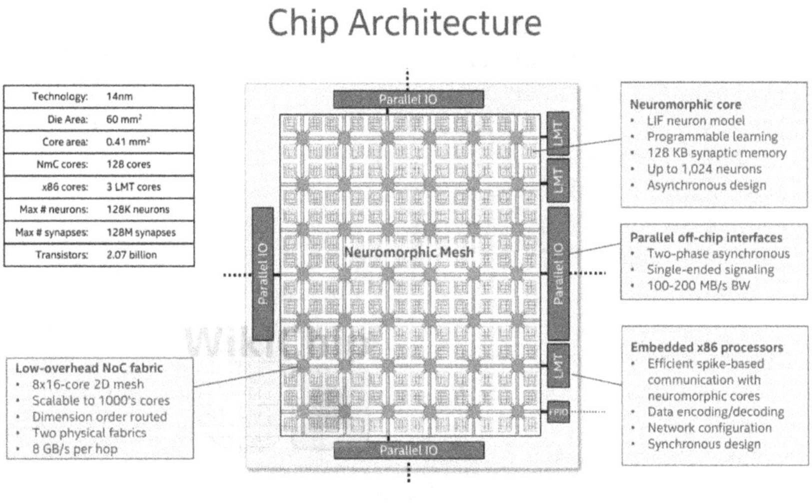

Figure 8.9. Architectural overview of Intel's 'Loihi' neuromorphic computing chip. Adapted from http://fuse.wikichip.org/news/2519/intel-labs-builds-a-neuromorphic-system-with-64-to-768-loihi-chips-8-million-to-100-million-neurons/.

parallel calculations accustomed to executing learning and induction with high effectiveness [12]. The chip is a 128 neuromorphic cores manycore IC, factory-made on Intel's 14 nm process, and notably exhibits a programmable computer code learning motor as an on-chip readymade SNN. The chip was formally exhibited at the 2018 Neuro Galvanised Machine Components (NICE) workshop [12].

8.2.2 Quantum-dot cellular automata architecture

In the 1940s, Stanislaw Ulam and John von Neumann introduced the concept of a cellular automaton. Neumann was working on self-replicating systems at that time, and initially his model was based on the notion of robots building other robots, but this idea failed as it was very difficult and expensive to build. To tackle this problem, a suggestion to use a discrete system in order to create a self-replicating model was made by Ulam. Ulam and Neumann worked together in the 1950s on the problem of calculating liquid motion. They used Ulam's suggestion of using a discrete system to create a self-reproducing model, and built the first ever cellular automata model. The definition and concept of the cellular automaton, as used by Neumann and Ulam, are explained in the following.

Discrete models are studied in complex science, physics, mathematics, computer science and many other growing fields. Cells are the basic building blocks of a cellular automaton, which is composed of a regular grid of such cells [13–15]. Each cell in the cellular automaton is in a state selected from some finite set of predefined states. ON and OFF are examples of such states. The dimensions of the grid are finite and the size can be set to any finite number. In cellular automata an initial state is defined at $t = 0$, and these states are updated by advancing it by 1 and by using a mathematical function that works on the neighbourhood of the cell, which is established in relation to itself [13].

    (i) Why cellular automata?

        As seen in earlier chapters, it has been established that current CMOS technology is failing in many respects and that its sustainability is being challenged due to many scaling and feature sizing issues. In response to these problems, much research has been undertaken and many solutions proposed. The most widely researched solutions are carbon nanotubes (or CNTs), SETs, and cellular automata based on quantum dots (or QCAs). Out of all these, most researchers believe that QCAs are the most best option to replace CMOS. There are many advantages of QCA technology: it is transistor-less, can achieve THz operating speeds and a device density of 10^{12} devices/cm^2. These advantages make QCAs very useful for devices that require high performance or low power [13–15].

        QCAs, first proposed in 1994, are different from previous computational techniques where information is transferred using electric current. In QCAs, the binary information is transferred from one place to another using the Coulombic force of interaction between QCA cells. Today, due to the requirements of global industry, circuits of smaller and smaller size need to be fabricated, but due to the limitations of conventional technologies like CMOS VLSI design it is difficult to meet these requirements; hence, QCAs are proposed for such circuits. The functioning of QCAs depends upon the charge configuration in the quantum dots of the cellular automata. As mentioned, the binary information is encoded by this charge configuration and propagated in the circuit by Coulombic forces of interaction between the QCA cells. One advantage is that, in a QCA's circuit, no power supply is needed for the individual cell and no current flows between the cells [13, 14].

(ii) Basics of QCA devices

Quantum dots are arranged such that they form a square. These quantum dots have a very small diameter, which makes their charging energy greater than $k_\mathrm{B}T$. In these cells, two mobile electrons are inserted; these electrons can travel between the four quantum dots of a square via electron tunnelling. This type of movement is represented through a path in figure 8.10. Due to Coulombic repulsion, the electrons can only take the opposing corners of the cells. Thus there are only two types of orientation possible, with each orientation associated with a polarisation, as shown in the figure. The figure also shows the maximum distance that the two electrons can acquire between them without escaping the cell. It is thought that electron tunnelling can be totally controlled by potential barriers placed underneath the cells. Capacitive plates can be used to control this potential barrier between QCA cells [13].

In QCAs, an isolated cell has two polarisation states, as explained above. These states are represented as $P = +1$ (figure 8.10(a)) and $P = -1$ (figure 8.10(b)). These polarisations store the binary data. The polarisation state $P = +1$ can be characterised as binary 1, whereas the polarisation $P = -1$ can be characterised as binary 0. The polarisation, i.e. P, of a QCA cell can be calculated using the formula [13, 14]

$$P = \frac{(P1 + P3) - (P2 + P4)}{P1 + P2 + P3 + P4} .$$ (8.1)

Here, P_1 is the top-right quantum dot and the numbering is done in clockwise order from that dot. Aside from these polarised states, there is also an unpolarised state for a QCA cell. In this state the potential barrier

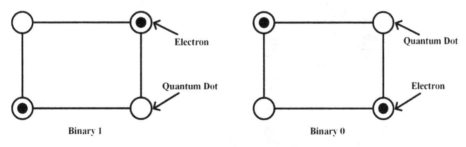

Figure 8.10. QCA cell polarisation and representation of binary 1 and 0. Adapted from [16], Copyright (2022), with permission from Elsevier.

between the quantum dots is lowered, which causes the electrons to occupy a single quantum dot. In such a state the cell exhibits a small amount of polarisation or no polarisation at all.

QCA cells can be implemented by using metallic dots, using molecular redox centres or by utilising bipolar nature of nanoparticles. Metal QCA cells consisting of four aluminium islands connected by aluminium oxide tunnel connections and capacitors has been successfully built using electron-beam lithography. In molecular QCAs, each molecule acts as a QCA cell in which redox centres acts as dots and tunnelling is provided by bridging ligands. Molecular QCAs can operate at higher frequency and at room temperature. Molecular QCAs also offer higher device density. In magnetic QCAs, interaction between nanoparticles guarantees that the structure is always bistable. As a result, information can be transferred through these nanoparticles. However, this does not offer great switching speed compared to current-generation computers. Basic logic gates can be designed using fundamental QCA logic units like wires, majority gates and inverters. In the following discussion, we will first discuss these fundamental building blocks and then some examples of logic gates built using QCAs.

(iii) The majority gate

The majority gate is the basic building block in a QCA [14]. They are used in nearly all circuits designed with QCAs. It is therefore beneficial to discuss this logic gate in detail while learning about QCA logic devices. A three-input gate is shown in figure 8.11. 'Cell 4' in the figure is called the device or the driver cell. The majority gate works by taking the device cell to its lowest energy state. The lowest energy state is attained by the driver cell when its polarisation is equal to the majority gate of the three input cells' polarisation. The input cell is defined as a cell whose polarisation changes

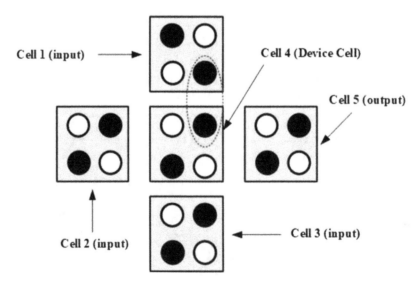

Figure 8.11. QCA majority gate. Adapted from [14]. Copyright (2015), with permission from Springer Nature.

based on a signal travelling towards the driver cell. The reason why the driver cell has minimum energy when it assumes the majority polarisation is that when it assumes the majority polarisation the electronic repulsion between the driver cell and the input cell is minimum [14].

Figure 8.11 is an example where the centre cell attains minimum energy while having majority polarity, i.e. $P = +1$. To understand this, let us look into the Coulombic interaction between all the input cells with the driver cell. In the figure, if we focus only on the driver cell and cell 1 then the driver cell should change its polarity to minimise repulsion, but to minimise repulsion with cells 2 and 3 the driver cell should have the same polarity.

Since the majority of the input cells have polarisation $P = +1$, in this case the driver cell will have the same polarity as these cells to minimise repulsion with the majority. The output of this gate can be characterised by the equation:

$$out = a \times b + b \times c + c \times a. \tag{8.2}$$

(iv) A '90-degree' QCA wire

Figure 8.12(a) represents a '90-degree' straight QCA wire and shows how data propagates in these wires [14, 15]. A 'wire' in QCA terms is a

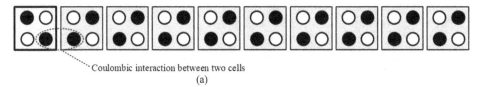

Coulombic interaction between two cells

(a)

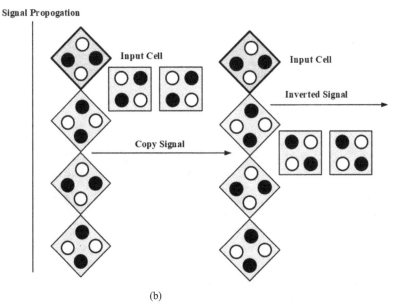

(b)

Figure 8.12. (a) QCA wire, (b) data propagation in QCA wire. Adapted from [14]. Copyright (2015), with permission from Springer Nature.

horizontal sequence of QCA cells. In figure 8.12(b), binary information passed to the wire transmits from left to right without changing because of Coulombic interactions between the cells. In the first part of the figure, the first cell has a polarity of −1 and the second has a polarisation of +1. Here, we assume that the charges in cell 1 are fixed, i.e. they are trapped in the quantum dot they are occupying, whereas the charges in the other cells are not fixed or trapped but are free to move between the quantum dots using quantum-mechanical tunnelling. This assumption saves us from the danger of charges travelling in the opposite direction from which they came. Here, the $P = -1$ travels across the wire because of Coulombic interaction. In this example, cell 1 changes the polarisation of cell 2 and the charges in cell 2 get trapped, which changes the polarisation of cell 3, and so on. This process continues down throughout the length of the wire, causing propagation of the data.

(v) QCA NOT gate

There are various implementations through which a NOT gate can be formed; we will discuss one such implementation here. A NOT gate is used to invert the input voltage; it gives the output 1 on input 0 and vice versa. The structure of a NOT gate using QCAs is shown in figure 8.13. The parallel wires contain the same polarisation state as the input, but the structure of the NOT gate inverts the polarisation state for the output wire.

(vi) Logic implementation using QCAs [14, 15]

Now, let us take a look at the implementation of AND and OR gates. The logical function is represented for a majority gate below:

$$Z = PQ + QR + PR. \tag{8.3}$$

By setting any of the input cells to 0, the majority gate works like an AND gate:

$$AND = PQ + Q \times 0 + P \times 0 = PQ. \tag{8.4}$$

Further, the OR function can be obtained by setting any input cell to 1:

$$OR = PQ + Q \times 1 + P \times 1 = P + Q. \tag{8.5}$$

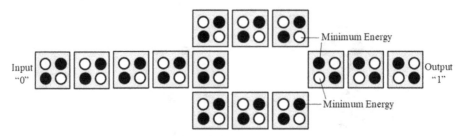

Figure 8.13. NOT gate using QCAs. Adapted from [14]. Copyright (2015), with permission from Springer Nature.

Using QCA cells AND, OR and NOT functionalities can be implemented. As such, QCAs can demonstrate a functionally complete logic set. Thus, using QCA cells any Boolean circuit can be implemented. To discuss one such circuit, we show the implementation of a 2×1 multiplexer in figure 8.14.

(vii) Clocking in QCA cells

Clocking is very necessary in QCAs to ensure that the information is transmitted correctly from the input cell to the output cell. In QCAs the clock is divided into four different periodic phases, switch, release, relax and hold, as shown in figure 8.15 [13]. A clock can be given to each QCA cell, but the QCA cells are divided into subarrays so that the advantages of multiphase and pipelining can be taken. Each QCA cell of a given subarray holds the same clock phase, which changes the polarisation state. In the switch phase of the clock, the QCA cell is in a nonpolarised state and changes to a polarised state. During this phase, the potential barrier is raised to high. During the hold phase, the polarisation state does not change and holds the previous phase; the potential barrier remains high. In the release phase, the transition again occurs; the potential barrier is lowered so that the cells lose their polarisation state and become nonpolarized. In the

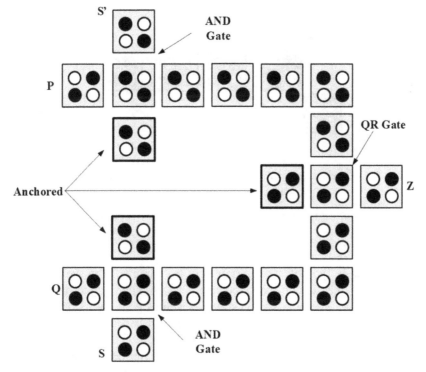

Figure 8.14. 2×1 QCA multiplexer with logical equation $Z = PS' + QS$. Adapted from [14]. Copyright (2015), with permission from Springer Nature.

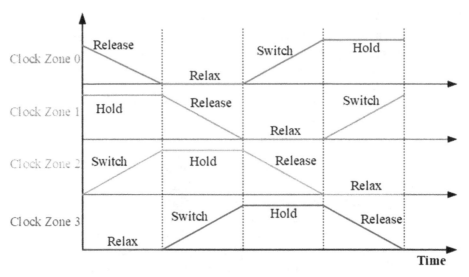

Figure 8.15. Clocking in QCA cells. Adapted from [13]. CC BY 3.0.

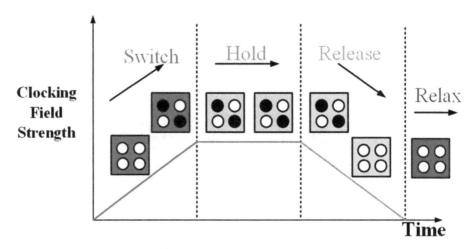

Figure 8.16. Stages of operation. Adapted from [13]. CC BY 3.0.

relax phase, the potential barrier does not change from the previous phase; thus, the cells remain in nonpolarized states. Then, again, the switch phase takes place. The polarisation states as well as potential barriers of the QCA cells during each phase are shown in figure 8.16.

(viii) Limitations of QCA technology

It can be seen that, with regards to contemporary technologies, one can either achieve a power-efficient device or a high-speed device. However,

QCA designs potentially allow for fast as well as low-power devices [17]. This is why QCAs are deemed to be strong candidates for the replacement of CMOS. QCA technology has demonstrated successful benchmarks of integrating around 10^6 devices down to a feature size of 10 nm, whilst operating at a speed of 100 GHz at 4 K and requiring a power supply of just 0.1 mV. However, the fabrication of QCAs is expensive. There are very few techniques available for lithography. Furthermore, accurate alignment of the cells is very difficult to achieve, and the cells malfunction and lose reliability. Moreover, for a circuit to work at room temperatures, the cell dimensions need to be of the order of a nanometre. Due to all these challenges, QCA technology is presently lagging very much behind the mainstream of CMOS technology.

8.2.3 Cortical architecture

Circuits based on cortical architecture is one of the approaches through which modelling based on the brain has been attempted. In a cortical architecture, the modelling of circuits is done by analogy to the neocortex of the human brain [18]. The neocortex is a complex brain structure responsible for higher-order functions in humans such as cognition, generation of motor commands, language, sensory perception, etc.

Our brain uses electrochemical impulses, or spikes, between neurons to transfer messages within itself and throughout the human body. These spikes hold information in them, and complex computation relies on how these spikes communicate in highly connected networks. Researchers hope to observe the performance of greater numbers of biological neurons in real time by modelling machines on the architecture of the brain. A full-scale supercomputer based on cortical architecture will help us understand even larger networks, which were previously thought to be out of reach. This will help in understanding the healthy and unhealthy functioning of the brain. Cortical architecture is a very new field and hence a lot of research is still ongoing. Researchers as well as big companies have all attempted to make circuits or products that come close to the efficiency of the neocortex. These have applications in digital signal processing, machine learning, computer vision, etc. In this section, we discuss some examples and try to understand cortical architecture through a few applications.

(i) SpiNNaker

SpiNNaker, or the Spiking Neural Network Architecture [19], as shown in figure 8.17, is the largest supercomputer in the world that is designed to work in a manner mimicking the human brain. It is a manycore supercomputer designed at the School of Computer Science, University of Manchester. On 14 October 2018, the Human Brain Project announced that the million-core milestone had been achieved [19].

Each of the chips used in the SpiNNaker machine consists of 100 million transistors, and can handle actions of the order of millions of millions per

Figure 8.17. The SpiNNaker million-core machine assembled at the University of Manchester. (This Spinn 1m pano.jpg image has been obtained by the author(s) from the Wikimedia website where it was made available under a CC BY-SA 4.0 licence. It is included within this book on that basis. It is attributed to Pabogdan.)

second. As mentioned, it replicates the biological neurons of the human brain. The network of chips used inside this machine are designed to imitate the neocortex of the brain. To this end, SpiNNaker is currently better than any other machine, and is capable of modelling huge numbers of biological neurons. Its electronic circuits are used to mimic electrochemical spikes, which are a characteristic of neuromorphic computing, and support parallel communication [19]. Small amounts of information are sent in parallel (in the order of billions) to thousands of different destinations. Real-time, high-level processing in a cluster of isolated brain networks is simulated using SpiNNaker. One such cluster is an 80 000 neuron model of a subpart of the cortex. As aforementioned, the cortex receives and processes information from the senses. The human brain is 1000 times bigger than the brain of a mouse, which itself consists of more than 100 million neurons; this gives an idea about the scale of information it processes. The fundamental use of this supercomputer is in aiding neuroscientists to better understand the working of the human brain. The cortical architecture used in SpiNNaker chips is illustrated in figure 8.18. There are various regions identified for utilisation in the SpiNNaker machine. The practical applications of this machine include simulation of the basal ganglia, a region of the brain that is vulnerable to Parkinson's disease [19]. This hints at the possibilities for potentially massive neurological breakthroughs in pharmaceutical testing and other sciences. SpiNNaker has also been recently used to control a robot, called the SpOmnibot. The robot interprets visual information in real-time and navigates towards certain objects while ignoring others using functionalities provided by SpiNNaker.

(ii) DARPA UPSIDE/Cortical processor study

The Defence Advanced Research Projects Agency (DARPA) [20] commenced developing so-called 'bio-inspired algorithms' for machine learning. Their main application was for digital image processing using principles from deep learning. Their work presents advancements in two aspects, one on the

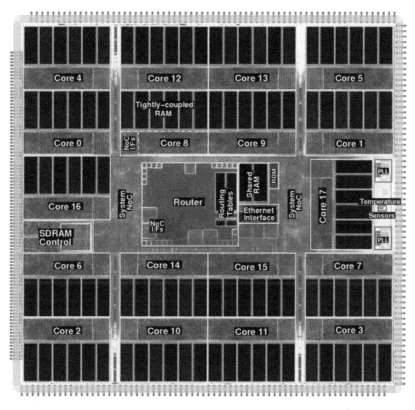

Figure 8.18. Architecture of the 18-core SpiNNaker chips (Cortical Architecture) [19]. Image courtesy of The University of Manchester.

software side and the other on the hardware side. Through their approach, DARPA was successful in exploiting the physics of emerging devices to perform extremely fast and low-power computation. Preprocessing, segmentation, tracking and classification of images was achieved at a rate 100 times better than conventional approaches. The power efficiency was also much better, with their hardware being 1000 times more power efficient than normal systems [20]. For sophisticated Department of Defence (DoD)-related applications, DARPA contributed to the Unconventional Processing Signals for Data Exploitation (UPSIDE) research program. The UPSIDE project generally consists of three tasks [20]. UPSIDE Task 1 consists of image-processing pipeline preparation and inference module development. UPSIDE Task 2 consists of implementation using mixed CMOS computational models. UPSIDE Task 3 consists of demonstrating image processing using next-generation devices with proper computer modelling. A block diagram of the DARPA UPSIDE image-processing system is illustrated in figure 8.19.

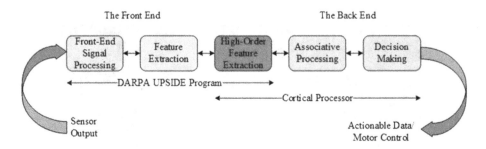

Figure 8.19. Block diagram for image processing using DARPA UPSIDE program. Adapted from [20]. Image stated to be in public domain.

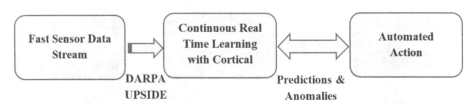

Figure 8.20. DARPA cortical processor concept. Adapted from [20]. Image stated to be in public domain.

Siance the algorithms used by **DARPA** are inspired by biology, the hardware has to be designed in a manner which closely resembles the human body, more specifically the cortex of the human brain. As ascertained in the previous secftion, it is difficult to mimic brain comprehensively, but enough knowledge with which to abstract some biological principles is available. This helps to modify machine-learning algorithms according to the needs of the application. The DoD is trying to create a novel computer architecture with suitable algorithms that can mimic specific human brain functions, including learning and pattern recognition, as well as addressing data recognition control and its challenges. Today, the most common approach is to use algorithms that are handcrafted and application specific. However, they require long computing time and high precision, which limits their ability to learn large datasets rapidly and does not work well in real-time applications [20].

A new approach is to use neural algorithms inspired by the brain that utilise a low-precision, temporal and hierarchical memory structure. These algorithms evolve with changing data and require shorter training times. However, only using better algorithms is not enough: the hardware has to be improved as well to meet require-ments. As seen in figure 8.20, a cortical processor is required for continuous real-time learning. High-performance and lower-power algorithms can be created using optimised silicon architectures for real-time applications. Reconfigurability, higher-connectivity, better local memory and parameter storage, and simple and low-precision computation are some of their advantages. A cortical architecture can boost complex embedded system operations as well as large-scale applications. Mapping cortical algorithms to hardware is of utmost significance [20]; that is, bio-inspired learning algorithms require matched hardware. Conventional processors are not

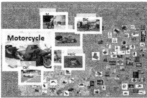

Figure 8.21. Object detection experiment by DARPA using cortical algorithm and architecture. (a) New object detection, (b) fixed training data for the algorithm, and (c) adaptive-learning environment for the algorithm. Adapted from [20]. Image stated to be in public domain.

suitable for cortical algorithms: they provide limited parallelism and have constrained processor/memory partition. Such processors consist of complex instruction sets, which make it difficult to build good programs for them. On the other hand, custom architectures are more suitable for bio-inspired approaches. For neuro architecture and computation, optimised versions of conventional CMOS fabrication has been suggested. They also eliminate the need to use high-risk components. While exploring the possibilities of cortical processor hardware, DARPA testing demonstrated the efficacy of cortical architecture when applied to object detection. One such example is shown in figure 8.21, in which the object used for detection is a motorcycle [20].

In their testing, the cortical processor exhibited much better performance. It runs in real time and exhibits online, on the go adaptation and learning. As figure 8.21(c) depicts, there is no problem when adding new objects, and they can be added continuously. The classification result comes out to be fixed-point single precision, and all nodes do not need to be active all the time. Hence, it can save a lot of power and demonstrate better efficiency, as well. It is possible to process 1000 images per second, with the processing of images taking just 0.0004 Joules per image. This is significantly lower than deep-learning approaches. Thus, in some respects, DARPA has overcome the challenges of Moore's law and Dennard scaling [20].

In summary, the functionality and performance of today's computing systems are increasingly dependent on the characteristics of the memory subsystem. There are many novel emerging memory devices and architectures, which however face challenges with regards to process compatibility, manufacturing, yield, performance variability and reliability. Although it is expected that, some day, the emerging memory architectures will replace existing memory technology, the timeline for this is not entirely clear. Emerging architectures are based on unconventional physical mechanisms, each of which offers different benefits. Their usage and maturity will be completely driven by the ability of industry to extract their benefits.

Questions

1. What architecture do present computing systems employ? Does it have any limitations? Why?
2. What are the considerations in designing an efficient architecture for future computing systems?
   Is it possible with present state-of-the-art technology?

3. Why are QCAs currently being investigated as an alternative to CMOS VLSI? With an example of a digital logic, compare the efficiency of QCAs in terms of area, delay and power to the CMOS counterpart.
4. What is the future of computer architecture?
5. What is cortical architecture? Is it feasible to develop this architecture using today's technology?

References

[1] Zhirnov L, Cavin R and Menzel S *et al* 2010 Memory devices: energy-space-time trade-offs *Proc. IEEE* **98** 2185–200

[2] Semiconductor Industry Association 2005 *International Technology Roadmap for Semiconductors (ITRS)* (https://semiconductors.org/resources/2005-international-technology-roadmap-for-semiconductors-itrs)

[3] Institute of Electrical and Electronics Engineers 2021 *International Roadmap for Devices and Systems (IRDS): Beyond CMOS* (irds.ieee.org)

[4] Semiconductor Industry Association 2011 *International Technology Roadmap for Semiconductors (ITRS)* (https://semiconductors.org/resources/2011-international-technology-roadmap-for-semiconductors-itrs)

[5] Definition: storage class memory (SCM), Techtarget https://www.techtarget.com/searchstorage/definition/storage-class-memory

[6] Semiconductor Industry Association 2007 *International Technology Roadmap for Semiconductors (ITRS)* (https://semiconductors.org/resources/2007-international-technology-roadmap-for-semiconductors-itrs)

[7] Meena J S, Sze S M, Chand U and Tseng Y T 2014 Overview of emerging non-volatile memory technologies *Nano Res. Lett.* **9** 526

[8] Covi E, Brivio S and Serb A *et al* 2016 HfO$_2$-based memristors for neuromorphic applications *2016 IEEE Int. Symp. on Circuits and Systems (ISCAS) (Montreal, QC, Canada)* pp 393–6

[9] Mead C 1990 Neuromorphic electronic systems *Proc. IEEE* **78** 1629–36

[10] Neuromorphics *Neuromorphics* Purdue Polytechnic Institute (https://polytechnic.purdue.edu/facilities/lobe/research/organic-neuromorphics)

[11] Cognitive Computer, Wikipedia (https://en.wikipedia.org/wiki/Cognitive_computer)

[12] Held J 2017 Leading the evolution of computing: neuromorphic and quantum computing *HPC Developer Conf. 2017*

[13] Laajimi R 2018 Nanoarchitecture of quantum-dot cellular automata (QCA) using small area for digital circuits *Advanced Electronic Circuits - Principles, Architectures and Applications on Emerging Technologies* ed M Niu (London: IntechOpen) ch 3

[14] Sridharan K and Vikramkumar P 2015 *Design of Arithmetic Circuits in Quantum Dot Cellular Automata Nanotechnology* (Cham: Springer International Publishing) Studies in Computational Intelligence vol 599

[15] Bernstein G, Amlani I, Orlov A, Lent C and Snider G 1999 Observation of switching in quantum-dot cellular automata cell nanotechnology *Nanotechnology* **10** 166–73

[16] Begum Y, Balaji M and Satyanarayana V 2022 Quantum dot cellular automata using a one-bit comparator for QCA gates *Proc. Materials Today* **66** 3539–46

[17] Pannase N P and Boke A 2016 Random number generator design using QCA logic for FPGA architecture *Int. J. Adv. Res. Comp. Commun. Eng.* **5** 432–6

[18] Wikipedia 2019 SpiNNaker (https://en.wikipedia.org/w/index.php?title=SpiNNaker& oldid=920391451)

[19] Advanced Processor Technologies Research Group 2023 SpiNNaker Home Page (http://apt.cs.manchester.ac.uk/projects/SpiNNaker/)

[20] Hammerstrom D 2023 UPSIDE/Cortical Processor Study (https://rebootingcomputing.ieee.org/images/files/pdf/RCS4HammerstromThu515.pdf)

IOP Publishing

Nanoelectronics
Physics, technology and applications
Rutu Parekh and Rasika Dhavse

Chapter 9

Nanosensors and transducers

Nanomaterials, with their extremely small size, accomplish explicit properties, thus greatly extending the materials science in different fields. The Universe works using the same laws as nanotechnology. The industry surrounding sensors technology today is not exceptional; nanosensors have been under research by numerous organisations for more than a decade. In this, various innovative contributions have been made to nanosensors for various applications. Although there are many present difficulties in their mass production, various approaches to their production have been already demonstrated. The day is surely not far away when nanosensors will see wide adoption and commercialisation. Nanosensors, due to their many virtues presented in this chapter, represent an advancement over current-generation technology.

9.1 Introduction to sensors science and technology

The process of detecting and perceiving various physical quantities is known as sensing. In everyday life, we encounter numerous activities in which sensors are used. More formally, the American National Standards Institute (ANSI) defines a sensor as 'a system [that] provides a useful output in response to a specific measurand, which can be a physical quantity, property, or event which is being measured' [1]. Here, what is meant by a useful output is an 'electrical quantity', which can be an electrical current, voltage, field, etc. A transducer is an electrical device that is utilised to change one type of energy into another. The words sensors and transducers have often been used interchangeably. The term transducer was preferred to sensor by the ANSI standard. However, the ANSI definition has not been widely adopted, and due to this sensor is the most commonly used term currently. It should be noted that the given definition does not define the general physical components of a sensor. In order to be more specific, terms such as sensing element, sensor and sensor systems are frequently used. The human organism itself has sensory organs like eyes, ears, tongue, the nose and skin. Further, electronic, mechanical, thermal, chemical and optical sensors are used extensively on a regular basis in cars, smartphones, consumer electronics items, energy equipment, home automation, portable healthcare equipment and many other personal gadgets and public systems. Figure 9.1 illustrates a more precise definition of a sensor based on its different physical components [1], several key terms of which are elaborated upon below:

- Sensing element(s): this component is the fundamental material of a sensor that converts one form of energy into another form. The sensor may have more than one type of sensing element. We define such a sensor as a compound sensor.
- Sensor: sensing elements with their physical packaging and external connection (s) constitutes the overall sensor component.
- Sensor system: this component involves a sensor and its associated signal-processing hardware, which can be analogue or digital and integrated in the same package or discrete from the sensor.

Various commercially available sensors include thermal sensors, colour sensors, alcohol sensors, gas sensors, smoke sensors, humidity sensors, touch sensors, infrared (IR) sensors, heartbeat sensors, glucose sensors, proximity sensors, etc. The list is endless. Some typical performance metrics for sensors and transducers are given below [2]:

(i) Range: the maximum and minimum value range of the parameter over which a sensor functions correctly. For example, a resistance temperature detector (or RTD) for the measurement of temperature has a range of $-200°C$ to $800°C$.

(ii) Accuracy: the error in measurement, defined as the deviation of the measured value from the true value. To extend this concept further,

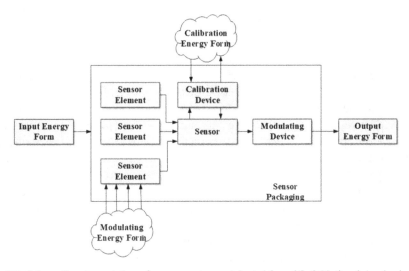

Figure 9.1. Schematic representation of sensor anatomy. Adapted from [1]. © National Academies Press.

absolute error is the measured value minus the true value, while relative error is the absolute value divided by the true value.

(iii) Precision: the similarity between successive readings while measuring under the same conditions. The values need not be close to the true value. However, when we say that the readings are accurate, it means all the readings are close to the true value and we can further say that the measurement is precise.

(iv) Sensitivity: the ratio of change in the output parameter to change in the input parameter.

(v) Resolution: the minimum change in the input parameter that can be detected by the sensor.

(vi) Linearity: a sensor is expected to behave the same throughout the practical range of the input. This is known as linear behaviour of the sensor. Linearity is the maximum deviation between the measured values of a sensor from an ideal curve, which is a specified straight line.

(vii) Hysteresis: an increasing or decreasing trend of the input sometimes affects the response of the sensor. Thus, the deviation in the sensor response during different trajectories of the input is termed as hysteresis. Ideally, it should be zero. However, engineers have found ways to utilise the hysteresis property of sensors.

(viii) Repeatability: the ability of a sensor consistently to produce the same output with the same input and physical and measurement conditions.

(ix) Response time: the time in which the output reaches a certain percentage of its terminating value.

Current sensor technology faces many issues and challenges. When a sensor interacts with the same environmental conditions, ideally it is supposed to react the same way every time. However, it is often the case that the sensing materials age with time, or are fouled by contaminants in the environment. Generally, current sensor system's response characteristics drift and degrade over time and thus are reduced in accuracy. Future nanosensors may solve this problem by using self-correcting materials or through the ability of the sensors to internally correct calibration drift at the least [3]. Further, as the surface-to-volume ratios for nanoscale materials are much higher than the macroscopic-scale sensors currently in use, their surface reactivity increases, and as a consequence nanosensors will be able to provide better sensitivity. Hence, in situations where the analyte is present in low concentrations, usually large samples are taken for proper measurement [3]. Secondly, integration of various components into sensor systems will be a significant technical challenge because, regardless of the quality, if a sensor is not integrable with the rest of the system, then the system will not perform satisfactorily [3]. It is no exaggeration to say that the field of sensor technology will be revolutionised by incorporating nanotechnology. All the sensor performance metrics, along with the power, performance, area and unit cost of the sensor and the interfacing electronics, can be improved manifold by scaling them to the nanolevel. Nanosensors can identify either tiny particles of an analyte or minimal traces of an external event: they can detect the presence of chemicals and nanoparticles in a given environment, or physical parameters such as temperature or pressure at the nanoscale. They boast small size, light weight and a huge reactive surface area. Thus, nanosensors are distinguished both by their physical characteristics and their sensing capabilities. Nanosensors find a special range of applications in medical diagnostics, food and water quality assessment, security and biometrics, integrated circuits (ICs), renewable energy sources, display technology, wireless sensor networks, etc [4]. Another advantage, working at the nanoscale, is that various emerging nanostructures such as quantum dots, quantum wires, quantum wells, films, coatings, monolayers, nanowires, porous materials, carbon nanotubes (CNTs), cantilever beams, diaphragms, solar cells, etc. can be used for sensing based on a variety of mechanisms. The response times of nanosensors are very small, making them the preferred tool over other experimental techniques currently in research. In this chapter, we give an overview of some generally popular nanosensors and transducers.

9.2 Nanosensors and transducers in food industry, healthcare and defence

The utility of nanosensors, in a practical sense, does not really have any limit. Including all of them in this book would be an impossible task. As such, we restrict ourselves to some of the most promising domains where consistently positive outcomes have been reported by numerous researchers. It is observed that the food industry, healthcare systems and military equipment in particular are finding the use of nanosensors and transducers to be vital. Typical applications of nanosensors and transducers include identification or detection of various chemicals in gases to detect pollution; medicinal diagnostics, wherein blood-borne sensors or lab-on-a-chip-type devices are used; monitoring physical parameters, such as in the accelerometers used in microelectromechanical (MEMS) devices like airbag sensors; monitoring plant signalling and metabolism to understand plant biology; and in the study of neurotransmitters in the brain to understand neurophysics.

9.2.1 Nanosensors and transducers in the food industry

The most important thing for human life is food. With the globalisation of the food supply chain, food quality and its safety have become primary concerns for human health. Many global food chains supply the same foodstuff worldwide, in which ingredients are procured at their source locations and then distributed around the world to local stores. To this end, many companies sell food that has been packaged. What, therefore, if the ingredients become inedible or unsafe during this process? To address this issue, food quality needs to be checked at all stages of the food supply chain, that is, its production, processing, packaging, marketing and distribution.

Food standards also have been increased following the globalisation of food. So, it is necessary for every food supply chain to check quality at every step of food production. Food safety can be quantitatively evaluated by different techniques, which include fine instrumental investigation and cell culture at the laboratory level. The primary disadvantages of those techniques is their long investigation times, ranging from few hours to days, ordinarily with various necessary pretreatment steps. These disadvantages, plus other novel difficulties in the food sector, have led to the development of new and fast analytical techniques. In this, nanotechnology incorporated with diagnostic devices presents a crucial solution for the advancement of new gadgets [5]. With the help of nanosensors and nanotechnology, it is possible to check food quality and safety quickly and in an efficient manner compared to biological and chemical methods. Currently, nanosensors are used only in food packaging and transport, and to detect impurities in food; other applications of nanosensors are still at the research level. Nanosensors as used in food analyses combine knowledge from chemistry, biology and nanotechnology, and are referred to as nanobiosensors. In modern life, ordinary sensory exposure to and the ability to check food is prevented by the food packaging, so consumers have to depend on the expiry dates given by the food producer, which are dependent on a set of ideal assumptions about food transportation methods and food storage. But what if these storage or transport conditions are violated before the packaged food reaches the consumer? Clearly, the quality of the food will deteriorate, which the consumer will not discover until the food package is opened or consumed. This disadvantage in food packaging can be solved by nanosensors through their unique electro-optical and chemical properties. Nanosensors can also detect pathogens, chemical contaminants, aromas, gases and even environmental conditions. Use of nanosensors thus ensures that consumers can purchase fresh and delicious food, as well as reducing the risk of food poisoning, which improves overall food safety [5]. To fulfill these tasks, a class of biosensor, referred to as aptasensors (which combine aptamers and nanomaterials) is used. An overview is shown in figure 9.2.

Aptamers are molecules of nucleic acid (DNA or RNA) of approximately 25–40 kDa dimension. Aptamers are target-specific elements, meaning they are highly selective and specific towards their target such as viruses, proteins, microns, toxins, ions, etc. because of their highly tailored structures. DNA aptamers are often robust and can be synthesised with a high grade of reproducibility and purity. They also make biosensor fabrication process easy. In contrast, RNA aptamer-based biosensors, which are easily degraded by endoribonuclease present in biological surroundings, are usually used for one-shot detection. Aptamers can be based on optical, electrochemical and mass transduction label-based or label-free techniques. Nanomaterials are used for signal enhancers or signal transducers. There are many different types of nanomaterials that can be used in aptasensors such as magnetic nanoparticles, carbon nanoparticles, semiconductor nanoparticles, metal nanoparticles, etc. The use of aptasensors is dependent on the kind of nanoparticles used to produce them. Aptasensors are classified into electrochemical and optical systems based on the detection method [6].

A numbers of nanosensors have been developed for the food industry, either for integration into packaging as a nanoscale tracer to show the history of food production or to recognise risks and their tolerance in the event of alleged food poisoning at any time. For instance, nanosensors in food packaging can be made to change colour when certain microorganisms grow beyond a threshold value. This can be used to track storage conditions to prevent food poisoning. In order to detect

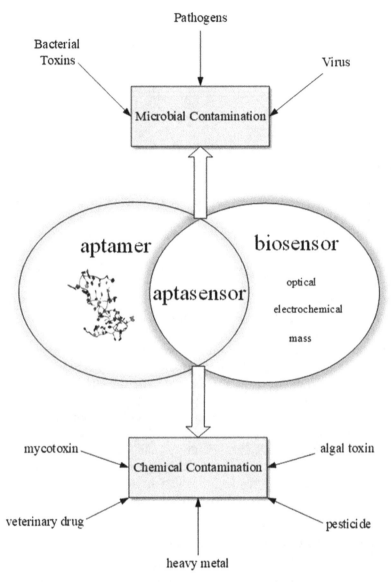

Figure 9.2. Overview of aptasensors.

chemicals like pesticides, gold nanoparticles with appropriate binder coatings can be used. In addition, nanosensors using Raman spectroscopy are ideal for food forensics, wherein a food's origin, contamination and adulteration are investigated. This highlights the unique strengths of nanosensor applications and methods, which enable various analyses of macrofood, preservatives, pigments, carbohydrates, proteins, dyes, etc. Nanosensors can provide quality assurance by detecting toxins, contamination and microorganisms throughout the food-processing chain, using data capture for automated control functions and

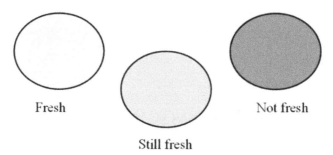

Figure 9.3. Example of a change in the colour of nanosensors in response to food spoilage.

documentation, among other uses; further, nanosensors to detect aqueous toxins are quite close to commercialisation [5].

As mentioned, nanosensors can communicate with food and its environment inside packaging and give an indication of the status of the food. They can respond to changes in the humidity, temperature and level of oxygen exposure. An illustration of the principle of a nanosensor that changes colour according to environmental changes and the condition of the food is given in figure 9.3 [7].

Oxygen detection in food packages using nano-sized particles such as SnO_2 and TiO_2 and methylene blue (redox dye) has also been developed using highly promising photo-activated indicator ink. As in the previous example, when nano-sensors come into contact with oxygen, they can be designed to change colour. This method is also used to detect leaks in packaging after production [5]. Smart sensor technology can be used to monitor grain quality, dairy products, fruits and vegetables in a storage and the sources and extent of spoilage, if any. Nano encapsulated flavour enhancers, nanotubes and nanoparticles can be used as viscosifying agents, and so on. In agriculture, nanosensors can be used to sense soil conditions like pH value, moisture, crop growth, heavy metals, etc., and environmental conditions like temperature and humidity. Nanoparticles can also be used to spray and disperse pesticides and fertilisers in a controlled way.

9.2.2 Nanosensors and transducers in healthcare

The healthcare sphere is probably the most likely to benefit from nanosensors and transducers. The intrinsic complexity of the human body, critical diseases and their sequelae, medical treatments with undesirable side effects, etc., often complicate the course of a patient's healthcare. Often, more research is undertaken to reduce the side effects of a strong medicine than into the medicine itself. As a further issue, the symptoms of certain diseases only manifest after a certain time, by which time the condition of the patient has progressed to a critical stage, making treatment complicated that could have been potentially easier had the disease been discovered earlier. For example, pancreatic cancer is typically only recognised after it has spread to other body parts. The same is true with a hip implant; when it is found to be infected, the only way to deal with it is to replace it. Nanotechnology can contribute to this field immensely as materials and structures behave altogether differently when it comes to the nanoscale, in that their physical, chemical and biological reactions can differ remarkably in contrast to bulk counterparts. The sensing capacity of nanosensors is so high that they can sense low concentrations of harmful viruses, for instance. Hence, healthcare diagnostics, monitoring and healing in a 'nano' way has huge potential to improve outcomes. The potential advantages of nanosensors can be leveraged if they are used and improved upon regularly. The purpose of these nanosensors is to gather information at a nuclear scale before transferring it for inspection. Some of the applications in the medical field are mentioned below.

(a) **pH sensing**

Fluorescent sensors contain at least one binding substrate that absorbs and releases some amount of energy when light is thrown onto it. According to any mismatches in the absorbed light and emitted light, a particular molecule can be sensed. The first nanosensor was a fluorescent sensor that measured pH value, for which polyacrylamide particles were used. The given phenomenon can be explained using the example of a receptor and an analyte, as illustrated in figure 9.4, where a receiver response for an analyte is defined by the change in the detected colour. These nanosensors can also be used for glucose monitoring in the body, by exploiting the use of nanoparticles on the skin [8].

Nanowire field-effect transistor (or NWFET) devices can also be used to detect particular species from liquid solution. The same concept is used in pH sensing (the species here is a H^+ ion) [9]. A p-type silicon nanowire can be modified such that its surface charge density becomes sensitive to the concentration of H^+ ions. Changes in the device conductance can be monitored and calibrated to obtain an accurate, linear pH sensor. The higher the conductance (i.e. the lower the surface charge), the higher the pH value.

(b) **Glucose monitoring**

Diabetes is one of the most prevalent diseases in the world. To aid with this, nanosensors offer a new solution to measure glucose. Nanotubes are one of the potential candidate to build such sensors due to their high surface-area-to-volume ratio. At the nanoscale, characteristics like quantum-mechanical

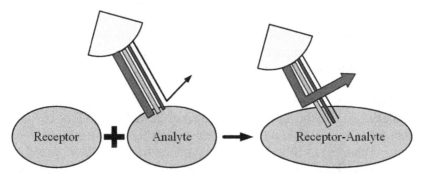

Figure 9.4. Conceptual schematic of luminance dye for intracellular sensing. Adapted from [8]. CC BY 4.0.

changes, optical properties and altered reactivity becomes prominent. The higher surface area improves catalytic activities, which in turn enhances the sensor's sensitivity, signal-to-noise ratio (SNR) and selectivity in measurement. During fabrication, specific substances can be introduced into these nanotubes at high temperature to react with the glucose. The optical properties can change and the fluorescent signal can be measured, which reflects the glucose level. Such sensors are transplanted under the skin and irradiated with a laser. Single-walled nanotubes (SWNTs) that have a large energy band gap absorb greater energy and hence more fluorescence. In addition to this, other options consist of biocompatible polymeric nano-sensors implanted under the skin and the use of quantum dots [9].

(c) **Detection of viruses and bacteria**

Although there are several existing methods to detect viruses and bacteria, like immunological assays, polymerase chain reaction (or PCR)-based testing and transmission electron microscopy, they do not show rapid detection if there are few viruses in the sample and they also require some manipulation (or preparation) of the sample. Modified nanowire arrays with influenza A can be used to achieve real-time virus detection. When a virus particle binds to an antibody receptor, the level of conductance changes from a baseline value; when it unbinds, it returns to the baseline value. Using this mechanism, a virus can be detected rapidly [8]. The quick detection of pathogenic bacteria is extremely important in the field of medicine to cure disease. The problem in the detection of bacteria is that existing methods give delayed results and also lack fine sensitivity. Thus, the need for nanosensors arises. Nanoparticles can be easily used to detect molecules such as antibodies due to their high fluorescence. A silicon nitride cantilever can be used to detect *Salmonella* bacteria by the change in surface stress. In the sensing of phage-triggered ion cascade (or SEPTIC) method, a nanowell device with two electrodes is used to identify bacteria through alternations in the electric field [8].

(d) **Asthma detection**

Nanobiosensors can be used to detect asthma attacks. They can detect an attack up to 3 weeks prior by using a hand-held tool to check the level of nitric oxide within the patients' breath. Patients can thus be warned if their levels are too high, or rising. This indicates the possibility of oncoming asthma attacks. A polymer-coated carbon nanotube field-effect transistor (or CNTFET) forms the base of the sensor and contains an arbitrary network of SWNTs between the source and drain [10].

(e) **Astronaut health and safety**

Astronauts spend large periods of time in space and work in space vehicles. Astronauts are also exposed to deadly amounts of potentially dangerous gases. Usually, hydrazine and similar gases are used as fuel by space vehicles. Even at a very low concentrations, these gases are very dangerous. Therefore, the detection and identification of these dangerous gases are very important. To detect these gases in real-time, nanosensors are used. Any early infection or damage due to radiation to astronauts can be observed by sensing biochemical changes in the body. This can be done using biosensors derived from synthetic polymers called dendrines. Their size is less than 5 nm and are created layer by layer. They are introduced though skin membranes to the white blood cells lymphocytes. The development of this type of sensor can also lead to the removal of the need for blood sampling and testing during space missions [11].

(f) **Nanorobotics in medicine**

Nanorobots can help protect our body from pathogens. A nanorobot is a type of nanodevice with many components such as actuators, controllers, sensors, etc. Surgical nanorobots inserted into the body can act as an on-site surgeon in the body, which are programmed or controlled by a human doctor. They can perform operations like finding pathogens, and diagnosing and conveying messages to the supervising doctor or medical staff through ultrasound signals [12]. Medical nanorobots can be used to monitor cells and microorganisms, for testing and diagnosis in the blood. They have the capability to note and report irregularities in immune system parameters such as pressure and the temperature in various body parts [12]. Nanorobots can also be used in the treatment of genetic diseases. This is done by relating proteins and DNA structure in the cell. The irregularity is thus corrected in the DNA and protein sequence. In place of cell correction, a chromosomal replacement can be efficiently used, which uses in-built repair vessels from the body to maintain genetics. The information stored in the database of a nanocomputer is put outside the nucleus for the purpose of comparison [12]. Nanorobots containing chemical biosensors can be used to measure epithelial cadherin signals' intensity for recognising tumours in their initial stage. They can be also used for drug delivery to obviate the side effects of chemotherapy [12]. Further, nanorobots can help efficiently in addressing dentistry problems. They can be helpful in repairing the tooth, fixing its appearance and in increasing its durability, as well as in oral anaesthesia and treating issues of tooth sensitivity [12].

(g) **Drug discovery and delivery**

The ability to discover drugs in human blood and their pharmaceutical formulation is very important for medical and pharmaceutical science. High selectivity, low sensitivity and minimal interference effects are needed for successful analysis [13]. Biosensors provide many benefits such as real-time analysis, cheap instrumentation, high sensitivity and short analysis time. Biosensors are categorised as a useful signal transduction mechanism and bioreceptor. The different classes of biosensors depending on signal transitions include optical, electrochemical and thermal. The most commonly used transducers for drug analysis are optical transducers and electrochemical transducers. Several biological components like antibodies and enzymes are used in the production of biosensors. The most used components for recognition are antibodies and aptamers, as in drug screening. Up until recently, several traditional techniques have been used for drug analysis [13]. These techniques offer a good range of detection but have some limitations, as well, such as lengthy analysis times, costly equipment and the need for skilled and experienced staff. Thus, low-cost analytical techniques using biosensors are in high demand, generated by interest in the field of drug science. Compared to traditional techniques, biosensors offer several unique advantages. For the task of drug determination, various biorecognition components and different electrode modification materials are used for the construction of the biosensors. Presently, many biosensor construction methods are being developed to amplify the performance of biosensors. In the years to come, advances in technology may provide low-price, selective and sensitive biosensors for drug analysis [13].

(h) **Remote-technology-based nanobiosensors**

Technologies for remote sensing are well developed, but applications of these technologies that utilise the noninvasive capabilities of wearable bioinstruments with wireless transmission have just began to appear in recent years. Among these, ring sensors are a very innovative example. A ring sensor is a wearable sensor that is used for observation and monitoring of essential signals. While measuring waveforms for blood volume and saturation of blood oxygen from a finger, it minimises motion artifacts. Ring sensors do not require any kind of slip be placed on the finger of the patient to observe key important signs like pulse rate. A stream of real-time data is fed to the operating system of the sensor to monitor the waveforms. If there is a reasonable amount of deviation in the waveforms, then the ring sensor senses it and can immediately alert a medical professional via mobile or personal digital assistant (or PDA) device by sending a warning [14].

(i) **Intelligent Healthcare Data Management Systems**

In the past few years, the rate of technological development has been very high. In this general progress, several healthcare-based technologies have been developed based on smartphones, nanosensors, smartwatches and many more. An Intelligent Healthcare Data Management System (IHDMS) is based on nanosensors. It permits a patient or sick person to use healthcare services like diagnostic services, monitoring services and emergency management services from any location and at any time. There are three components to a IHDMS [15]. The first component comprises wearable sensors, worn by the patient. Every nanosensor can identify, represent and process one or more signals, i.e. brain electrical activity monitoring by an electroencephalogram (or EEG) nanosensor, muscle activity observation by an electromyogram (or EMG) nanosensor, heart rate observation by an electrocardiogram (or ECG) nanosensor, blood pressure recording by a blood pressure nanosensor, etc. The second component includes a personal application (or app) that runs on a smartphone. This app accounts for significant commitment. The application is utilised as an interactive interface between the patient's health data, gathered using sensors, and a remote healthcare server to which the patient's data can be transferred. The third component involves the healthcare server or medical server. This service is accessed via the Internet and is also able to contact other servers such as those of emergency services, healthcare providers and informal caregivers. There is thus a communication channel between the patient's app and the healthcare server. The server collects data from the app and consolidates it into the medical records of the patient. If reports appear to indicate an inconsistent state, the service may allow suggestions or notifications to be pushed to the user [15].

9.2.3 Application of nanosensors and transducers in defence

The present integration of nanosensors and defence into the new realm of 'nano-defence' is expected to deliver advancements in broader areas that will change militaries and play an important role in maintaining national security. There is a belief that nanotechnology can be utilised in two vital ways by military personnel. The first is the scaling down of existing gadgets to enable it to be smaller, lighter, more energy efficient and functionally prompter. The second is to create and adjust new materials for military purposes. Current nanosensing applications in the military use surface coating, nanomaterials and fabrication methods to increase desired capabilities. For example, there are possibilities to detoxify poisonous regions; to detect the beginning of infection in a region presented to military medics; to detect poisons or radioactive material; to secure electronic, data and correspondence systems; to protect human lives and troop safety through the use of nanotextures and related materials; to spy and accumulate intelligence both more accurately and discreetly, etc [16]. In the coming years, nanoscience, nanodesign and nanotechnology will likewise provide lighter, more proficient and increasingly viable military apparatuses; nano-robots for nanoscale devices and frameworks; invisible suits based on metamaterials that are both lighter and tougher; adaptive nanoscale sensors for brain and body sensing; virtual tracking systems for nanoinformation hardware [16]; as well as many other advancements in the defence realm. Some example applications are illustrated below.

(a) **Nanomechanical cantilever sensors in defence**

Microcantilever sensors have numerous applications in the locating of different analytes in a fluid or vapour. These sensors are characterised by their high affectability, need for minimal stimuli, quick reaction and extra-ordinary unambiguity. They detect a biochemical response occurring on a surface by approximating a mechanical bending-like event [14].

Explosives such as trinitrotoluene (TNT), dinitrotoluene, pentaerythritol tetranitrate and hexahydro-1,3,5-trinitroazine (RDX) are all substances that pose a threat to public safety. Further, their detection is very hard because they contain a complicated mixture of chemicals, and are revealed only at low air pressures. To address this, by covering a sensor surface with metallic self-assembled monolayers (SAMs), it can detect explosive substances. To detect the particular components of an explosive, cantilever arrays can be utilised. As they feature many reversible receptors, they can detect if any volatile component is present, even in small quantities. To remove any kind of noise from the environment, instead of using a single cantilever beam an array is set up, as shown in figures 9.5(a) and (b), respectively, which further improves the reliability of the system.

In a second approach, the microcantilever beam's bimetallic effect is used, which probes the IR spectrum of a given substance on the cantilever's surface. It was demonstrated that several nanograms of RDX and TNT on a gold-coated cantilever sensor show a response to 3–5 μm of IR illumination and induce a characteristic bending spectrum, as shown in figure 9.6.

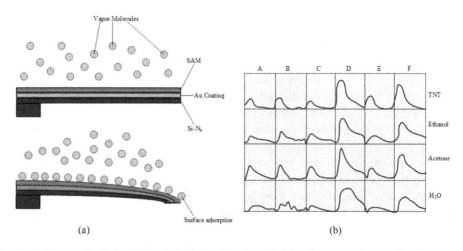

Figure 9.5. Nanomechanical cantilever in explosive detection. (a) A bending-mechanism cantilever sensor, (b) the unique response when an array of six microcantilevers is exposed to water vapour, acetone, ethanol and TNT. [14], adapted by permission of the publisher (Taylor & Francis Ltd, http://www.tandfonline.com).

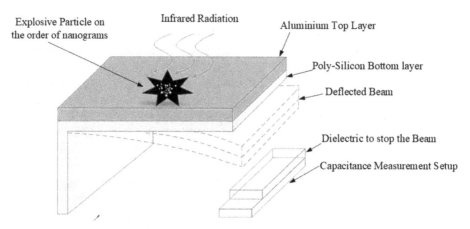

Figure 9.6. Schematic of a microcantilever-based detector to trace explosive particles. [14], adapted by permission of the publisher (Taylor & Francis Ltd, http://www.tandfonline.com).

(b) **Communications in defence**

Currently, it is not clear how novel nanosensors will communicate with each other. There are two ways to aid communication at the nanoscale: molecular communication and nanoelectromagnetic communication, as shown in figure 9.7. Molecular communication is referred to as the exchange of information contained in molecular form. Because of their small size and wide operating range, molecular transceivers will be simple to include into nano-devices. These transceivers have the ability to respond to particular molecules and release different molecules in response to internal commands or by following some sort of processing.

The propagation of the released molecules occurs either by active carriers that carry the molecules along predetermined paths, or through spontaneous diffusion of the molecules in a fluid media. However, novel channel models, network topologies, and communication protocols are required by this profoundly altered communication paradigm [17]. By contrast, nanoelectromagnetic communication is characterised as the transmission and gathering of electromagnetic radiation from parts dependent on novel nanomaterials. Ongoing advancements in molecular and carbon devices have opened a pathway to another age of electronic nanocomponents, e.g. nanobatteries, logical hardware and nanomemory at the nanoscale, and even nanoradio wires.

From a communications point of view, certain properties observed in novel nanomaterials will allow them to operate at particular bandwidths and to discharge electromagnetic radiation. Again, however, all these involve an essential change in the present state-of-the-art of logical channel models, network structures and communication protocols [17].

(c) **Chemical, biological, radiological and nuclear defence using nanosensors**

Today, in the world of technology, weapons development has advanced to an extremely dangerous level. The most dangerous are chemical, biological, nuclear and radiological weapons, which together are abbreviated as CBRN. In the field of CBRN weapons, the ongoing exponential improvement of nanosensors is driving the advancement of efficient instruments for the

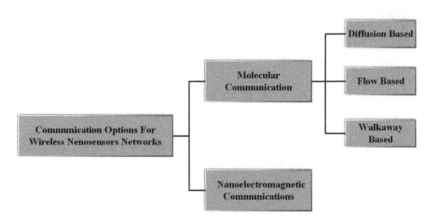

Figure 9.7. Nanoscale communications in defence.

recognition, observation and reduction of CBRN specialists. During the twentieth century, due to the improvement of safety devices, for example activated carbon gas masks, polymer-based securing suits and cleaning arrangements or powders, the potential effects of the use of chemical or biological weapons on the battlefield have been greatly reduced. Nevertheless, with a view towards the future, specific considerations need to be made to assure against new and unconventional weapons. To this end, nanosensors are helping to develop extraordinary strategies, instruments and hardware to neutralise efficiently and successfully the dangers posed by CBRN. In this way, if the pace of advancement in defence technology is high, then the dangers of weapons of mass destruction can be counteracted and the risk to people lessened [18].

Nanoparticles are perfect for heterogeneous catalysis. The uncovered dynamic sites of a solid catalyst must cooperate efficiently with substrate particles and must advance their change. Thus, most heterogeneous catalysts depend on nanoparticles, specifically inorganic oxide nanoparticles as well as metal nanoparticles. Nanosystems with reactant properties would thus be able to assume a significant role in anticipating or limiting the negative impacts on substance, organic and, to a lesser degree, radiological operators. Nano-sized reactant frameworks can find application not just in the location of but also in the security, purification and reduction steps during operations with a hazardous agent. Particularly in the expulsion and destruction of CBRN operators, nanocatalysts have shown great promise [18].

(d) **Military transportation**

Military and guard systems incorporate drones, rockets, shuttle, military vehicles, marine frameworks, ships, satellites and rockets. These frameworks work in the harshest of conditions during battle situations. Interior and outside security systems depend on smart and wise sensor innovation for observation, insight and tactical tasks. In this way, sensors have become a necessary component of military frameworks and in meeting the performance needs of every type of military vehicle, hardware and their related frameworks. Sensors are utilised in in-flight controls, impetus, monitoring the environment, weapon controls, indicators and communications, etc. Today, one can find sensors in various security frameworks, explosives-detection systems, chemical warfare, crime-detection systems, intrusion-recognition systems, and so on. Additionally, sensors are utilised in warzone surveillance systems. These can be deployed on land, in the air, in space and underwater, for concentrated monitoring of a particular zone. Electronic hardware including multiple sensors is a fundamental component of rocket design. With present-day innovation in sensor systems, missiles are considerably more advanced, accurate and powerful as compared to previous generations.

(e) **Futuristic nanosensor applications in military services**

Scientists are gaining confidence in metamaterials to make objects imperceptible. Metamaterials are man-made materials with properties that cannot be found in nature. An optical negative index material (or NIM) can curve light in a manner unlike traditional materials [19]. This can be used to manufacture an invisibility-cloak-like fabric with seemingly magical properties. Nanotechnology research has aimed to upgrade the characteristics of military attire and its various materials. Future smart uniforms will have nanosensors attached to them, and which will have modified properties to help heal injured soldiers. Further, such technology will make it easier to locate soldiers and to know their condition. Such suits will be able to perform ideal tasks in certain situations. For instance, if a fighter is injured and/or falls unconscious, it might be able to shut itself [19]. As another example, in space elevators the important element in the structure of the airship is the weight-to-power ratio. Making the plane lighter and smaller makes it far less expensive. CNTs show different electrical properties based on the angle or amount of bending, which can be tailored. CNT materials can help to scale down electronics, reduce structural mass and reduce power consumption due to atomically precise materials [19], and thus may lead to better airship weight-to-power ratio.

9.3 Metal nanoparticles and quantum-dots-based sensors

Metal nanoparticles are defined as sub-100 nm entities that are made using metals such as gold, cerium, platinum, zinc, silver, thallium, iron and titanium in their pure form, or in compound forms of hydroxides, oxides, sulphides, phosphates, chlorides and fluorides. Gold and silver metal nanoparticles are examples of single-element metal nanoparticles, with their most common uses being found in medical imagery, while metal oxides are widely used as stabilisers or in paint formulations. The large surface-area-to-volume ratio of nanoparticles make them strong in comparison to bulk materials, and they also exhibit higher surface energies. Nanoparticles have a specific electronic structure owing to the transition between molecular and metallic states, also known as the local density of states (or LDOS) [20]. Plasmon excitation, quantum excitement, short-range ordering, a higher number of kinks and an increased number of low-coordination sites like corners and edges lead them to have very high numbers of dangling bonds and the capability to store extra electrons. A range of metals can be used, but the industry typically prefers using gold metal nanoparticles for a number of reasons. These particles show exceptional properties at the nanoscale, although many applications for them existed years before this discovery. For example, gold is a rare element used in jewellery, coins, medicine as well as electronics. Chemically, gold is considered to be inert as it does not corrode and has a high electrical conductivity. As gold nanoparticles are reduced smaller and smaller in size, they become highly efficient for targeted drug delivery in medicine, and also act as a great catalyst in helping certain chemical reactions. They can also be substituted for iron nanoparticles as they are nontoxic [20]. Further, they can be functionalised easily to develop an affinity towards diseased regions of the body and to help in curing cancer tumours and other such diseases. Their optical properties vary as per their size and are interdependent on their electronic properties. One more special property that gold nanoparticles possess is that they can convert certain wavelengths of light into heat, because of which the free electrons in the metal can easily move throughout the metal. Thereafter, depending on the size and shape of the gold metal, the gold nanoparticles resonate with the correct wavelength on its surface. Gold nanorods and nanospheres are more efficient in all these functionalities. Gold colloids are widely used to put their optical properties to use in a number of analytical methods, as well as in numerical methods that are used for calculating extinction and scattering across cross sections and other related topics having random shapes as well.

9.3.1 Gold-nanoparticles-based biosensors

Owing to the aforementioned unique properties of gold nanoparticles such as their excellent biocompatibility, strong catalytic properties and high surface-to-volume ratio, they are used extensively in the construction of biosensors. The mechanisms of gold nanoparticles being very efficient, they can be used to improve the analytical performance of sensors. A biosensor based on gold nanoparticles can be designed to interact with an analyte within the body, and the physical/chemical change converted into a signal. Here, the transducing element recognises the analyte and, because of its contact with it, their interaction leads to some physical or chemical change. This change is then represented in the form of a signal [21]. The history of biosensors dates back to 1962. They function using two basic principles: first, the elements that participate in biosensing are actually biological structures; and, second, the sensors measure these biological or physical processes. Gold-nano-particle-based sensors are highly sensitive in comparison to other conventional biosensors. Essentially, three basic types of biosensor design can be considered: optical biosensors, electrochemical biosensors and piezoelectric biosensors. In these biosensors, different working principles are utilised for the nanoparticles wherein the gold nanoparticles either change their optical properties, electrical characteristics or undergo a change in their mass, respectively.

Optical biosensors are used to measure a change in light output. Surface plasmon resonance (SPR) is the most widely used technique for optical sensing. In SPR, an optical phenomenon takes place because of the interaction between the free conducting electrons and an electromagnetic wave; this is further explained in a later part of this section. A coherent oscillation takes place owing to the electro-magnetic radiation, which further excites the electrons present on the surface of the conductor. As a result, a change in the laser light from the surface of the metal film after reflection as a change in the dielectric constant is induced [21].

Figure 9.8 shows a schematic of a SPR detection unit. The SPR signals are amplified by the gold nanoparticles; it has been proved experimentally that the sensitivity of an optical fiber coated with a layer of gold nanoparticles is very high as compared to one without any coating. A number of researchers have devised new techniques to correlate the peak intensity along with the position of the gold nanoparticles' SPR with the surrounding medium's refractive index. This procedure is known as as localised surface plasmon resonance (or LSPR) [21].

Figure 9.9 shows the signal amplification mechanism of gold nanoparticles in a biosensor. The bioreceptors and analytes immobilise the gold nanoparticles. Owing

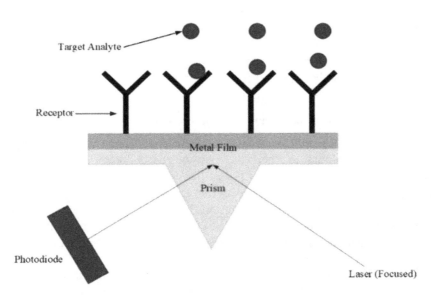

Figure 9.8. Schematic of surface plasmon resonance detection unit. Adapted from [21]. Copyright (2010), with permission from Springer Nature.

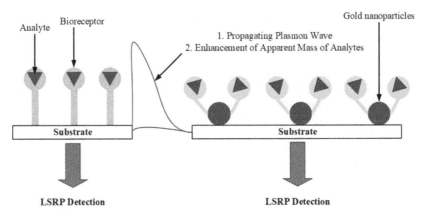

Figure 9.9. Propagating plasmon wave where the gold nanoparticles, bioreceptors and analytes interact with each other. Adapted from [21]. Copyright (2010), with permission from Springer Nature.

to the high density and high molecular weight of these gold particles, the mass of the immobilised analytes on the surface is increased. These sensors can also be structured in the form of arrays to increase throughput screening of the interactions between the biomolecules.

Electrochemical sensors can be used to obtain useful electrical signals by converting the interactions between biological elements; these are referred to as electrochemical biosensors. For this, gold nanoparticles have a number of advantageous properties such as their larger area, better biocompatibility and conductivity; thus, they are widely used to analyse electrochemical signals within the body. Alternatively, materials like quartz, tourmaline, topaz, etc. are said to exhibit a piezoelectric effect. Piezoelectric biosensors are used to analyse the biological recognition processes taking place and thereafter by measuring changes in mass. A combination of gold nanoparticles with other materials is a key factor in the design of piezoelectric biosensors.

9.3.2 Metal-nanoparticles-based colorimetric sensors

At the nanoscale, gold particle solutions have a specific absorption spectrum and display different colours to their bulk counterparts based on their sizes and interparticle distances. This strategy of detection using gold nanoparticles is effective, easy and fast, and hence it is used in the detection of hazardous chemicals. Gold nanoparticles with particle distributions from 1 nm to 100 nm are referred to as colloidal gold. With alterations in the size of the particle and interparticle separation between the gold nanoparticles, the characteristic peak absorption wavelength shifts from ultraviolet (UV) to the visible spectrum due to SPR.

This mechanism facilitates its use in environmental sensing tasks. Figure 9.10 explains the process whereby gold nanoparticles can be used to detect target matter using its optical properties. Optical sensing technologies based on gold nanoparticles can be deployed by using the integration or disintegration of the gold nanoparticles with the targeted element. This will cause the solution of gold nanoparticles to change its colour from red wine to blue, corresponding to its surface plasmon property discussed earlier. This shifts the absorption spectrum related to this property from 523 nm to 610~670 nm. Optical sensors intensively use this SPR property of gold nanoparticles. It arises when there is vibration among the free electron, giving rise to a certain frequency, which in turn resonates with the frequency of the light that is incident on the solution. This causes the solution to shift the frequency spectrum of the maximum possible wavelength from one spectral

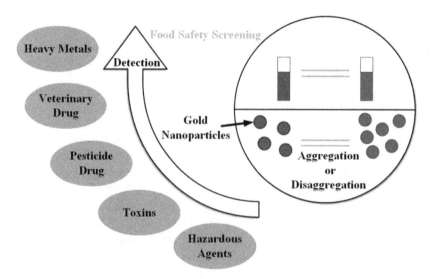

Figure 9.10. Gold nanoparticles responding differently to different wavelengths. Adapted from [22]. CC BY 4.0.

region to another depending upon the frequency that was incident on the solution. This shift is caused due to a change in the size of the particles or by changing interparticle distance between the particles in the solution due to the resonance frequency. Figure 9.11 shows a basic experiment, conducted by Liu *et al* [23], to sense the presence of atrazine using a cysteamine-gold nanoparticle solution. The experiment illustrates that the solution changes its absorption spectra with varying proportions of atrazine.

There are two main determining factors for the colorimetric sensors that determine their response time, sensitivity and SNR [22, 23]. One of these factors helps in determining the specific response or correspondence with the intended substance. The other is the conduction factor, which converts the change in colour to the spectral detection and, hence, simulates the sensitivity effect. This can be used in the detection of toxins from food samples; as aforementioned, the problem of food contamination is one of the major problems faced by supply chains worldwide, and means of early and effective detection are required. In this, different kinds and sizes of pathogens are present in food. Therefore, it is very difficult to detect them in the food samples taken. An atomic absorption spectroscopy technique based on gold nanoparticles, as described above, is very sensitive and selective and hence can be used for this purpose. Fluorescence is the most widely used method for the detection and quantification of biomolecules.

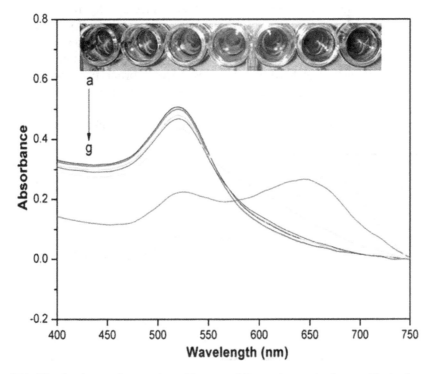

Figure 9.11. The absorbance of cysteamine-gold nanoparticles can be seen to change with atrazine concentration. Reproduced from [23] with permission from the Royal Society of Chemistry.

Nanomaterials, then, offer truly inimitable materials for the development of new methods in chemical and biological sciences. In addition to these, the effects of Raman scattering and ground plasmon resonance can be utilised. With this, rapid steps in the domain of biosensors can be made, and portable instruments based on these biosensors can be constructed. Nevertheless, in view of the versatility of electronic instrumentation for signal conditioning, it is important to concentrate further efforts on electrical sensors. In addition, it should be noted that the use of gold nanoparticles as a chemical and biochemical sensing agent and as a building block for nano-optical devices is not completely understood as of yet. They remain a great area for research.

9.3.3 Quantum-dots-based sensors

Quantum dots have zero dimensions, with high density as compared to the structure of high dimension. For this reason they are widely used in the optical and transport domains. They are suited for sensory applications in the fields of biology, diode lasers and amplifiers. They also play a major role in the detection of ions and small molecules and in indicating pH level. Semiconductor quantum dots have a photoluminescent property, which has an adjustable wavelength. Quantum dots, when illuminated with light, become excited, resulting in energy release as light when the electron falls back to the valence band. The dot sizes can be adjusted to generate different emitted light from identical material, as depicted in figure 9.12.

Quantum dots also possess very high surface-to-volume ratios. As a consequence, they display special electrical properties. Due to these unique optical and electrical properties, they have started to be used in the production of light-emitting diodes (LEDS) and solar cells. Furthermore, these responses to property changes have been utilised to develop sensors. Quantum dots can play a major role in sensors because of their ideal and tunable characteristics. Quantum dots are used in sensors because of their unique properties such as their surficial chemistry, photophysical

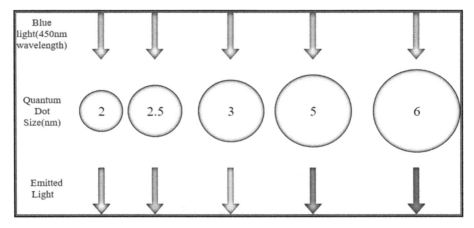

Figure 9.12. Each quantum dot produces a different colour according to its size. Adapted from http://www.nanowerk.com/what_are_quantum_dots.php.

characteristics and binding capacity. They have the capacity to bind with or attach to different molecules while maintaining their fluorescence and photophysical properties. For instance, quantum-dots-based photoluminescence chemosensors have been developed. A quantum-dot chemosensor is a molecular sensor that is used for detecting ions within an analyte by producing a detectable change.

Transition metal ions are used in many processes. However, they pose serious threats to human health due to their toxic properties, and are harmful to the environment. Subsequently, researchers have been trying to develop new methods for their detection. The main principle in the development of these sensors relies upon the functional properties of quantum dots with selective ion reception. There are various methods through which this process can take place. Quantum dots can aggregate due to the removal of receptor on itself; this occurs when the photoluminescence of quantum dots is quenched due to specific collisions among the receptors and ions. Charge transfer can also cause quenching [24]. In the example shown in figure 9.13, cadmium telluride (CdTe) quantum dots and zinc selenide (ZnSe) quantum dots are capped with receptors for the detection of Cu_2+ analyte. The CdTe is coated with mercaptoacetic acid (or MAA) and glutathione (or GSH), which act as receptors and attract the copper ions. The photoluminescence of the coated quantum dots is selectively quenched when the Cu_2+ ions bind themselves with the receptors. In this way, quantum dots can also be used for the simultaneous determination of multiple analytes.

Without the receptor capping, quantum dots can still be used for the determination of Cu_2+ ions. However, apart from Cu_2+ ions, ions such as Ag_2+ and Hg_2+ have shown the capacity of replacing Cd_2+. To prevent these competing ions, a coating of thiosulfate is used. The coating of thiosulphate prevents competing ions, making way for Cu_2+ detection.

Detection of physiological process regulators such as peptides and amino acids can be achieved using quantum-dot sensors. Molecularly imprinted polymers (or MIPs) and attached receptors on quantum dots are the two generally used methods. After the combination of receptor and quantum dot is selected, aspects of the photoluminescence property can be tuned [24]. Nickel is coordinated with dopamine-coated quantum dots and manganese is coordinated with GSH-coated

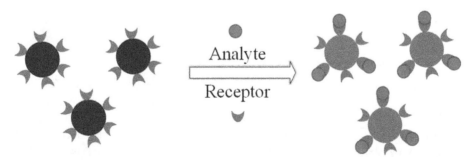

Figure 9.13. Detection of analyte using receptor-coated quantum dots. Adapted from [24] with permission from the Royal Society of Chemistry.

quantum dots. Because of the binding of amino acids and metal ions, recovered photoluminescence is seen. Quantum dots can be covalently linked to MIPs. For instance, GSH, a tripeptide, is overexpressed in cancer cells. Quantum dots can thus be used to build sensitive sensors, thus showing better analysis in cancer studies.

The utilisation of quantum dots for biosensors is yet more interesting due to their optical properties, such as high quantum yields, wide absorption spectra, tight and symmetrically shaped tunable photoluminescence emission spectra, long periods of fluorescence, and outstanding photostability with a solid protection from photo-bleaching. Indeed, quantum dots have high photostability in comparison to organic fluorescent dyes, and hence they are more suitable for usage in biomedical imaging and related applications. The surfaces/tops of quantum dots can be coated with amphiphilic, hydrophobic and/or hydrophilic types of ligands. These quantum dots can then be linked with proteins, antibodies and/or drugs. As aforementioned, the emission spectrum can be tuned by regulating their size. Quantum dots conjugated with biomolecules can exhibit and 'turn-on' photoluminescence upon interaction with receptors. pH level detection can also be achieved using quantum dots. The process of functionalisation is used in this. This means the process of attaching or interchanging an attached chemical molecule present on the exterior of the quantum dots. Here, the top of a quantum dot can be attached with an organic ligand to impart a sensation of pH. On placing this modified quantum dot in a base, it forms a substance that causes luminance, whereas in an acid it results in a substance that does not absorb in the visible region and is also a poor electron donor. Hence, there is a visible change between the two, which is further a reversible one, making quantum dots a good potential sensor of pH.

In future, quantum dots will surely be very helpful in the domains of research and development. They will play a major role in biological and pharmaceutical research, in the study of tumour detection, stem cells, lymphocytes and embryogenesis. Scientists also believe that, in the future, quantum dots will have a significant contribution to fluorescence spectroscopy. The fluorescent nature of quantum dots may be incorporated into devices. In view of improvements in sensors based on quantum dots' fluorescence, one could move on to reducing the sizes of the same. Scaling in nanoelectronics has been increasing, which could eventually help create a pathway to mobile sensing fluorescent platforms using quantum dots. In summary, there is great capacity for further experiment in the integration of quantum dots on chips with the help of miniaturisation.

9.4 Carbon-nanotubes-based sensors

Because of the high surface-area-to-volume ratio, the sensitivity of CNTs is very high due to their very large contact interfaces in a small volume. As such, even if some target gas is present in a proportion of parts per million, it can be detected. Further properties such as their high tensile strength and high electrical and thermal conductivity can be used to make sensors for specific applications. Also, CNTs are highly sensitive to charge transfer, and so are suitable for making chemical sensors for industrial purposes. CNT-based sensors can be used for various applications such as monitoring environmental pollution, improving diagnostics in the medical field, chemical detection, quality checking in the food industry, in gas sensors detecting harmful gases in mines, as pressure sensors for medical purposes and in warning systems for military bases.

(a) **Temperature sensors**

Temperature sensors are commonly used in manufacturing industries. CNTs show good electric response with respect to temperature changes, and so considerable sensitivity can be achieved with the help of CNT-based temperature sensors. A spray deposition method is used to deposit CNT films for temperature sensing. For effective sensing, this deposition should be uniform. However, due to van der Waals forces, CNTs stick to each other and so achieving a uniform distribution is difficult. To overcome this, they are dispersed in an aqueous solution of the cellulose derivative sodium carboxymethyl cellulose (CMC). During fabrication, silicon wafers with thermally grown oxides are used as substrates. Photolithography is carried out to define an interdigitated electrodes (or IDE) structure on the samples. The CNTs are deposited by an automated spray system. Before testing, earlier deposited CMC dispersant should be removed, otherwise the CNTs will not be able to touch each other, forming a percolating layer. CMC is removed with the help of HNO3. CNTs are highly sensitive to oxygen, so they are encapsulated using UV-curable epoxy and glass [25].

(b) **Sensors for pressure and strain measurements**

In the medical field, a microsensor has been designed using SU-8 and polyimide film [26]. SWNT-based resistors are best for this purpose because of their metallic and semiconducting properties as well as compact size. Observations show that this type of microsensor has higher sensitivity compared to conventional piezoresistive sensors. Silicon pressure sensors based on MEMS have a major drawback in that they are not flexible and biocompatible. As such, these types of sensors cannot be used in medical implants. SU-8 is biocompatible and can function in higher thermal conditions. As shown in figure 9.14, two configurations are possible for fabricating pressure and strain sensors, flat-design sensors and cavity-design sensors.

Further, figure 9.15(a) shows the fabrication procedure, and figure 9.15(b) shows a prototype of a fabricated sensor [26]. Figure 9.16 illustrates how

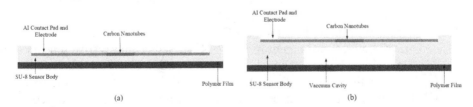

Figure 9.14. Designs of CNT-based thermal sensors. (a) Flat-design sensor prototype, (b) cavity-design sensor prototype. Adapted from [26]. CC BY 4.0.

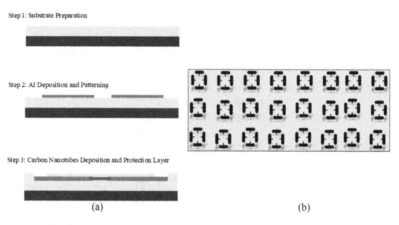

Figure 9.15. Pressure/strain sensors. (a) Process flow, (b) graphical representation of flexible flat-design sensor array. Adapted from [26]. CC BY 4.0.

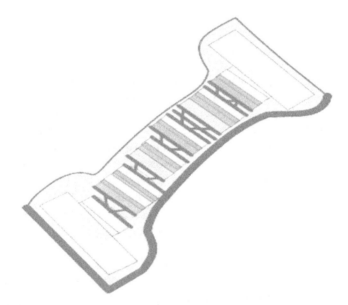

Figure 9.16. Schematic of CNT-based pressure sensor to monitor knee movement. Adapted from [27].

pressure and strain sensors can be used to manufacture a knee belt with which to sense knee pressure in daily life [27].

(c) **Chemiresistive sensors based on CNTs**

One of the most prominent uses of CNT-based chemiresistive sensors is in the form of conducting channels. A demonstration of chemiresistive sensors is shown in figure 9.17. Two main fabrication methods are used to align CNTs between two electrodes. CNT networks are placed between the electrodes with the help of various techniques such as printing, spraying, solid transfer or chemical vapour deposition (or CVD). Conductance

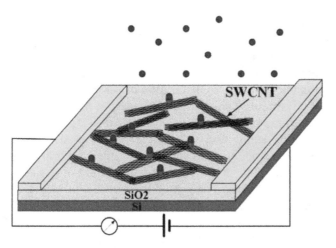

Figure 9.17. Structural configuration of chemiresistive sensors. Adapted from [28]. © IOP Publishing Ltd. All rights reserved.

between the two electrodes can then be measured to check the range and scope of sensitivity. As CNTs mainly consist of surface atoms, a minute change in the chemical configuration of the surrounding environment will result in a change in conductance between the electrodes. The absorption of analytes on the surface of CNTs can result in a change in conductance. For instance, CNTs possess characteristics of p-type semiconductors. As electrons are donated to the valence band, due to the absorption of analytes the number of holes decreases, which will result in a decrease in conductance. Similarly, if electrons are withdrawn from p-type CNTs, the hole concentration will increase, which will increase conductance. The introduction of scattering sites can decrease the charge mobility in CNTs and so the conductance also changes. Further, if the CNT–metal interface absorbs the analytes, then the conductance can also change due to modifications in the Schottky barrier. If an intertube junction absorbs an analyte, then the conductance of CNTs is again modified [28]. Conducting channels are fabricated by the connection of many CNTs, owing to the fact that single CNT units are not long enough to form conducting channels.

(d) **Carbon-nanotubes-based** biosensors

CNTs can be conjugated with several entities, which in turn can potentially enhance their recognition capacity. This also provides a means for them to become multi-functional. Owing to their very high capacity to conduct electricity as compared to that of copper wires, CNTs are beneficial for the transduction of electrical signals generated by the identification of a target. The strength of CNTs is a hundredfold higher as compared to that of steel. Moreover, their thermal conductivity is also much higher than that of diamond. Furthermore, the performance of CNTs in cross-biological membranes makes them very suitable for use in living processes, and that too with minimal invasiveness. CNTs have been shown to be favourable materials to enhance electron transfer, owing to their electrochemical and electrical properties. This makes them appropriate for fusing with electrochemical biosensors. In order to detect protein biomarkers, metabolites, and so on, diverse electrochemical CNT biosensors have been synthesised [29].

(e) **Gas sensors for detecting the expiration of perishable food**

The main reason behind the degradation of food is the activity of microorganisms, which produce gases like CO_2. The amount of CO_2 produced is very small, and as such the sensors used to detect the presence of CO_2 need to be highly sensitive. The challenge faced in the synthesis of this type of gas sensor is the high synthesis temperature of CNTs, which is 600°C. This exceeds the maximum permitted temperature of MEMS/CMOS devices, which makes it difficult to integrate CMOS with CNTs on a single chip. To overcome this, a new technique has been developed found in which CNTs are integrated with silicon-based circuits. A microheater consisting of two suspended silicon bridges is used to achieve high temperatures on the chip, as shown in figure 9.18. Thermal annealing is carried out to convert the iron film into nanoparticles at 700°C. The iron nanoparticles work as a catalyst for the CNTs to grow. After this, the CNTs are obtained at 900°C. The thus formed Si–CNT–Si structure acts as a sensor, and also shows the characteristics of a diode with a breakdown voltage of 2.3 V. It should be noted that the detection of CO_2 in an argon environment is relatively simple in comparison to detection in the atmosphere [30].

Power consumption has also been a major hurdle for sensor industries. In this, the development of power-efficient CNT-based sensors has a huge market scope. Miniaturisation is another future goal towards which CNT-based sensors can surely contribute.

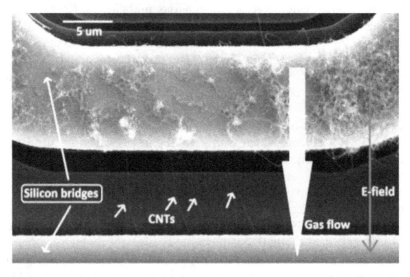

Figure 9.18. Microheater used for the fabrication of CNTs. © 2013 IEEE. Reproduced, with permission, from [30].

9.5 Electronic skin based on nanotechnology

The human skin is the largest organ of the body, with an area of up to 20 square feet. It acts effectively as a very large sensor consisting of nerves. This sensor helps to keep the brain in touch with the outside world by responding to external stimuli. The skin also allows the regulation body temperature with the help of its blood supply, while permitting the sensations of heat, cold and touch. The skin possesses several types of sensory receptors such as free nerve endings, which can detect pain and damage to tissue, and thermal receptors, which can detect a change in temperature. Due to the versatility of the human skin, it is very desirable to imitate its many functionalities in such a way that they can be adapted for applications. Any such 'electronic skin' should be built in such a way that it does not require high temperature or UV rays to be activated. For this, plastic nickel, a conductive metal with the ability to repeatedly self-repair, can be used. The electronic skin can also be sensitive to pressure and highly flexible [25]. Lighting electronic skin is based upon pressure sensors that are fit onto the electronic skin. When pressure is applied to the skin, the sensor's resistance is changed, and with it the electricity flow changes depending on the pressure, hence giving different colours of light. This design consists of a synthetic rubber and plastic composite material, which is thinner than a slice of paper. Organic LEDs are kept between layers and then lit by CNTs and a spot of conductive silver ink. There even is an 'sweating' electronic skin, wherein a soft and flexible device is used that sticks to the skin and can detect changes in the body due to sweat/exercise. This has been developed keeping in mind the possible applications in the field of disease diagnosis. It can also help users to detect how much water (i.e. hydration) is required in the body [25]. Furthermore, in certain areas of medicine, surgeons have to depend extensively on tactile sensation during operations to locate arteries and tissues. This represents a shortcoming for surgical robots as they lack tactile sensation. To overcome this obstacle, the development of electronic skin has proven to be a blessing as it has the potential to provide tactical sensation in order to differentiate between different kinds of tissues. Tactile properties such as elasticity are used to detect unusual stiffness or to discover the positions of tumours. In modern-day health monitoring, the machines used to perform physiological measurements are connected to the patient using wires and cables through complicated wiring, which can cause inconvenience and suffering to patients as well as caretakers. Therefore, it might be beneficial to develop miniaturised versions of these machines that can be attached to the body of the patient directly. As such instruments are attached to the skin of the patient, it would also allow one to collect physiological data from the patient constantly, and might also be helpful in monitoring the effect of a given treatment on the patient. To develop an electronic skin for this purpose, a layer of this skin must have mechanical properties that are similar to those of the human skin in order not to cause any discomfort while wearing it for longer durations. Electronic skin is also useful in the field of smart fabrics. This is an application based on common fabrics and is divided into three categories: sensing, actuating and adapting. Sensing means sensing motion in the environment. Actuating means reacting to sensed objects. Adapting means being ready to detect, respond and adjust conduct to given conditions. A textile structure can be built up with the help of smart materials by using techniques such as embroidering, knitting, weaving, etc. Sensors give a sensory system the ability to distinguish signals; consequently, in a passive keen material, the presence of sensors is basic. The actuators follow up on the recognised sign either independently or from a focal control unit. Creating mixtures of these source materials can result in an endless scope of novel materials. However, now and then, the business yield is spoken to by pieces of clothing that contain traditional links, scaled-down electronic segments and exceptional connectors. Nowadays, people prefer to wear soft and flexible materials instead of donning rigid structures (i.e. gadgets), so textile materials are the first choice for performing the functions of electronics. Threads can be arranged in such a way (e.g. multiple crossings) to exhibit electrical properties. Also, these properties can be embedded into yarn and a circuit-like element can be made. In this, the core part of the yarn is treated as a gate, as shown in figure 9.19, and the rest can be built up using metals. Such devices are in demand nowadays due to their cheap manufacture, flexibility, large enough surface area for sensing and ease of integration [28].

There are some simple logical insights to the material aspects of electronic skin. A silicon element is used as a substrate for new development. The ability to conform to an uneven surface is an advantage of elastomers in e-skin applications, which also makes it easier to distribute sensors. Specific material properties like tensile strength and biodegradability should also be considered [26]. Polydimethylsiloxane is one of the most important elastomers, which can be used both as a substrate and also as a

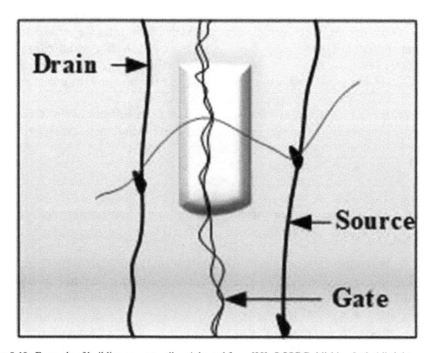

Figure 9.19. Example of building smart textiles. Adapted from [28]. © IOP Publishing Ltd. All rights reserved.

dielectric. It is used to build pressure sensors. The microstructure of this substrate is such that by applying an external force or pressure, it deforms elastically, thus showing its viscoelastic behaviour. Further, its response time is also very fast, so as to be comparable to the response time of real skin. One of the most important components for flexible electronics is a high capacitance for good performance and low-temperature processability [26]. Another likelihood in electronic skin applications is the utilisation of nanowires. Nanowires bring a phenomenal range of possibilities for electronic skin applications, essentially due to their high bearer versatility. Inorganic nanowires are an appealing decision for acknowledging hardware in automated applications. Specifically, semiconducting nanowires have fascinating electrochemical, mechanical, electrochemical and electrical properties, which are ideally suited for various uses, e.g. nanoelectronics, nanotube-based sensors, optoelectronics and photovoltaics [26].

CNTs and graphene are two widely contemplated compounds due to their interesting behaviour. Their intrinsic material properties, for example close to ballistic transport and exceptional mechanical properties, offer another avenue of innovation to imrove detection as compared to ultra-slim substrates. In the same way, as with any newly invented material, the effective use of these materials for e-skin and related methodologies depends entirely on the ability to invent reliable fabrication techniques with ease and versatility. Arrangement process procedures, for example turn coating, shower covering and inkfly printingm, are only a small portion of the strategies that could help these materials in the advancement of large-scale gadgets. In addition to CNTs, graphene is another contender for the advance-ment of e-skin parts for mechanical technology and comparable applications. Since the first segregation of graphene, incredible advancements have been made in the amalgamation of its enormous territory. In accordance with these improvements, the coordination of graphene sheets onto ultra-slender, adaptable and delicate substrates could unlock an assortment of applications in mechanical skin, ranging from movement detection to show applications. For instance, the use graphene sheets of enormous territory as adaptable and straightforward anodes in an automated skin would respect propelled detecting of numerous natural parameters because of their high transporter versatility and high surface inclusion. On the other hand, utilisation of graphene-based adaptable electrochromic gadgets in addition to graphene/nanotube-based keen windows could give skin like showcase boards over the attachments of a human-like device to show data regarding the present status of the framework [26].

Natural semiconductors have helped underlying improvements in the domain of stretchable and adaptable gadgets. Despite the fact that these materials have poor portability in comparison with inorganic semiconductors, the ease of their manu-facture and enormous surface regions are some of their relative strong points. Natural materials have colossal planned applications in electronic skin applications. Organic semiconductors are widely used in the improvement of adaptable hardware. Some example natural semiconductors utilised in the improvement of electronic skin incorporate rubrene and pentacene [26]. Polymers, if used as individual components in the domain of flexible electronics, can also help due to their good mechanical properties, though they also have the drawback of poor conductivity. To overcome this, combining conductive fillers in a polymer can result in a good conductor. Such a combined structure makes a conductive network [26].

Thus, it can be seen that the future scope of electronic skin is based on the capacity of energy storage devices and how reliable the sources of energy are. Material strategies are in research that could lead to the successful integration of the human body with skin-inspired electronics. The inherent limitations to these materials include low stretchability, mechanical stability and areal density, due to which bottom-up approaches will emerge in the near future.

9.6 Microelectromechanical/nanoelectromechanical sensors

MEMS/NEMS sensors have more advanced functionalities because of the addition of mechanical components on the electronic chip, resulting in moving parts being integrated with semiconductor ICs. Typically, integration of the electronic and mechanical functionalities in one single device makes it more compact [31]. Because of their structural properties, MEMS/NEMS are sensitive to a wide range of stimuli such as temperature, pressure, etc. MEMS have countless applications in sensing fields such as airbag accelerometers, gyroscopes, inertial MEMS for automotives, automotive pressure sensors, magnetic field sensors, and so on; as do NEMS devices, being a miniaturised version of MEMS. NEMS have two basic properties: (i) they either deflect or vibrate in response to an applied force, and (ii) they convert mechanical energy to electrical energy and vice versa. These properties allow them to be used as devices by which to measure static or time-varying forces [31]. The biggest future application of MEMS/NEMS technology is in fabricating the smallest possible sensor systems such that they can be applied to nearly any device and used nearly anywhere, in every possible situation. The main goal, then, is to design a device that is able to make full use of the transduction mechanism at these small scales, both at low power and with high sensitivity. Such small-range devices will be able to find use in nearly every man-made mechanism. They also provide high sensitivity because of their structure. NEMS offer even smaller mass and very high surface-to-volume ratio compared to MEMS. Because of this, NEMS are more suitable for applications such as high-frequency resonators or ultra-sensitive sensors. Three of the most widely used types of NEMS sensors are cantilever sensors, graphene nanoelectromechanical (GNEMS) resonators and accelerometers.

9.6.1 Cantilever sensors

Cantilever sensors are used to sense biochemical reactions taking place on their surface by measuring a nanomechanical reaction. There are two basic working principles to such sensors: first, that cantilever bending can be observed due to specific interactions between molecules on the sensor surface and molecules in the analyte; second, that a change in the resonant frequency of the sensor can be observed due to a mass change by molecular adsorption on the surface of the sensor. Two methods exist to measure these changes, optical and piezoresistive, which are described further below. When a periodic function is applied to a system, the

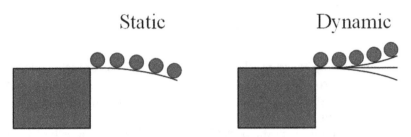

Static   Dynamic

Figure 9.20. The two modes of cantilever sensors, static and dynamic. [14], adapted by permission of the publisher (Taylor & Francis Ltd, http://www.tandfonline.com).

amplitude varies. The amplitude attains a maximum at a certain frequency, known as the resonant frequency of the system. Much research is ongoing to produce devices using cantilever sensors for biomedical applications. However, the biggest challenge in this aspect is the need for fast functionalisation and reliable output, along with high sensitivity and reproducibility.

As illustrated in figure 9.20, two modes of operation are defined for cantilever sensors [14]. One is a static mode, where bending of the surface is measured; the other is a dynamic mode, where change of the resonant frequency is measured.

(a) **Static mode**

When exposed to the substrate, we can observe some surface stress building up due to interactions on the surface. This stress results in the bending of the surface of the sensor. In static mode, we measure this bending of the surface. Equation (9.1), known as the Stoney equation, shows the relation between stress and the bending of a surface:

$$\Delta z = \frac{3(1-v)L^2}{Et^2}\Delta\sigma, \qquad (9.1)$$

where Δz is the cantilever bending amplitude, $\Delta\sigma$ represents the surface stress change, v is Poisson's ratio, E is Young's Modulus, and L and t are the length and thickness of the cantilever sensor, respectively.

(b) **Dynamic mode**

Mass change can be observed due to adsorption of the molecules of the substrate we seek to detect on the surface of the sensor. Due to this mass change, the resonant frequency of the system changes. In dynamic mode, we measure this change by observing the resonant frequency of the sensor before and after its exposure to the substrate we want to detect by using equation (9.2):

$$f_0 = \frac{1}{2\pi}\sqrt{k/m}, \qquad (9.2)$$

where k is the spring constant and m represents the suspended mass. The change in frequency can be calculated by equation (9.3):

$$\Delta f = f - f_0 = -\frac{\Delta m}{2m} f_0. \tag{9.3}$$

Many readout techniques are available, such as optical, piezoresistive/piezoelectric, capacitive, and electron tunnelling, to measure the real-time response of molecular interaction on the surface of the sensor. The most popular are optical readout and piezoresistive readout.

As shown in figure 9.21, in the optical readout technique a laser beam is focused on a surface and the reflected beam is then read by a position-sensitive detector. The deflection in the beam is proportional to the bending of the surface. Hence, the bend can be measured by calculating the deflection angle of the reflected beams before and after performing the experiment.

The most attractive feature of this method is that it can detect deflection in the nanometre range. This method is also very reliable and its setup is quite easy. One of the major disadvantages of this method is that it cannot be used in liquid media as parameters like temperature control, diffraction, etc. are taken into consideration. Further, it is difficult to use this method when using an array of sensors as it requires precise alignment and multiple laser diodes.

Piezoresistive materials have a property wherein their resistivity changes when a strain is applied to them. This can easily be measured by connecting it to an electrical circuit as a resistor. Generally, a Wheatstone bridge circuit, as shown in figure 9.22, is used for this purpose. This constitutes the piezoresistive readout technique.

The main advantage of this method is that the external electrical components can easily be fabricated on the same chip as contains the cantilever sensor, thus making this method more practical to use in the case of sensor arrays. But, when talking

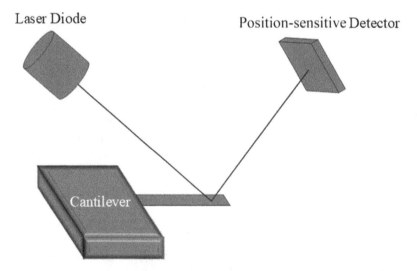

Figure 9.21. Optical readout technique for cantilever sensor. [14], adapted by permission of the publisher (Taylor & Francis Ltd, http://www.tandfonline.com).

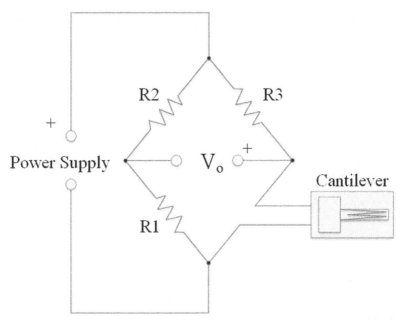

Figure 9.22. Piezoresistive readout technique for cantilever sensors. [14], adapted by permission of the publisher (Taylor & Francis Ltd, http://www.tandfonline.com).

about electrical circuits, we have to consider built-in noise, which affects the precision of the output by this method. Also, as there is an electrical connection on the chip, there is a need to take extra care while using this method in liquid media. There is also a technological limit to this method, as everything is fabricated on a thin, small chip.

Fritz was first to propose the method of using an array of cantilevers to detect DNA hybridisation [14]. Because of the molecular interaction between DNA molecules and the complementary-oligo-coated surface of a sensor, surface stress can be observed and thus surface bending can be observed. To detect a single mismatch between two DNA molecules, different deflection angles can be measured between sensors coated with different oligo-substrates. Hansen revealed that the deflection angle and beam's direction is dependent on the number of mismatches between the DNA molecules, because as mismatches increase, repulsion forces increase [14]. Meanwhile, research into the application of cantilever sensors in protein identification has been progressing rapidly due to the need to discover disease biomarkers and subsequently to produce healthcare devices that are low cost and easy to use (a disease biomarker is an indicator of the presence of a disease). Moulin observed that different proteins can bend a cantilever surface in different directions and at different angles [14]. In addition, specific antigen–antibody interactions can be observed for disease biomarkers by coating the cantilever surface with a protein-recognition layer. In order to make use of these sensors in the medical field for biomarkers, the sensors should be able to detect protein concentrations and antibody–antigen interactions in a biological environment (i.e. in the blood). To detect different biochemical molecules, a recognition material needs to be coated on the surface of the cantilever. For example, when coated with phosphoric acid, cantilever sensors can behave as a humidity sensor. The phosphoric acid absorbs water molecules and, due to the mass change, the resonant frequency changes. When coated with gold, mercury can be detected using cantilevers sensors. As mercury is attracted to gold, the resonant frequency changes due to mass change. When detecting large molecules, the dynamic mode of operation is applied as the mass change is greater. One of the most important applications in the detection of gases and vapours that does not require coating on the surface is photoacoustic spectroscopy (or PAS). As shown in figure 9.23, an IR radiation beam is focused through a window. The gas is absorbed by the IR radiation, which in turn produces sound waves. A cantilever microphone senses changes in pressure in the sound waves, and bending can be observed accordingly.

As mentioned earlier when discussing defence applications, there are many explosive substances that are a complicated mixture of chemical compounds which exhibit very low pressure, thus detecting them is a difficult task. To this end, many explosives-detection materials are available such as metals, polymers, etc. that can be coated on cantilever surfaces for the detection of these substances. Cantilever arrays can also be used to detect explosive components in these substances. This method gives very accurate and sensitive readings. Furthermore, as it is not easy to find a substance that is reactive to only one specific explosive substance, by using array setups different coatings can be applied and responses taken from each sensor to determine the exact substance.

9.6.2 Graphene nanoelectromechanical resonators

All sensors, whether developed using top-down or bottom-up techniques, have extremely small structures that are designed to oscillate in response to some mechanical stimuli for their application. In order to sense some particular stimuli, this oscillation is essential. To enable oscillation, these structures require an elastic and inertial element that is situated upon a substrate with appropriate precision.

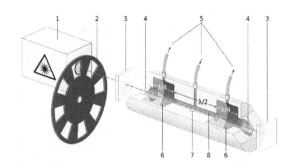

1. Light source (e. g. Laser)
2. Chopper wheel
3. Semipermeable mirrors form an optical resonance system
4. Cell windows (tilted for polarisation)
5. Gas flow inlet and outlets
6. Buffer gas volumes
7. Acoustic sensor (Microphone or transducer)
8. Resonator

Figure 9.23. Photoacoustic spectroscopy. (This Photoacoustic spectroscope.svg image has been obtained by the author(s) from the Wikimedia website where it was made available under a CC BY-SA 3.0 licence. It is included within this book on that basis. It is attributed to Hk kng.)

GNEMS resonators are one such structure. To speed up the identification of protein and disease biomarkers, a device that is able to characterise proteomes at the single-cell or single-molecule level is needed. The sensing mechanism of a mechanical mass resonator involves shifts in resonant frequency. Therefore, the basic idea to improve mass sensing resolution is to increase the resolution frequency of the device. With high resonant frequencies, a small fraction of change should also absolutely be large enough to detect easily. The resonance frequency is proportional to $h/L^2 \times \sqrt{E/\rho}$, where h is the thickness, L is the length, E is Young's modulus and ρ is the mass density of the resonator [32, 33]. Based on this, there are two ways to achieve high resonant frequency: (i) by reducing the structural dimension, which makes h/L^2 larger and also the fraction change in mass comparatively large in small-dimensional resonators; and (ii) by using a material with high $\sqrt{E/\rho}$ such as graphene [32]. For NEMS resonators, materials with a high-value quality factor (Q) is used. The quality factor denotes how slowly the oscillation of the oscillator dampens; a high-value Q factor shows a narrow resonance peak in the frequency domain. Thus, a high-value Q factor is desired to optimise the performance of a NEMS resonator acting as a sensor. NEMS resonators have recently been shown to be capable of detecting a mass of 1 dalton. A graphene NEMS resonator can be fabricated using silicon wafers with a silicon dioxide layer on top. After that, graphene layers are suspended from the SiO_2 substrate through van der Waals forces. These graphene sheets are exfoliated mechanically over long narrow ditches that are etched into the silicon dioxide substrate through dry plasma etching. This is shown in figure 9.24. The metallic electrodes can be made through a photolithographic process. They are often made of gold. Vibrations are applied across the graphene layer with fundamental frequencies and detected using a selected readout technique to gain the resonance frequency [33].

Just as in acoustic-wave-based sensors, NEMS resonators are capable of detecting an addition to the mass in a system because of the change in frequency at which the graphene sheets oscillate. The mass of an analyte and its position on the resonator are the main factors that determine these shifts in resonant frequency. As shown in figure 9.25, there is an increase in frequency shift as more and more mass is added.

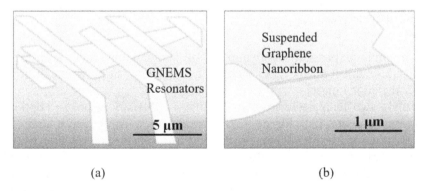

(a)   (b)

Figure 9.24. Graphical representation of (a) GNEMS resonators, and (b) a suspended graphene nanoribbon. Adapted from [34] with permission.

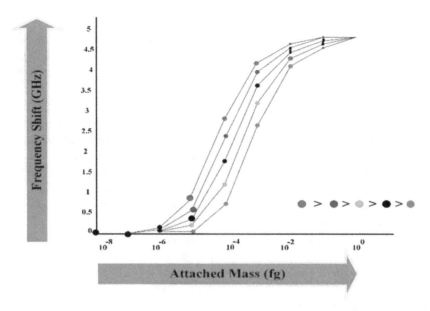

Figure 9.25. Frequency change in accordance with the addition of mass: red dots denote highest mass and pink dots denote lowest mass. Adapted from [33]. Copyright (2013), with permission from Springer Nature.

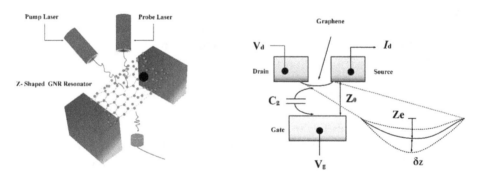

Figure 9.26. Readout techniques for a graphene resonator. (a) Optical, and (b) electrostatic actuation. Adapted from [34] with permission.

There are infinite possible combinations of analyte mass and position for a particular resonant frequency. As such, if the analyte position is known it is easy to detect the added mass.

There are two main techniques to actuate a GNEMS resonator: optical and electrostatic actuation [34]. Figure 9.26 illustrates these readout techniques. In the optical readout technique, a laser is focused on a sheet to modulate it at drive frequency. This modulation results in temperature modulation of the graphene, which causes the periodic shrinking and expansion of a layer. This continuous periodic action results in motion of the sheet. Another laser beam reflects from the suspended graphene as well as passes through it and reflects from the substrate under it, creating an interference pattern. The intensity of this reflected signal is sensitive to the position of the graphene. Thus, by tracking the intensity modulation of a reflected signal using a fast photodiode, the motion of the graphene can be detected. In the electrostatic readout technique, a graphene sheet is connected to source and drain electrodes, and this structure is situated on a gate electrode. When a DC voltage is applied to the gate, because of the uniform electrostatic force the sheet is statically deflected towards the gate electrode and creates a parabolic shape. By applying an additional radio frequency (RF) signal of frequency W at the gate, resonant motion is created upon the graphene sheet, which leads to RF force. When the RF is applied, the graphene vibrates around its static parabolic shape with a sinusoidal-mode shape. Parts with small deflection (parts near the source and drain electrodes), the difference between exact-mode shape and parabolic shape is infinitesimal and thus ignorable. Therefore, detection of the motion of the graphene using only a midpoint (with maximum deflection) is possible. As this midpoint oscillates, it acts like an actual spring with a static equilibrium position at the midpoint of a parabola. Now, this mechanical motion can be transduced to a time-varying current using graphene's charge-dependent conductance, G. By measuring changes in the current, changes in the resonant frequency can be measured.

9.6.3 Single-chip-based nano-optomechanical accelerometer

Inertial sensors like accelerometers and gyroscopes have applications in the area of maneuver modern vehicles including motorcycles, missiles, aircraft, including unmanned aerial vehicles (UAVs), among many others, and spacecraft, including satellites and landers. The accelerometers used in the past were formed with capacitive, piezoresistive and piezoelectric outputs. Optical sensors have an advantage due to their increased sensitivity. Basically, an accelerometer works by movement of a proof mass with respect to some fixed focal point, by compression or stretching of a piezoelectric material, and then by measuring current. Measuring acceleration with light is a little more complex. Such devices are made of two layers. The first layer contains a proof mass that can move in the vertical direction, as shown in figure 9.27. The second layer on top of it has a hemisphere mirror, as shown in figure 9.28. Together they form a cavity. Most of the light incident on the device is reflected back except the light matching the resonant frequency of the cavity. When there is an acceleration in the device, the cavity changes, and thus the resonant frequency changes. It is then possible to continuously match the intensity to the resonant frequency of the cavity and to calculate the acceleration of the device.

To summarise, in this chapter we have given a glimpse of the fantastic world of nanosensors and transducers. Suffice to say, the contents of this chapter are like a poppy seed as compared to ocean. However, for a student/researcher embarking on the journey of nanoscale sensor design and applications, it will surely pave a solid pathway. A vast literature is available on these and similar topics, and we leave it to the interest of the reader to explore them further. Only one statement can truly express the potential of the world of nanosensors: the sky is the limit. But, again, this is perhaps true of every aspect of nanoelectronics!

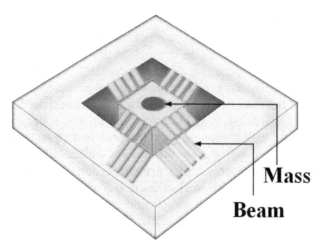

Figure 9.27. First layer of an optomechanical accelerometer. Adapted from [35]. Image stated to be in public domain.

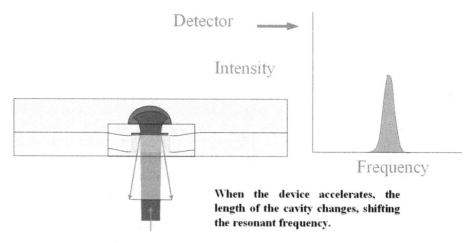

Figure 9.28. Function of an optomechanical accelerometer. Adapted from [35]. Image stated to be in public domain.

Questions

1. What advantages do sensors at the nanoscale offer and why?
2. Differentiate between a nanosensor and a transducer with an example.
3. With proper physics, explain two future nanosensor designs and their characteristics.
4. Do you think nanosensors have the capability to change warfare? Why and how?
5. Explain how graphene NEMS resonators work on the principle of resonance frequency? Give two applications of such a sensor.

6. Mention the challenges faced in developing and implementing nanosensor-based applications.
7. Name two commonly used nanosensors in today's state-of-the-art and describe how they are fabricated.

References

[1] National Research Council 1995 *Expanding the Vision of Sensor Materials* (Washington, DC: National Academies Press)
[2] Patel B C, Sinha G R and Naveen G 2020 Introduction to sensors *Advances in Modern Sensors* (Bristol: IOP Publishing) pp 1-1–1-21
[3] Heather H 2009 *Nanotechnology-Enabled Sensing* (Arlington, VA: National Nanotechnology Initiative Workshop)
[4] AZoNanos 2007 Nanosensors: Definition, Applications and How They Work (https://azonano.com/article.aspx?ArticleID=184)
[5] Omanović-Mikličanina E and Maksimovićb M 2016 Nanosensors applications in agriculture and food industry *Bull. Chem. Technol. Bosnia Herzegovina* **47** 59–70
[6] Ning Y, Hu J and Lu F 2020 Aptamers used for biosensors and targeted therapy *Biomed. Pharmacother.* **132** 110902
[7] Kuswandi B 2016 Nanotechnology in food packaging *Nanoscience in Food and Agriculture* ed S Ranjan, N Dasgupta and E Lichtfouse (Cham: Springer) pp 151–83
[8] Mousavi S M, Hashemi S A, Zarei M, Amani A M and Babapoor A 2012 Nanosensors for chemical and biological and medical applications *Med. Chem.* **8** 8
[9] Zehe T and Ramirez 2016 Nanotechnology tackles problems with noninvasive glucose monitoring *Eur. Pharm. Rev.* **2**
[10] Tallury P, Malhotra A, Byrne L M and Santra S 2010 Nanobioimaging and sensing of infectious diseases *Adv. Drug Delivery Rev.* **64** 424–37
[11] Visakh P M (ed) 2022 *Nanomaterials and Nanotechnology in Medicine* (New York: John Wiley & Sons) https://onlinelibrary.wiley.com/doi/book/10.1002/9781119558026
[12] Mazumder S, Biswas G R and Majee S B 2020 Applications of nanorobots in medical techniques *Int. J. Pharm. Sci. Res.* **11** 3138–47
[13] Yu D, Blankert B, Viré J C and Kauffmann J M 2005 Biosensors in drug discovery and drug analysis *Anal. Lett.* **38** 1687–701
[14] Lim T-C 2011 *Nanosensors: Theory and Applications in Industry, Healthcare, and Defense* (Singapore: SIM University) pp 70–88
[15] Kim M, Dorj U-O and Lee M *et al* 2017 The intelligent healthcare data management system using nanosensors *Hindawi J. Sens.* **2017** 1–9
[16] Pandya J 2012 *From Nanotechnology to Nano-Defense* Report ISN, Center for Security Studies (CSS), ETH Zurich
[17] Akyildiz I F and Jornet J M 2010 Electromagnetic wireless nanosensor networks *Nano Communication Networks* **1** 3–19
[18] Guidotti M, Rossodivita A and Ranghieri M 2012 Nano-structured solids and heterogeneous catalysts: powerful tools for the reduction of CBRN threats *Technological Innovations in Sensing and Detection of Chemical, Biological, Radiological, Nuclear Threats and Ecological Terrorism* ed A Vaseashta, E Braman and P Susmann (Dordrecht: Springer) pp 89–97

[19] Kharat D K, Muthurajan H and Praveenkumar B 2006 Present and futuristic military applications of nanodevices *Synth. React. Inorgan., Metal-Organ., Nano-Metal Chem.* **36** 231–5

[20] Arregui F J 2009 *Sensors Based on Nanostructured Materials* (New York: Springer)

[21] Li Y, Schluesener H J and Xu S 2010 Gold nanoparticle-based biosensors *Gold Bull.* **43** 29–41

[22] Liu G, Lu M, Huang X, Li T and Xu D 2018 Application of gold-nanoparticle colorimetric sensing to rapid food safety screening *Sensors* **18** 4166

[23] Liu G, Wang S, Yang X, Li T, She Y and Wang J *et al* 2016 Colorimetric sensing of atrazine in rice samples using cysteamine functionalized gold nanoparticles after solid phase extraction *Anal. Methods* **8** 52–6

[24] Cui L, He X P and Chen G R 2015 Recent progress in quantum-dot-based sensors *RSC Adv.* **5** 26644–53

[25] Verma S 2018 Electronic Skin: Advancements and Opportunities (https://electronicsforu. com/technology-trends/tech-focus/electronic-skin-advancement-opportunity)

[26] Yogeswaran N, Dang W, Navaraj T, Shakthivel D, Khan S, Ozan E, Gupta S, Heidari H, Kaboli M and Lorenzelli L 2015 New materials and advances in making electronic skin for interactive robots *Adv. Robot.* **29** 1359–73

[27] Yamada T, Hayamizu Y, Yamamoto Y, Yomogida Y, Izadi-Najafabadi A, Futaba D N and Hata K 2011 A stretchable carbon nanotube strain sensor for human-motion detection *Nature Nanotech.* **6** 296–301

[28] Castano L and Flatau A 2014 Smart fabric sensors and e-textile technologies: a review *Smart Mater. Struct.* **23** 053001

[29] Tîlmaciu C-M and Morris M C 2015 Carbon nanotube biosensors *Front. Chem., Sec. Chemical Biology* **3** 59

[30] Nguyen H Q, Ta B Q, Hoivik N, Halvorsen E and Aasmundtveit K E 2013 Carbon nanotube based gas sensor for expiration detection of perishable food *2013 13th IEEE Int. Conf. on Nanotechnology (IEEE-NANO 2013)* pp 675–8

[31] Srivastav S, Bhardwaj P and Marichi S 2011 Fabrication, sensing and application of MEMS/NEMS technology *Int. J. Core Eng. Manage.* **12** 57–60

[32] Zhang Y and Zhao Y P 2015 Mass and force sensing of an adsorbate on a beam resonator sensor *Sensors* **15** 14871–86

[33] Zhang J X and Hoshino K 2013 *Molecular Sensors and Nanodevices: Principles, Designs and Applications in Biomedical Engineering* (New York: Academic Press)

[34] Chen C 2013 Graphene nanoelectromechanical resonators and oscillators *PhD Thesis* Columbia University

[35] LeBrun T W and Gorman J J 2019 NIST On A Chip: Optomechanical Accelerometers (https://www.nist.gov/noac/technology/mass-force-and-acceleration/optomechanical-accelerometers)